IEE CIRCUITS, DEVICES AND SYSTEMS SERIES 14

Series Editors: Dr D. G. Haigh
Dr R. S. Soin
Dr J. Wood

Design of high frequency integrated analogue filters

Other volumes in the Circuits, Devices and Systems series:

Volume 1 **GaAs technology and its impact on circuits and systems**
D. G. Haigh and J. Everard (Editors)

Volume 2 **Analogue IC design: the current-mode approach**
C. Toumazou, F. J. Lidgey and D. G. Haigh (Editors)

Volume 3 **Analogue-digital ASICs**
R. S. Soin, F. Maloberti and J. Franca (Editors)

Volume 4 **Algorithmic and knowledge-based CAD for VLSI**
G. E. Taylor and G. Russell (Editors)

Volume 5 **Switched-currents: an analogue technique for digital technology**
C. Toumazou, J. B. Hughes and N. C. Battersby (Editors)

Volume 6 **High frequency circuits**
F. Nibier and co-authors

Volume 7 **MMIC design**
I. D. Robertson (Editor)

Volume 8 **Low-power HF microelectronics**
G. A. S. Machado (Editor)

Volume 9 **VLSI testing: digital and mixed analogue/digital techniques**
S. L. Hurst

Volume 10 **Distributed feedback semiconductor lasers**
J. Carroll, J. Whiteaway and R. Plumb

Volume 11 **Selected topics in advanced solid state and fibre optic sensors**
S. M. Vaezi-Nejad (Editor)

Volume 12 **Strained silicon heterostructures: materials and devices**
C. K. Maiti, N. B. Chakrabarti and S. K. Ray (Editors)

Volume 13 **RFIC and MMIC design and technology**
I. D. Robertson and S. Lucyszyn (Editors)

Design of high frequency integrated analogue filters

Edited by
Yichuang Sun

The Institution of Electrical Engineers

Published by: The Institution of Electrical Engineers, London,
United Kingdom

British Library Cataloguing in Publication Data

Design of high frequency integrated analogue filters
(IEE circuits and systems series; no. 14)
1. Linear integrated circuits 2. Electric filters — Design
I. Sun, Yichuang. II. Institution of Electrical Engineers
621.3' 8412

ISBN 0 85296 976 7

Typeset by Newgen Imaging Systems (P) Ltd., India
Printed in England by MPG Books Ltd., Bodmin, Cornwall

To Xiaohui, Bo and Lucy

Contents

Preface

Analogue filters have been an active topic in electronics, circuits, systems and communications for many years. Passive RLC, active RC and sampled data SC filters have been very well established. They can be found in many excellent textbooks (for example, [1, 2]) and thus will not be the subject of this book.

Motivated by the exploding mobile and wireless communications market, fully integrated analogue filters for high frequency applications have recently received world-wide interest. Continuous-time OTA/g_m-C filters and MOSFET-C filters have been extensively investigated and widely utilised. Research has also been conducted into active LC filters, log domain filters, sampled-data switched current (SI) filters and analogue adaptive filters. Many new architectures and designs of different types of filter have been reported. Crucial issues such as automatic on-chip tuning have been addressed. Many novel implementations have been realised in different IC technologies such as bipolar, CMOS, BiCMOS, GaAs, SiGe, etc. Filter performances have been enhanced rapidly in terms of high operation frequency, low supply voltage, low power consumption, low sensitivity, low parasitic effects, low noise and large dynamic range. Successful applications have been seen in many important areas such as video signal processing, communications systems, computer systems, telephone circuitry, broadcasting systems, and control and instrumentation systems. To date, however, there have been no books in the public domain which have systematically treated all these new topics in depth, although some on-chip tuning techniques are covered in [1] and a comprehensive treatment of OTA/g_m-C filters in both voltage and current domains is given in [2]. Whilst an excellent collection of papers on continuous-time filters, mainly MOSFET-C and OTA/g_m-C filters, published before 1993 has been published in a single volume [3], for more recent developments, engineers, researchers and senior students still rely on papers published in very diversified sources such as journals and conferences.

This book aims to systematically introduce the fundamentals and most recent advances of modern high frequency integrated analogue filters and to identify the most promising future directions. The book contains seven chapters covering different types of filter and design technique. Chapters 1 to 6 are concerned with OTA/g_m-C filters, MOSFET-C filters, active LC filters, log domain filters, switched current filters and analogue adaptive filters, respectively. Automatic on-chip filter tuning is dealt with in Chapter 7. All chapters have been written by internationally leading experts in the subject areas. Many practical issues have been discussed which include the system-on-chip, mixed-signal, low voltage/power and RF. The majority of the material has not been covered by published books. Some topics such as active filters using integrated inductors, log domain filters and analogue adaptive filters have been, for the first time, included in a filter textbook in a systematic way.

The book is oriented to any practitioners in electronic, electrical, communication, information, control and instrumentation systems. A must for modern analogue filter designers, the book can be used as a reference book by researchers and engineers and as a textbook for graduate students and senior undergraduate students in the form of either a single advanced course of analogue filters or as part of other courses related to analogue, mixed-signal and RF electronics, circuits and systems. It can also be used as a training/short course text for companies and government agencies.

Motivation for this book was partly due to my Guest Editorship of the Special Issue on High-Frequency Integrated Analogue Filters for IEE Proceedings: Circuits, Devices and Systems [4], and encouragement from David Haigh of UCL, IEE Circuits, Devices and Systems Series Editor, and Robin Mellors-Bourne of IEE, Director of Publishing Department. I must thank all the chapter authors for their great effort and high speed in contributing such high quality chapters, especially given that they are all extremely busy people. In the preparation of the book, David Haigh, Robin Mellors-Bourne, Eric Willner (previous Commissioning Editor at IEE) and Roland Harwood (current Commissioning Editor) have given much and detailed help, which is gratefully acknowledged. I would also like to thank Professor Rolf Schaumann of Portland State University for his useful communication. Last, but not least, I am greatly indebted to my lovely family, Xiaohui, Bo and Lucy. Their patience, understanding and support during the whole period of writing and editing are most appreciated.

References

1 R. Schaumann, M. S. Ghausi and K. Laker, Design of Analog Filters: Passive, Active-RC and Switched Capacitor, Prentice-Hall, NJ, 1990.
2 T. Deliyannis, Y. Sun and J. K. Fidler, Continuous-Time Active Filter Design, CRC Press, USA, January 1999.
3 Y. Tsividis and J. O. Voorman, Editors, Integrated Continuous-Time Filters, IEEE Press, 1993.
4 Y. Sun, Guest Editor, Special Issue on High-frequency Integrated Analogue Filters, IEE Proceedings: Circuits, Devices and Systems, Vol.147, No.1, February 2000.

Professor Yichuang Sun
Department of Electronic, Communication and Electrical Engineering
University of Hertfordshire
Hatfield, Hertfordshire
United Kingdom

Contributors

Yichuang Sun
Department of Electronic, Communication and Electrical Engineering
University of Hertfordshire
Hatfield, Hertfordshire
United Kingdom

Mihai Banu
Dandan Li
Agere Systems
Murray Hill, New Jersey
USA

Yannis Tsividis
Department of Electrical Engineering
Columbia University
New York
USA

Douglas Frey
Department of Electrical Engineering and Computer Sciences
Lehigh University
Bethlehem, PA
USA

John B. Hughes
Philips Research Laboratories
Redhill, Surrey
United Kingdom

Apisak Worapishet
Rungsimant Sitdhikorn
Department of Electronic Engineering
Mahanakorn University of Technology
Bangkok
Thailand

Anthony Carusone
David A. Johns
Department of Electrical Engineering
University of Toronto
Toronto
Canada

Rolf Schaumann
Department of Electrical Engineering
Portland State University
Portland, Oregon
USA

Aydin I. Karsilayan
Department of Electrical Engineering
Texas A&M University
College Station, Texas
USA

Chapter 1

Architectures and design of OTA/g_m-C filters

Yichuang Sun

1.1 Introduction

Fully integrated analogue filters for high frequency applications have received world-wide attention since the mid-1980s [1–78]. Most recent advances in this area have been reflected by a special issue in *IEE Proceedings-Circuits, Devices and Systems* [1]. Continuous-time (CT) filters have advantages over sampled-data and digital filters in terms of high speed and low power dissipation. In fact, CT filters can sometimes be the only alternative since the clock feedthrough problem in sampled data filters escalates at high speeds, and digital filters can be power hungry [2]. Among various CT filter techniques, operational transconductance amplifier/transconductor and capacitor (OTA-C or g_m-C) filters have been the most widely investigated [1–74] and used in many application areas such as video signal processing circuitry, computer hard disk drive systems [51–58], and transceivers for wireless communications. This chapter addresses these filters. The terms OTA-C and g_m-C will be used interchangeably in this chapter as both are popular in the literature.

Designing high frequency integrated continuous-time filters is a complex task which involves optimisation of sensitivity, frequency performance, parasitic effects, dynamic range, noise, low voltage operation, power consumption and chip area [1–4]. The performance of a filter depends mainly on the constituent devices, filter structure and design methods, and IC technologies, among others. In particular, different architectures can present very different performances for the chosen device and fabrication technology. A choice of more filter structures means more flexibility in design. Also filter structures at the building block level have generality, since they may be implemented in any suitable IC technologies.

In g_m-C filter design, the key component is the OTA [5–13]. An ideal operational transconductance amplifier is a voltage controlled current source, with infinite input and output impedances and constant transconductance. The OTA has tunable transconductance and can work at high frequencies, which make it most attractive for fully integrated high frequency filter design. It is used as an open loop amplifier in g_m-C filter design. Practical OTAs will, however, have finite input and output impedances. At very high frequencies, the OTA transconductance will be frequency-dependent. These nonideal characteristics will degrade frequency performances of OTA-C filters. Practical OTAs also exhibit nonlinearity for large signals and have noise, which will affect the dynamic range of OTA-C filters.

Many IC technologies have been utilized in OTA-C filter implementation. These include bipolar [52, 53], CMOS, BiCMOS, etc. To date, OTAs have been mainly implemented in CMOS [5–13, 27–30, 33, 39–51, 54–56, 60–63, 70, 73, 74], although more recently BiCMOS has also been widely used [5, 37, 38, 57, 58, 70]. From the system on a chip viewpoint, CMOS is preferred for analogue filter design, as digital circuits are fabricated in CMOS.

Methods for structure generation and design of active filters have been well established in general, and include the biquad cascade, LC ladder simulation and multiple loop feedback [3, 4]. These methods have also been used in OTA-C filter design. A large number of OTA-C filter structures have been published in the literature and their performances such as sensitivity, parasitic effects, dynamic range, noise have also been evaluated. OTA-C filters have been thoroughly included in a recent textbook [3].

In fully integrated high frequency g_m-C filter design, automatic tuning circuitry is usually included on the same chip to overcome the effects of parasitics, temperature and environment. Several tuning methods have been proposed in the literature including the master-slave and adaptive methods [4, 59]. In a mixed-signal environment the analogue part should be immune to the noises from the digital part and power supply. To reduce such noises, fully balanced structures are normally used in modern filter design [2]. Balanced structures can also eliminate even-order harmonic distortion.

This chapter is concerned with structure generation, performance analysis, design methods, on-chip automatic tuning and IC implementation of CMOS g_m-C filters. In the References, many recent publications on these topics are also given.

1.2 Two integrator loop g_m-C filter structures

The most popular method for high order filter design is the cascade method due to its modularity of structure and simplicity of design and tuning [3]. The cascade method is general in that arbitrary transmission zeros can be realised. The sensitivity of the method is however high compared with the LC ladder-based and multiple loop feedback filter configurations that will be investigated in the later sections of this chapter.

Second-order filters are the basic sections in the cascaded structures. A large number of papers have been published on second-order g_m-C filters [15–19]. A systematic and comprehensive treatment of these low order g_m-C filters has been given in Reference 3. In this section, we briefly discuss two integrator loop OTA-C filters and their performances.

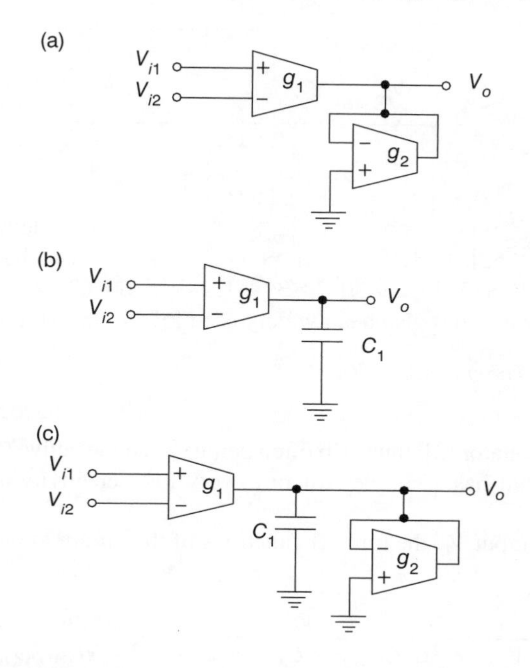

Figure 1.1 *(a) Amplifier; (b) ideal integrator; (c) lossy integrator*

1.2.1 g_m-C building blocks

The basic building blocks are shown in Figure 1.1

The amplifier in Figure 1.1(a) has gain

$$k = \frac{V_o}{V_{i1} - V_{i2}} = \frac{g_1}{g_2} \tag{1.1}$$

The ideal integrator in Figure 1.1(b) has the characteristic of

$$H(s) = \frac{V_o}{V_{i1} - V_{i2}} = \frac{1}{s(C_1/g_1)} \tag{1.2}$$

A lossy integrator is given in Figure 1.1(c), which has the lowpass function

$$H(s) = \frac{V_o}{V_{i1} - V_{i2}} = \frac{g_1}{sC_1 + g_2} \tag{1.3}$$

1.2.2 Tow-Thomas (TT) g_m-C biquad

The OTA-C architecture in Figure 1.2 consists of an ideal integrator and a lossy integrator in a single loop. This biquad is the single most popular biquad in practice. It can be considered to be the OTA-C equivalent of the Tow-Thomas (TT) active-RC biquad [3]. It can also be generated by OTA-C simulation of resistors and inductors of

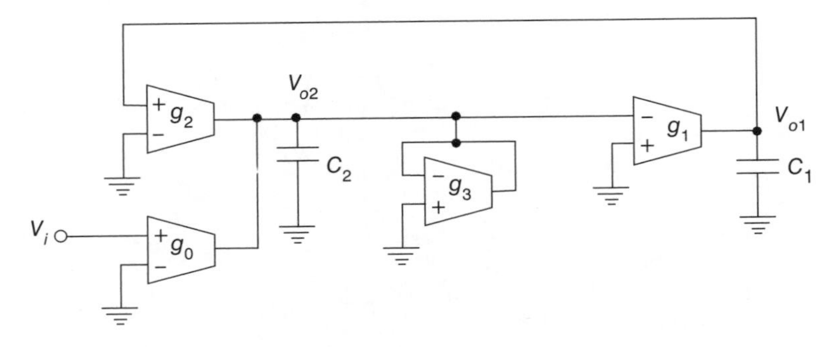

Figure 1.2 Tow-Thomas g_m-C filter

a passive RLC resonator [3], thus also often being called the active OTA-C resonator. The TT filter circuit has a simple structure, very low sensitivity and low parasitic effects.

For the given input V_i, the transfer functions of the circuit at output V_{o1} and V_{o2} can be derived as

$$H_{LP}(s) = \frac{V_{o1}}{V_i} = \frac{-g_0 g_1}{s^2 C_1 C_2 + s g_3 C_1 + g_1 g_2} = \frac{-K_{LP}\omega_0^2}{s^2 + s(\omega_0/Q) + \omega_0^2} \tag{1.4}$$

$$H_{BP}(s) = \frac{V_{o2}}{V_i} = \frac{s g_0 C_1}{s^2 C_1 C_2 + s g_3 C_1 + g_1 g_2} = \frac{K_{BP}(\omega_0/Q)s}{s^2 + s(\omega_0/Q) + \omega_0^2} \tag{1.5}$$

$$\omega_0 = \sqrt{\frac{g_1 g_2}{C_1 C_2}} \tag{1.6}$$

$$Q = \frac{1}{g_3}\sqrt{\frac{g_1 g_2 C_2}{C_1}} \tag{1.7}$$

$$K_{LP} = \frac{g_0}{g_2} \tag{1.8}$$

$$K_{BP} = \frac{g_0}{g_3} \tag{1.9}$$

As can be seen from Eqns. (1.4) and (1.5), the TT biquad has a lowpass function from V_{o1} and a bandpass from V_{o2}. Eqns. (1.6–1.9) also show that filter gain can be tuned independently of the quality factor and frequency and the quality factor can be tuned independently of the frequency. If the capacitors are equal, then orthogonal tuning can be achieved.

1.2.3 Effects and reduction of OTA nonidealities

Nonideal parameters of the CMOS OTA include the input capacitance C_i, output capacitance C_o, output conductance G_o, and frequency dependent transconductance

$$g(s) = \frac{g}{1 + s/\omega_b} \cong g e^{-j\varphi_e} \tag{1.10}$$

where ω_b is the bandwidth of the OTA, $\varphi_e = \omega/\omega_b$ is the excess phase and g is the DC transconductance. The input conductance of the CMOS OTA is normally very small and thus can be ignored. These nonideal parameters will limit the performance of OTA-C filters. The OTA frequency dependent transconductance and input and output capacitances will affect mainly the high frequency characteristic, and the finite output resistance will affect especially the low frequency response of OTA-C filters.

OTA parasitic capacitances added to design capacitances will reduce the filter working frequency. For the TT biquad, all parasitic capacitances are referred to the ground due to the use of only single ended input OTAs, and these parasitic capacitances can be absorbed into the grounded circuit capacitances, which make the filter particularly suitable for high frequency applications. The absorption approach determines the real component values by subtracting the parasitic induced increments from the nominal values. This requires that the nominal values should be larger than the total of the related parasitic capacitances. It should be noted that at very high frequencies, this may not always be possible. Careful design is thus needed to handle the parasitic effects.

We may apply the input voltage to the input terminal of the OTAs inside the loop, but this may cause feedthrough effects, that is, extra zeros due to finite OTA differential input capacitances. Applying input voltage through a capacitor is not desirable for several reasons. It will result in the floating capacitor, which itself has large parasitic capacitances and occupies a large chip area. Also, the floating capacitor cannot absorb the OTA finite capacitances. Furthermore, capacitor injection requires an ideal input voltage source, otherwise the internal resistance will change the filter performance. It is not suitable for high order cascade design as the low filter input impedance will cause loading effects. Thus capacitor injection must be rejected.

Transconductance frequency dependence or excess phase has the Q enhancement effect. Suppose that all transconductances in Figure 1.2 are equal; we can derive the changed Q' of TT biquad as [7]

$$Q' = \frac{Q}{1 + 2(1/gR_0 - \varphi_e)Q} \tag{1.11}$$

As can be seen clearly from Eqn. (1.11), the excess phase increases the quality factor. This effect may be compensated for by connecting a resistor in series with the capacitor [27] as shown in Figure 1.3. In IC design, this resistor is normally realised using a MOSFET in the ohmic region. The equivalent resistance value can be adjusted by voltage V_B. From the circuit in Figure 1.3 we can write

$$H(s) = \frac{g/sC}{(1 + sRC)/(1 + s/\omega_b)} \tag{1.12}$$

It is clear from Eqn. (1.12) that, setting $R = 1/\omega_b C$, the circuit in Figure 1.3 will become an ideal integrator. The effect of finite bandwidth is thus compensated for.

The finite OTA output resistance R_o or finite OTA DC voltage gain gR_o will reduce the filter quality factor as shown in Eqn. (1.11) and filter gain. In high-Q applications this effect may need to be eliminated by using negative resistance [3, 39].

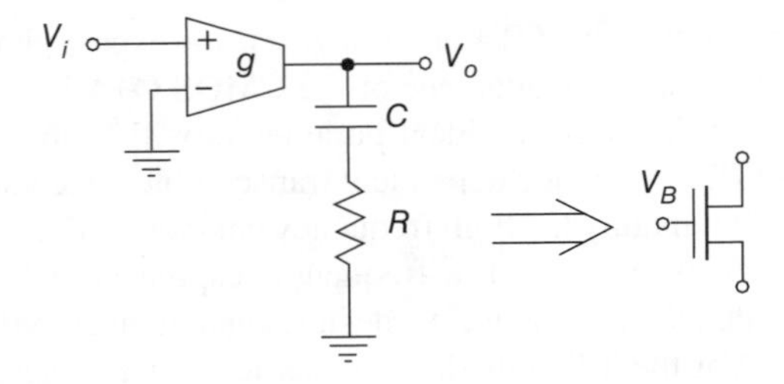

Figure 1.3 Compensation of finite bandwidth effects

The dynamic range of a filter can be analysed by calculating the ratio of the maximum signal magnitude to the noise level at either the input or the output nodes of the filter [3]. Generally, given the desired filter transfer function, the dynamic range of a filter is dependent on the dynamic ranges of the network elements, especially the active devices, and the filter network architecture. The limited dynamic range of the OTA is confined by its linear input range and noise level, which restricts the dynamic range of the filter. Several publications have dealt with noise performance analysis and large-signal capability of OTA-C filters [3, 32–36].

1.2.4 Balanced filter structures

Balanced structures are widely utilised in continuous-time integrated filter design [2]. This is because balanced structures can increase the common-mode rejection ratio, eliminate the even-order harmonic distortion components and reduce the effects of power supply noise. In mixed-signal or system on chip design, balanced structures are especially important for reducing interference/noise from digital circuits on the same chip. Balanced configurations can be obtained from single ended structures. The single ended to balanced conversion can be generally achieved by first mirroring the whole single ended circuit at ground (duplicating all the components and changing the terminal polarities of all mirrored active elements) and then combining each original amplifier and its mirrored counterpart into a balanced differential input-differential output device with inverting-noninverting gains [3, 4]. As an example, the balanced TT OTA-C biquad is shown in Figure 1.4.

Note that the OTA with a pair of fully differential inputs and outputs is used. Note also that inverting is achieved by a simple cross-connection. Since the construction of balanced structures is a simple matter once one has single ended circuits and also single ended circuits are more straightforward to handle, therefore single ended structures will be used in the relevant sections of the chapter. It should be noted that balanced structures are usually more complex than their single ended versions, which may occupy a bigger chip area and consume more power. Common-mode feedback circuits are usually required to preserve the balanced operation.

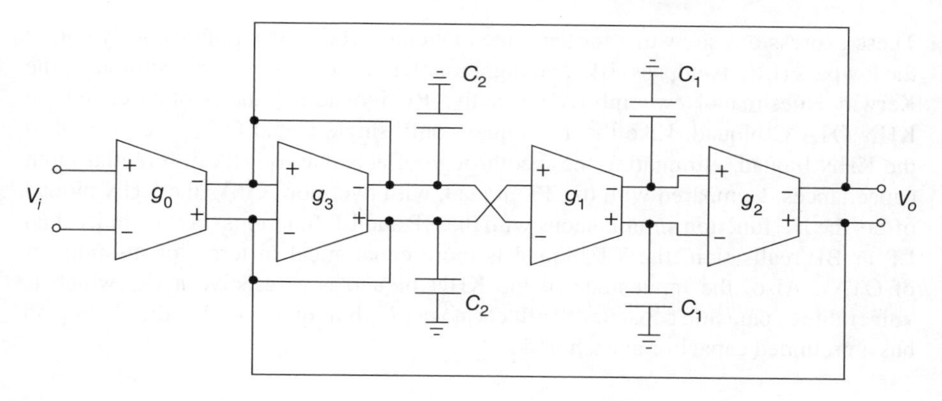

Figure 1.4 Balanced structure of TT bandpass OTA-C filter

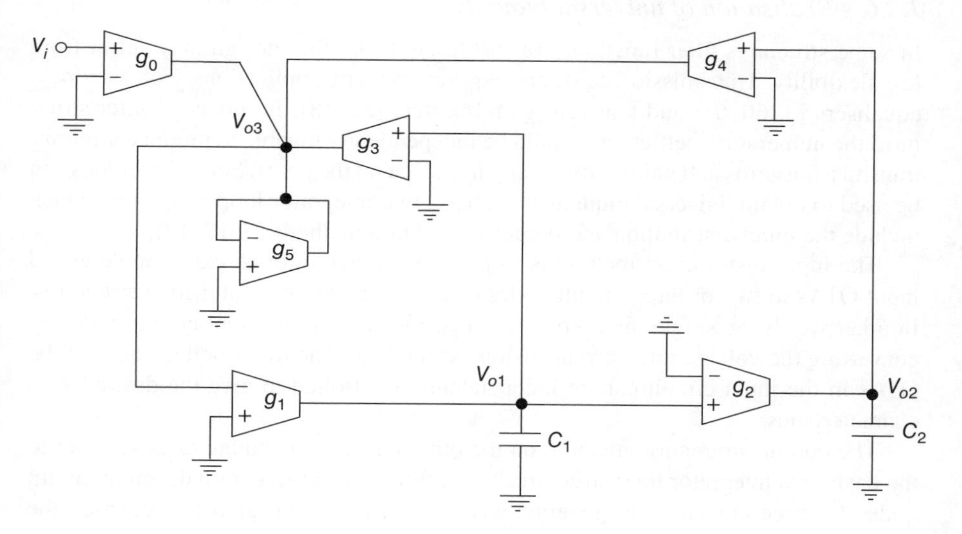

Figure 1.5 KHN OTA-C biquad

1.2.5 KHN g_m-C filters

Figure 1.5 [3, 16] shows another two integrator loop OTA-C filter. With $\tau_1 = C_1/g_1$, $\tau_2 = C_2/g_2$, $k_{11} = g_3/g_5$, $k_{12} = g_4/g_5$ and $K = g_0/g_5$ we can derive

$$\frac{V_{o1}}{V_i} = \frac{-K\tau_2 s}{\tau_1\tau_2 s^2 + k_{11}\tau_2 s + k_{12}} \tag{1.13}$$

$$\frac{V_{o2}}{V_i} = \frac{-K}{\tau_1\tau_2 s^2 + k_{11}\tau_2 s + k_{12}} \tag{1.14}$$

$$\frac{V_{o3}}{V_i} = \frac{K\tau_1\tau_2 s^2}{\tau_1\tau_2 s^2 + k_{11}\tau_2 s + k_{12}} \tag{1.15}$$

These expressions show that for the same input the circuit can simultaneously output the lowpass (LP), bandpass (BP) and highpass (HP) functions, which is similar to the Kerwin–Hueslman–Newcomb (KHN) active-RC biquad [3] and is often called the KHN OTA-C biquad. Like the TT biquad, only single ended OTAs are utilised in the KHN biquad, eliminating the feedthrough effects due to OTA differential input capacitances. Compared with the TT biquad, with two more OTAs the KHN biquad offers the HP function simultaneous with the BP and LP functions. For the individual LP or BP realisation, the TT biquad is more economical in terms of the number of OTAs. Also, the input node of the KHN biquad is a resistive node, which is vulnerable to parasitic capacitance effects at very high frequencies, but the TT biquad has a grounded capacitor at each node.

1.2.6 Realisation of universal biquads

In some structures filter functions are still limited and filter design may suffer from less flexibility. Transmission zeros are required in many applications such as allpass equalisers [3, 60–63] and real zero gain boosters [52–58]. In universal filter structures the numerator coefficients should be independently tunable to produce arbitrary transmission zeros and gains without any influence on the poles. Several methods can be used to obtain universal biquads from basic two integrator loop structures, which include the input distribution and output summation methods [3, 17–19].

The input distribution method is to apply a voltage through extra single ended input OTAs to two or three circuit nodes of the basic two integrator loop structures. In other words the single input voltage is distributed onto different circuit nodes by converting the voltage into currents using extra OTAs. The transconductances of the OTAs in the input circuit can be independently controlled to give the desired zero characteristics.

The output summation method, on the other hand, is to add the OTA summer to the basic two integrator loop structures to combine the outputs from different circuit nodes for a certain input to generate any desired zeros and gain by adjusting the transconductances of the OTAs in the output circuit.

The input distribution method will not introduce any extra circuit nodes. The summation method will, however, introduce an additional resistive node, the overall output node. This node will produce a pole at high frequencies due to parasitic capacitances, which should be dealt with carefully. The summation method also needs one more OTA (for voltage output) than the distribution method. In practice, what kind of method should be used will also depend on other considerations, for instance, whether a particular filter realisation of the method would need highly sensitive difference nulling of some numerator coefficient. In most cases the filter performance will depend on both the input/output methods and basic two integrator loop structures.

Finally we stress again that for high order filter design by cascade, to avoid the loading effects, OTA-C sections should be designed to drive high impedance nodes such as the inputs of other OTAs. Therefore only those biquads with the filter input applied to the OTA input terminals are practical for high order cascade design.

1.3 g_m-C filters based on LC ladders

The cascade method is simple in design and tuning. However, it has a very high sensitivity to component variations. It is well known that resistively terminated lossless LC filters have very low passband sensitivity [3, 4]. To achieve very low sensitivity, OTA-C filters are thus normally designed by simulating passive LC filters. The design methods, structures and performances of OTA-C filters based on passive LC ladder simulation have been presented systematically in [3]. The methods discussed in [3] include inductor substitution, Bruton transformation, admittance simulation, signal simulation, matrix simulation and coupled biquad approaches. Many typical filter structures have been given for different simulation methods. The conditional equivalence of the admittance and signal simulation OTA-C architectures has been demonstrated. Interesting ideas such as the partial floating admittance concept for the admittance and signal simulation structures have been given and some alternative leapfrog OTA-C structures have also been discussed. In this section we focus on component substitution and signal simulation methods.

1.3.1 g_m-C filters based on component substitution

OTA-C filter design based on a passive LC ladder can be conducted by substituting inductors by OTA-C counterparts [3, 20]. Such an OTA-C circuit has as low sensitivity as the passive counterpart. We consider OTA-C simulation of the inductor. The OTA gyrator, when terminated by a capacitor, will produce a simulated OTA-C inductor with the inductance being given by $L = C/g_1g_2$ as depicted in Figure 1.6(a). A floating OTA-C inductor based on two gyrators connected back-to-back may be reduced to the three OTA architecture shown in Figure 1.6(b). It can be shown that when $g_2 = g_3$ the equivalent inductance is given by $L = C/g_1g_2$.

A grounded inductor requires two OTAs and one capacitor, while a floating inductor needs three OTAs and one capacitor. The inductor substitution technique described above leads to a realisation that has the same topology as the original passive ladder

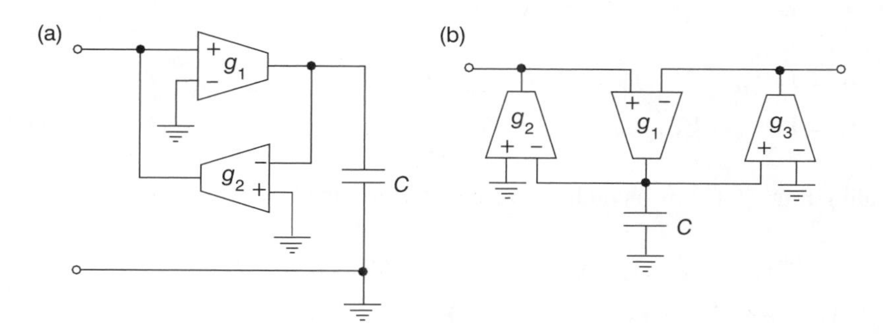

Figure 1.6 g_m-C simulation of inductors: (a) grounded; (b) floating

network. The difference is that each inductor is replaced by a circuit using OTAs and the capacitor.

It is well known that the inductor substitution method is most economical for the simulation of highpass LC ladders, as the inductors in highpass ladders are grounded. In all-pole LC ladders all capacitors are grounded, but others contain floating capacitors. In IC implementation, the grounded capacitor is simpler to implement in technology. The floating capacitor has substantial parasitic capacitances. The floating capacitor can be simulated using five OTAs and a grounded capacitor [3]. However, the large number of OTAs may cause other problems such as extra noise and power consumption. It should also be pointed out that the floating to grounded capacitor converter circuit has an internal node, which does not have any component to ground, which we call the suspending node. The suspending node with parasitic capacitances will produce an extra pole, and this parasitic pole will influence the filter responses at high frequencies.

1.3.2 g_m-C filters based on signal simulation

The signal simulation method is based on circuit equations that describe the passive ladder filter. Such equations are of the mixed current and voltage type, as the simulated signals are normally capacitor voltages and inductor currents. We can convert these equations to their voltage-only counterparts by scaling and simulate them using OTAs and capacitors [3, 21]. The resulting filters are of the leapfrog structure.

1.3.2.1 Finite-zero lowpass g_m-C filters without floating capacitors: Consider a fifth-order lowpass finite zero passive LC ladder, as shown in Figure 1.7(a). Denoting admittances in the series arms and impedances in the parallel arms as

$$\begin{aligned} &Y_1 = 1/R_1,\ Z_2 = 1/sC_2,\ Y_3 = sC_3 + 1/sL_3,\ Z_4 = 1/sC_4, \\ &Y_5 = sC_5 + 1/sL_5,\ Z_6 = 1/(sC_6 + 1/R_6) \end{aligned} \tag{1.16}$$

we can derive the equations relating the currents flowing in the series arms, I_j, and the voltages across parallel arms, V_j, as

$$\begin{aligned} &I_1 = Y_1(V_{in} - V_2),\ V_2 = Z_2(I_1 - I_3),\ I_3 = Y_3(V_2 - V_4),\ V_4 = Z_4(I_3 - I_5), \\ &I_5 = Y_5(V_4 - V_6),\ V_{out} = V_6 = Z_6 I_5 \end{aligned} \tag{1.17}$$

Scaling Eqn. (1.17) by a conductance g and denoting $V_j{}' = I_j/g$, we have

$$\begin{aligned} &V_1{}' = Y_1/g(V_{in} - V_2),\ V_2 = gZ_2(V_1{}' - V_3{}'),\ V_3{}' = Y_3/g(V_2 - V_4), \\ &V_4 = gZ_4(V_3{}' - V_5{}'),\ V_5{}' = Y_5/g(V_4 - V_6),\ V_{out} = V_6 = gZ_6V_5{}' \end{aligned} \tag{1.18}$$

where Y_j/g and gZ_i are voltage transfer functions.

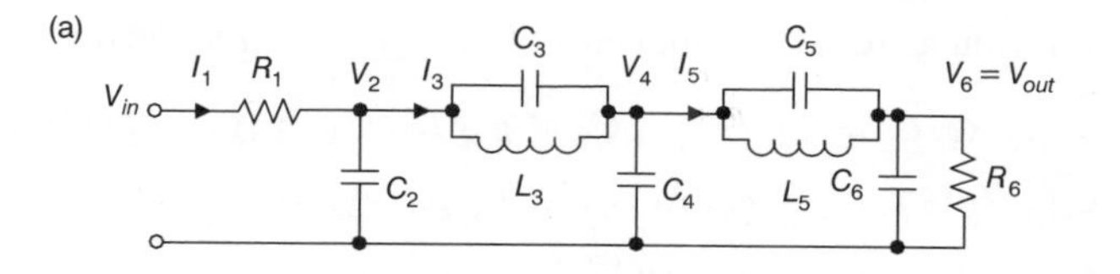

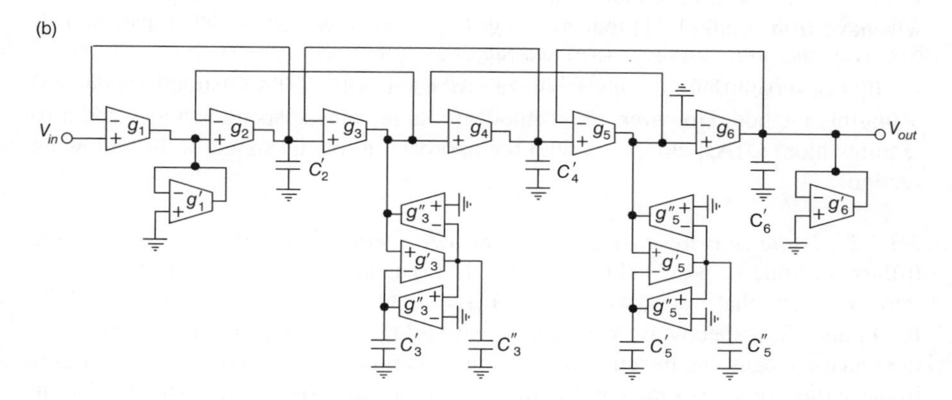

Figure 1.7 g_m-C simulation without floating capacitors: (a) fifth-order finite zero LC ladder and (b) leapfrog g_m-C simulation structure

Eqn. (1.18) can be realised in a leapfrog form using OTAs with transconductance g_j and grounded impedances $Z_j{}'$,

$$\begin{aligned} Z_1{}' &= Y_1/g_1g, \quad Z_2{}' = Z_2g/g_2, \quad Z_3{}' = Y_3/g_3g, \\ Z_4{}' &= Z_4g/g_4, \quad Z_5{}' = Y_5/g_5g, \quad Z_6{}' = Z_6g/g_6 \end{aligned} \tag{1.19}$$

From Eqn. (1.19) we can see that, besides the general scaling by g, each new grounded impedance has a separate transconductance which can be used to adjust the impedance level.

Now we further simulate the grounded impedances using OTAs and/or capacitors as shown in Figure 1.7(b). For the arms with a single component, this is very straightforward. For the cases such as Y_3 and Y_5 which are a combination of the admittances of two components, we can use inductor substitution. Taking Y_3 as an example, the corresponding grounded impedance can be written as

$$Z_3{}' = Y_3/g_3g = sL_3{}' + 1/sL_3g_3g \tag{1.20}$$

where $L_3{}' = C_3/g_3g$. The second term in the equation represents a capacitance of value $C_3{}' = L_3g_3g$. But the first term is equivalent to an inductor. This should then be further replaced by an OTA-C inductor with $L_3{}' = C_3{}''/g_3{}'g_3{}''$. Combining the two steps we can also obtain $C_3{}''$ in terms of C_3.

The design formulas for all components of the g_m-C filter are given below:

$$g_1' = g_1 g R_1,\ C_2' = (g_2/g)C_2,\ C_3' = g_3 g L_3,\ C_3'' = (g_3' g_3''/g_3 g)C_3,$$
$$C_4' = (g_4/g)C_4,\ C_5' = g_5 g L_5,\ C_5'' = (g_5' g_5''/g_5 g)C_5, \quad (1.21)$$
$$C_6' = (g_6/g)C_6,\ g_6' = (g_6/g)(1/R_6)$$

where g is the scaling conductance. If all transconductances are equal to g, we will have from Eqn. (1.21) that $g_1{}' = g^2 R_1, g_6{}' = 1/R_6, C_j{}' = C_j$ for even $j = 2, 4, 6$, $C_j{}' = g^2 L_j$ and $C_j{}'' = C_j$ for odd $j = 3, 5$.

It is observed that all capacitors are grounded in the OTA-C structures derived using this method. However, the method results in some suspended nodes and also requires more OTAs, compared with the approach to be discussed in the following section.

1.3.2.2 Finite-zero lowpass g_m-C filters with floating capacitors: Consider the fifth-order finite zero LC ladder in Figure 1.7(a) again. This time, we split Y_3 and Y_5 into two parts, that is, $Y_{31} = sC_3$ and $Y_{32} = 1/sL_3$, $Y_{51} = sC_5$ and $Y_{52} = 1/sL_5$ for Y_3 and Y_5, respectively. We want to leave admittances Y_{j1} floating and simulate the nodal voltages and the currents flowing in admittances Y_{j2} (not the total currents in the series arms), as shown in Figure 1.8(a). The equations for the whole ladder in Figure 1.8(a) can be written as

$$I_1 = Y_1(V_{in} - V_2),\ V_2 = Z_2[(I_1 - I_3) - Y_{31}(V_2 - V_4)],\ I_3 = Y_{32}(V_2 - V_4),$$
$$V_4 = Z_4[(I_3 - I_5) + Y_{31}(V_2 - V_4) - Y_{51}(V_4 - V_6)],\ I_5 = Y_{52}(V_4 - V_6), \quad (1.22)$$
$$V_{out} = V_6 = Z_6[I_5 + Y_{51}(V_4 - V_6)]$$

Scaling Eqn. (1.22) by conductance g and denoting $V_j{}' = I_j/g$ we can obtain

$$V_1{}' = Y_1/g(V_{in} - V_2),\ V_2 = gZ_2[(V_1{}' - V_3{}') - Y_{31}/g(V_2 - V_4)],$$
$$V_3{}' = Y_{32}/g(V_2 - V_4), \quad (1.23)$$
$$V_4 = gZ_4[(V_3{}' - V_5{}') + Y_{31}/g(V_2 - V_4) - Y_{51}/g(V_4 - V_6)],$$
$$V_5{}' = Y_{52}/g(V_4 - V_6),\ V_{out} = V_6 = gZ_6[V_5{}' + Y_{51}/g(V_4 - V_6)]$$

Based on Eqn. (1.23), we can obtain the corresponding OTA-C LF structure with floating capacitors as shown in Figure 1.8(b), where g_j is the OTA transconductance and $g_2 = g_4 = g_6$. The design formulas for the grounded impedances Z_j', $j = 1, 2, \ldots, 6$, are the same as those in Eqn. (1.19) in form, except that Y_3 and Y_5 should be replaced by Y_{32} and Y_{52}, respectively, and $g_2 = g_4 = g_6$. The design formulas for the floating admittances $Y_{31}{}'$ and $Y_{51}{}'$ are given below:

$$Y_{31}{}' = g_4/gY_{31},\ Y_{51}{}' = g_6/gY_{51} \quad (1.24)$$

The values for $g_1{}'$, $C_j{}'$ and $g_6{}'$ can be calculated using Eqn. (1.21). The values for the two floating capacitances can be computed from Eqn. (1.24), given by $C_3{}'' = (g_4/g)C_3$ and $C_5{}'' = (g_6/g)C_5$. Note again that there should be $g_2 = g_4 = g_6$.

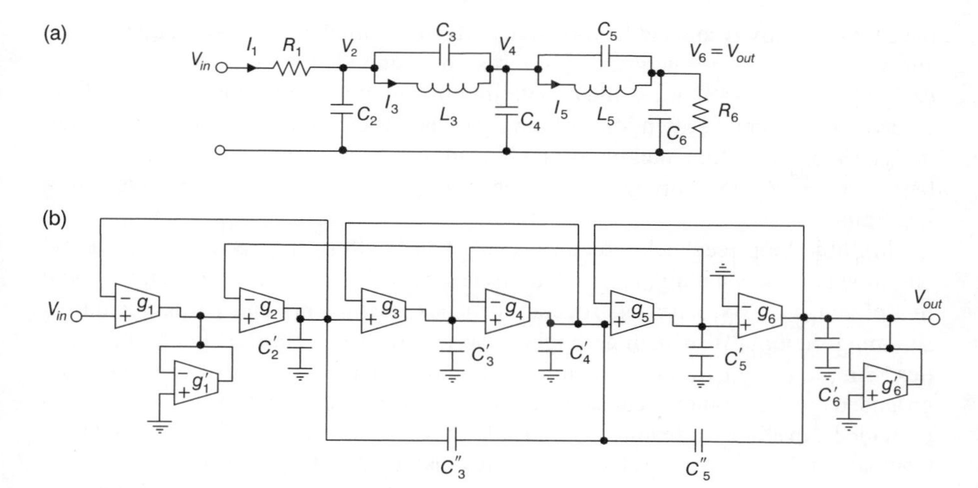

Figure 1.8 g_m-C simulation with floating capacitors

This approach results in the OTA-C structure whose every node has a grounded capacitor and requires fewer OTAs, at the expense of using floating capacitors. In practice, the structure in Figure 1.8 has been used more often than the one in Figure 1.7.

Both the component substitution method and the signal simulation method can ensure that all but one, or at most two, transconductances are identical. Identical transconductance values can make on-chip tuning, design and layout much easier, since a standard transconductance cell can be used throughout the circuit [4].

It can be proved that the component substitution and signal simulation methods are equivalent in OTA-C filter design under certain conditions [3, 22]. These two methods are thus equally important for OTA-C filter design. However, the signal simulation method has the possibility of scaling the component values such that the circuits have the maximum possible dynamic range. This scaling is not normally available in the component substitution design.

To finish this section we draw the reader's attention to recent progress in direct integration of inductors on silicon for RF applications [75]. Active filters using integrated inductors will be covered in Chapter 3.

1.4 Multiple loop feedback g_m-C filters

Filter design is concerned with the synthesis of a transfer function of the form:

$$H_d(s) = \frac{A_n s^n + A_{n-1} s^{n-1} + \cdots + A_1 s + A_0}{B_n s^n + B_{n-1} s^{n-1} + \cdots + B_1 s + 1} \tag{1.25}$$

Multiple loop feedback (MLF) active filters [3, 4] have the advantage of both low sensitivity and arbitrary transmission zeros, compared with cascade structures that have high sensitivity and ladder topologies which can implement only imaginary axis zeros. Note that non-imaginary-axis zeros are required, for example, in equaliser design [29, 30, 52–58]. g_m-C filters with multiple loop feedback configurations have received considerable attention in high frequency integrated continuous-time filter design [3, 23–33]. Both integrators and biquads can be used in the design. Integrator-based techniques are more general, as biquads themselves can be constructed from integrators.

Multiple loop feedback structures consist of feedforward, feedback, input and output networks generally. A detailed discussion of MLF g_m-C filters can be found in Reference 3. By setting particular feedback coefficients and particular output summing or input distributing weights, we can realise arbitrary filter characteristics. Simple design formulae require no complex optimisation. All capacitors are grounded, and canonical realisations can guarantee that all internal nodes have a grounded capacitor – a feature useful for high frequency integrated implementation. Comparison has shown that these structures may be advantageous over the cascade and ladder simulation architectures in generality of function, simplicity of design and structure, and insensitivity to OTA nonidealities.

1.4.1 All-pole MLF g_m-C filters

We consider the multiple loop feedback g_m-C configurations and all-pole filter design. The basic building blocks are voltage integrators and amplifiers. The general multiple loop feedback all-pole g_m-C model with all capacitors being grounded is shown in Figure 1.9, which consists of n OTA-capacitor integrators connected in cascade and a feedback network that may contain OTA-based voltage amplifiers and/or pure wire connections.

The relationship between the overall circuit input voltage V_{in} and the integrator output voltages $V_o = [V_{o1} V_{o2} \ldots V_{on}]^t$ including the overall circuit output voltage $V_{out} (= V_{on})$ can be obtained as

$$A(s)V_o = BV_{in} \tag{1.26}$$

where $B = [1\ 0 \ldots 0]^t$ is the input related vector. The superscript t stands for transpose. $A(s)$ is the system coefficient matrix, given by

$$A(s) = M(s) + F \tag{1.27}$$

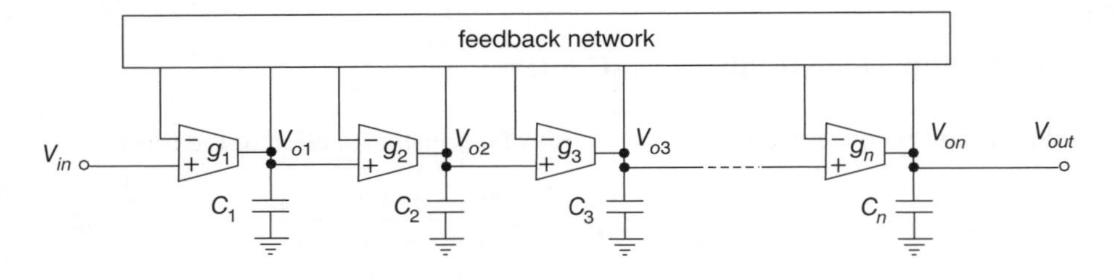

Figure 1.9 Multiple loop feedback all-pole g_m-C filter structure

where $M(s)$ is related to the integrator part, which is given by

$$M(s) = \begin{bmatrix} s\tau_1 & & & \\ -1 & s\tau_2 & & \\ & & \ddots & \\ & & -1 & s\tau_n \end{bmatrix} \tag{1.28}$$

in which $\tau_j = C_j/g_j$ is the time constant of integrator j and s is the complex frequency. $F = \left[f_{ij}\right]_{n \times n}$ is the feedback coefficient matrix. The element f_{ij} is the voltage feedback coefficient from the output of integrator j to the input of integrator i. The coefficient f_{ij} can be realised with an OTA-based voltage amplifier.

From Eqn. (1.26) the system transfer function can be derived as

$$H(s) = \frac{V_{out}}{V_{in}} = \frac{1}{|A(s)|} \tag{1.29}$$

where $|A(s)|$ represents the determinant of the system coefficient matrix $A(s)$. This transfer function usually has the form of an all-pole characteristic. The structure can therefore be utilised to design all-pole lowpass filters.

If there are only pure wire connections (unity feedback) in the feedback network, the whole system will have the minimum number of components, n OTAs and n capacitors for an nth-order system. Along with various features due to the minimum number of components, canonical structures also have the property that each node contains a grounded capacitor, and therefore all the finite OTA output capacitances can be absorbed. Noncanonical realisations with more components may offer some design flexibility and result in more architectures. However, there are some nonintegrating nodes in the feedback network, and any parasitic capacitances including OTA input and output capacitances on these nodes will thus cause extra parasitic poles. More OTAs may also increase the effects of the OTA excess phase, the chip area and power consumption.

The one-to-one correspondence between the feedback matrix F and the circuit configuration is very useful for filter structure generation, as different Fs will result in different circuit structures. For any order and any particular F we can easily draw the associated structure, obtain the corresponding transfer function, and calculate the component values. For fourth-order filters, for example, when $f_{12} = f_{23} = f_{34} = f_{44} = 1$ and other $f_{ij} = 0$, the inverse-follow-the-leader feedback (IFLF) structure can be obtained as shown in Figure 1.10(a). With $\tau_j = C_j/g_j$ denoting the time constant of integrator j, the transfer function is derived as

$$H(s) = \frac{1}{\tau_1\tau_2\tau_3\tau_4 s^4 + \tau_1\tau_2\tau_3 s^3 + \tau_1\tau_2 s^2 + \tau_1 s + 1} \tag{1.30}$$

The LF structure will result from $f_{14} = f_{24} = f_{34} = f_{44} = 1$ and other $f_{ij} = 0$, shown in Figure 1.10(b). The transfer function is given by

$$H(s) = \frac{1}{\tau_1\tau_2\tau_3\tau_4 s^4 + \tau_1\tau_2\tau_3 s^3 + (\tau_1\tau_2 + \tau_1\tau_4 + \tau_3\tau_4)s^2 + (\tau_1 + \tau_3)s + 1} \tag{1.31}$$

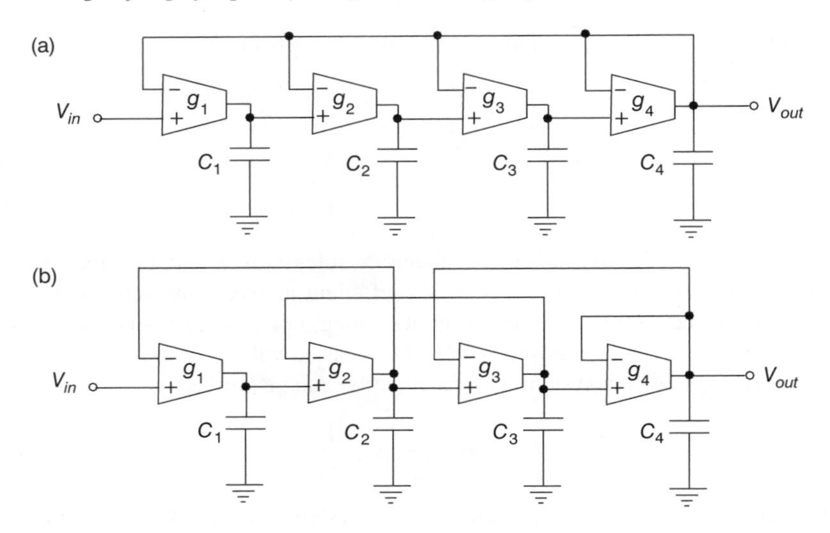

Figure 1.10 Fourth-order all-pole canonical OTA-C filters: (a) IFLF; (b) LF

In both cases explicit design formulae for τ_j can be obtained by comparison with Eqn. (1.25) for a given unity gain fourth-order all-pole lowpass characteristic. The formulae are:

$$\tau_1 = B_1, \tau_2 = \frac{B_2}{B_1}, \quad \tau_3 = \frac{B_3}{B_2}, \quad \tau_4 = \frac{B_4}{B_3} \tag{1.32}$$

for the IFLF structure and

$$\tau_1 = B_1 - \frac{B_3}{B}, \tau_2 = \frac{B}{B_1 - \frac{B_3}{B}}, \quad \tau_3 = \frac{B_3}{B}, \quad \tau_4 = \frac{B_4}{B_3}, \quad B = B_2 - \frac{B_1 B_4}{B_3} \tag{1.33}$$

for the LF structure.

As can be seen from Eqn. (1.27) the network performance is a function of F. Different F matrices will lead to different transfer characteristics. F is also linked with filter structures, and different F matrices will correspond to different architectures. Thus the relationship between the performance and the structure is established through the feedback matrix.

Note that both filter structures in Figure 1.10 have the same number of differential OTAs and grounded capacitors. Thus, they have the same power consumption and chip area for the same OTA. But other performances will be different due to the difference in their feedback connections.

Analysis and comparisons of magnitude sensitivity, maximum input signal, noise and OTA nonideality effects of the IFLF and LF all-pole OTA-C filters have been carried out for the realisation of the fourth-order unity gain Butterworth characteristic with a cut-off frequency of 2 MHz using a double crossed quad CMOS OTA [32, 33]. Both theoretical and transistor level simulation results show that the LF filter is superior to the IFLF filter in all four performances. Comparison with the cascade architecture has also been conducted [30, 33]. Advantages and disadvantages of the LF, IFLF and cascade structures have been identified, again with the LF configuration showing the overall best performance.

1.4.2 g_m-C synthesis of transmission zeros

The synthesis of the general transfer function requires that the numerator coefficients are controllable separately from each other, which ensures that any zeros can be achieved. The numerator should also be adjustable separately from the denominator which is determined by the n integrators, to keep the poles unchanged while tuning zeros.

For a given input we may combine the different node outputs with a summation OTA network to give the overall circuit output, or for a fixed output distribute the overall input onto different circuit nodes using an OTA distribution network. A general transfer function can thus be obtained. Then by properly selecting the summation or distribution weights for respective cases one may attain any filter characteristics. In the following we will introduce the distribution method.

The voltage signal is applied to circuit nodes by an input OTA network as shown in Figure 1.11. Using the relations in Eqn. (1.26) and noting that now the input related vector becomes $B = [\beta_1\beta_2 \ldots \beta_n]^t$, where $\beta_j = g_{aj}/g_j$, $j = 1, 2, \ldots, n$, we can derive the transfer function as [25]

$$H(s) = \frac{V_{out}}{V_{in}} = \frac{1}{|A(s)|}\sum_{j=1}^{n} \beta_j A_{jn}(s) \tag{1.34}$$

where A_{jn} (s) represent the cofactors of $A(s)$.

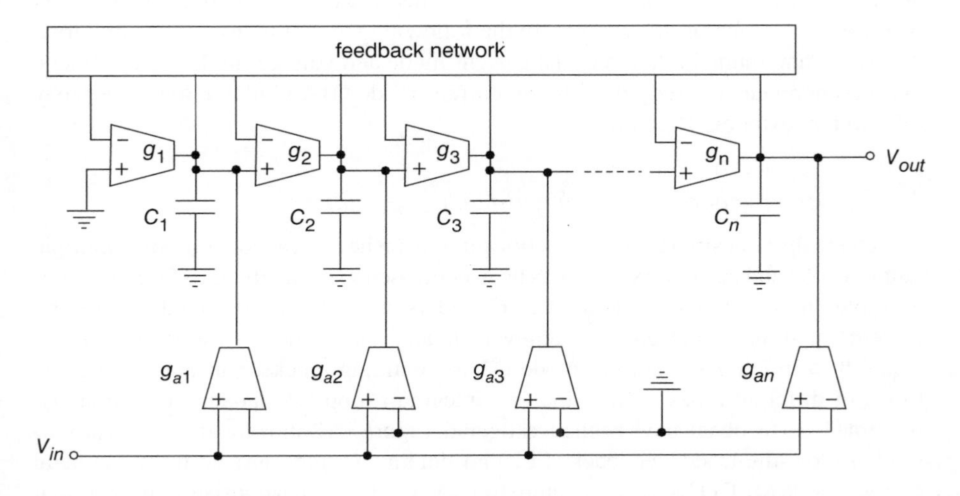

Figure 1.11 MLF OTA-C filters with input distribution OTA network

Transmission zeros can be realised by adjusting β_j, that is, g_{aj}. If a specific function is realised, the structures and design formulae can be simplified. In particular, canonical feedback may be used and some of the distribution OTAs may be removed. Also, the method may be advantageous over other methods in many cases. For example, for the odd-order elliptic filter based on the input distribution method, canonical IFLF or LF configurations can be used and only those distribution OTAs with the odd number subscripts are needed. Thus the total number of OTAs required will decrease to $n + (n + 1)/2$, which is smaller than the numbers of OTAs required by the cascade and ladder simulation methods [31, 67]. As another example, an nth-order linear phase filter with two real zeros' gain boost used in computer hard disk drive systems [28, 29] can be realised with only $n + 2$ OTAs in total, but such filters cannot be realised by simulating lossless LC ladders, which can only realise imaginary axis zeros. The MLF structures with their poles determined by τ_j and f_{ij} and zeros controlled arbitrarily by g_{aj} in β_j may be particularly useful for adaptive continuous-time equalisation applications [56, 58, 59] (see Chapter 6). Also, these structures use only grounded capacitors and have a grounded capacitor on each node, thus having reduced parasitic effects. The ladder simulation method either contains floating capacitors or a very large number of OTAs and suspending nodes.

1.5 Current-mode g_m-C filters

In the synthesis of conventional OTA-C filters, signals at the building block level are treated as electrical voltages, and therefore voltage integrators and voltage amplifiers are involved and feedback is also voltage type. Current-mode OTA-C realisations of filter functions are based on current integrators, current amplifiers and current feedback, with signals in the circuit also being currents. Current-mode filters have received substantial attention [3, 62–72]. Detailed treatment of current-mode filters is beyond the scope of this chapter. In the following we will focus on the similarity, difference and mutual relation of the current-mode and voltage-mode OTA-C filters. For a comprehensive study of different current-mode OTA-C filters, the reader may refer to the textbook, Reference 3.

1.5.1 Direct generation of current-mode g_m-C filters

One can realise a desired transfer function directly in the current domain using multiple output OTAs and capacitors. This has been addressed extensively in Reference 3. For example, in the simulation of passive LC ladders, we can convert mixed voltage and current equations to current-only equivalents and realise the corresponding current signal flow graph using current-mode OTA-C building blocks. Current-mode g_m-C filters of different types such as the two integrator loop, LC ladder simulation and multiple loop feedback have been investigated. Figure 1.12 shows a 4th-order current-mode follow-the-leader feedback (FLF) all-pole g_m-C filter derived from a general current-mode MLF OTA-C architecture [66]. As can be seen, it consists of four current integrators with current input, unity current feedback and current output. Balanced

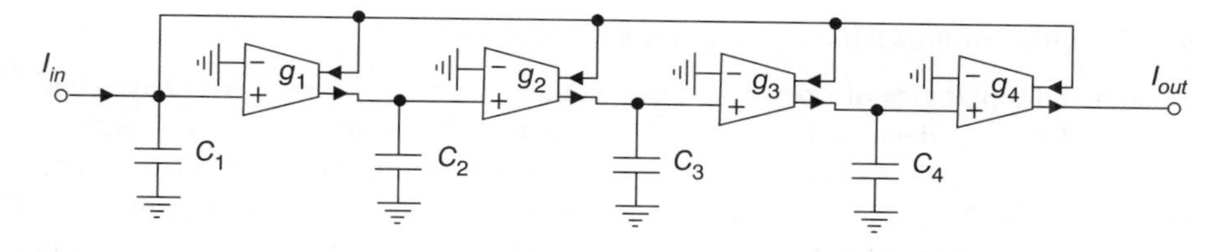

Figure 1.12 Fourth-order current-mode FLF all-pole g_m-C filter

structures of current-mode g_m-C filters require two-input four-output OTAs and/or two-input two-output OTAs.

1.5.2 Adjoint method for current-mode g_m-C filter design

Current-mode filters can be derived from voltage-mode OTA-C filters using the adjoint circuit concept [71, 72]. The former is the adjoint of the latter and vice versa. The process for obtaining the adjoint circuit, that is the current-mode OTA-C filter, of a known voltage-mode OTA-C filter from the circuit viewpoint can be conducted in the following way: leave all capacitors unchanged, reverse all OTAs, but retain the transconductance, interchange the circuit input and output ports, and turn port signals to their duals, that is, from voltage to current, with all circuit parameters gs and Cs being kept unchanged. For example, the current-mode FLF filter in Figure 1.12 is the adjoint circuit of the voltage-mode IFLF OTA-C filter in Figure 1.10(a).

Clearly an m-input n-output OTA in the current-mode circuit will correspond to an n-input m-output OTA in the original voltage-mode circuit. Current integrators and amplifiers are associated with voltage equivalents. Structurally, for example, the current-mode FLF OTA-C structure corresponds to the voltage-mode IFLF OTA-C filter. Voltage-mode OTA-C filters are more convenient for voltage signal processing with voltage inputs to OTA input terminals and voltage outputs from circuit nodes, while current-mode OTA-C structures complementarily are more straightforward for current processing with current inputs to circuit nodes and current outputs from OTA output terminals. Realising the same transfer function, the current-mode filter and its voltage-mode counterpart have identical sensitivities, but may have differences in other performance aspects such as high frequency capability and dynamic range [36, 72].

1.6 Design examples of g_m-C filters

In this section a high performance fully differential OTA is first described. Two filter design examples are then presented based on passive LC ladder simulation and IFLF structures. The simulations were conducted assuming use of a 1.2 μm CMOS technology.

1.6.1 Operational transconductance amplifiers

Design and topologies of operational transconductance amplifiers (OTA) have been discussed in some detail in [5–7]. The design of high frequency g_m-C filters requires high performance OTAs. The integrator is the main building block of active filters. One of the major problems in high frequency applications is the phase error of the integrator. In order to keep the phase as close as possible to -90 degrees, a wideband OTA with sufficiently high DC gain is required. If this can be achieved, the Q-tuning circuit for filters will not be required. To improve the CMRR, PSRR and dynamic range of integrated filters, fully differential structures are used. This requires multiple input multiple output fully differential transconductors. In filter design, use of multiple input OTAs can also reduce the number of active components, chip area and power consumption.

A CMOS fully differential multiple input OTA with high DC gain and wide bandwidth is shown in Figure 1.13 [9]. This OTA combines the linear cross-coupled quad input stage with the enhanced folded-cascode circuit to increase the output resistance of the amplifier. DC gain enhancement is obtained without significant bandwidth limitation.

The input stage of the OTA consisting of transistors M1a–M4a and M1b–M4b in Figure 1.13 is in fact a linear V-I converter. Using the standard square-law model for

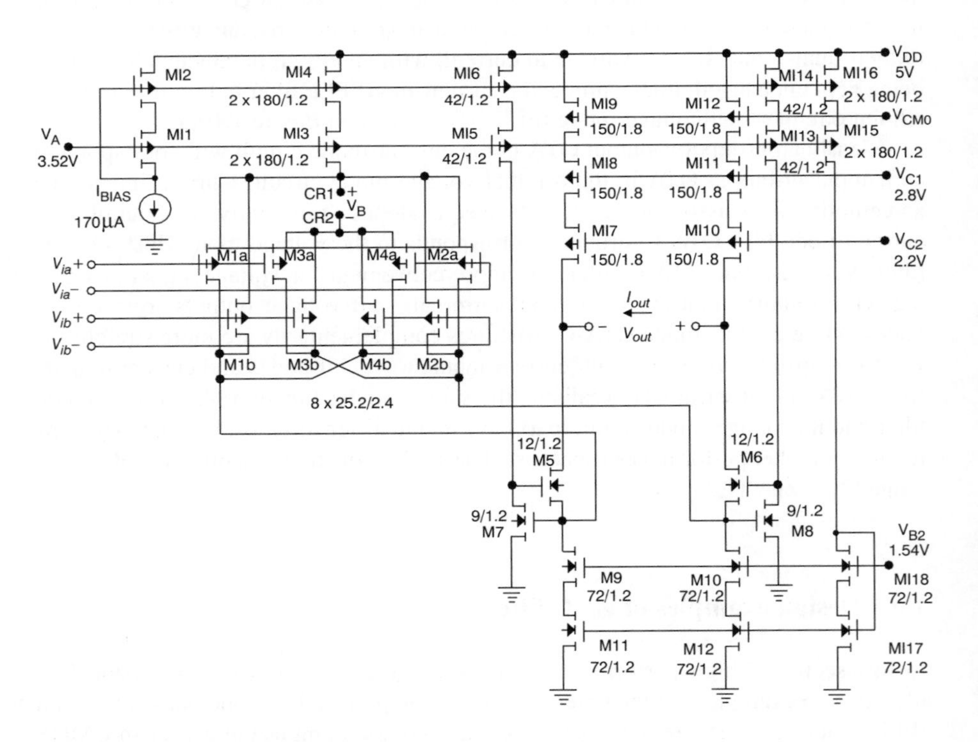

Figure 1.13 Fully differential multiple input OTA

MOS transistors in their saturation region and assuming that the transistors M1a–M4a and M1b–M4b have the same dimensions W and L, the differential output current, I_{out}, can be expressed as:

$$I_{out} = kV_B \left[(V_{ia+} - V_{ia-}) + (V_{ib+} - V_{ib-}) \right] \qquad (1.35)$$

where $k = 0.5\,\mu_o C_{ox}(W/L)$ is the transconductance parameter, and μ_o, C_{ox}, W and L are the mobility, oxide capacitance per unit area, and channel width and length, respectively. V_B is the voltage of the floating DC voltage source connected between points C_{R1} and C_{R2} and, according to Figure 1.13, $V_{ida} = (V_{ia+} - V_{ia-})$ and $V_{idb} = (V_{ib+} - V_{ib-})$ are the differential input voltages.

This circuit can be considered as two identical transconductance stages with output nodes connected in parallel and exhibits a perfectly linear transconductance of value $g_m = kV_B$. The V-I converter is tunable by varying bias voltage V_B, with detailed analysis being given in Reference 10. The control of the voltage V_B is actually achieved by varying a DC current, I_{CF}, in the floating voltage source circuit [9].

The output stage employs an enhanced folded-cascode circuit composed of transistors M5–M12. Transistors MI17 and MI18 form the improved cascode current mirror. Negative current-shunt feedback is used for M5/M6 to increase the output resistance. DC gain enhancement by current feedback in the output cascode has no significant effect on the high frequency response. The common-mode voltage at the output nodes is stabilised by a typical common-mode feedback loop [9]. The desired value is 2.5 V, i.e. a half of the power supply voltage V_{DD}. V_{CM0} in Figure 1.13 is connected to the common-mode circuit.

A significant improvement in linearity and distortion reduction of the OTA in Figure 1.13 is obtained by cross-coupling differential MOS pairs. However, cross-coupling increases the noise of the OTA due to the effect of the g_m subtraction. There is trade-off between noise performance and large-signal handling capability at the input of the cross-coupled OTA [10].

Mobility reduction, body effects and channel length modulation can cause distortion in the linear transfer function in Eqn. (1.35). An analysis of the influence of second-order effects on a similar transconductance stage has been reported in Reference 10. A scaling condition concerning mobility and body effects can be derived under which the input stage remains linear [9]. The condition is a function of tuning voltage V_B, hence the scaling is possible only for a given transconductance g_m of the OTA. The channel length modulation effect of transistors M1a–M4b can be neglected due to the application of current-shunt feedback in the output stage.

The OTA was simulated using SPICE Level 3 transistor models for the 1.2 μm AMI ABN CMOS process. The bias current flowing through the input stage is equal to 340 μA. All p- and n-channel transistors have their bulks connected to V_{DD} and GND, respectively. The main parameters of the OTA in Figure 1.13 obtained from the simulation are presented in Table 1.1 [9]. As can be seen, the transconductance may be tuned over a decade. Implementation of the enhanced folded cascode in the output stage results in an increase of the output resistance by 11 times, which increases the DC voltage gain significantly. The worst-case PSRR and worst-case CMRR were

Table 1.1 Simulated results of the OTA

Parameters	Value at $V_{DD} = 5\,V$
Transconductance range	5–50 μS for I_{CF} 9 – 70 μA
DC voltage gain	50–70 dB for I_{CF} 9 – 70 μA
THD at 10 MHz, with $V_{id} = 0.5$ V (magnitude)	< -47.1 dB for I_{CF} 9–70 μA
CMRR	>46 dB
PSRR	>62 dB
Equivalent input capacitance C_i	0.048 pF
Equivalent output capacitance C_o	0.033 pF
Power consumption	7.65 mW

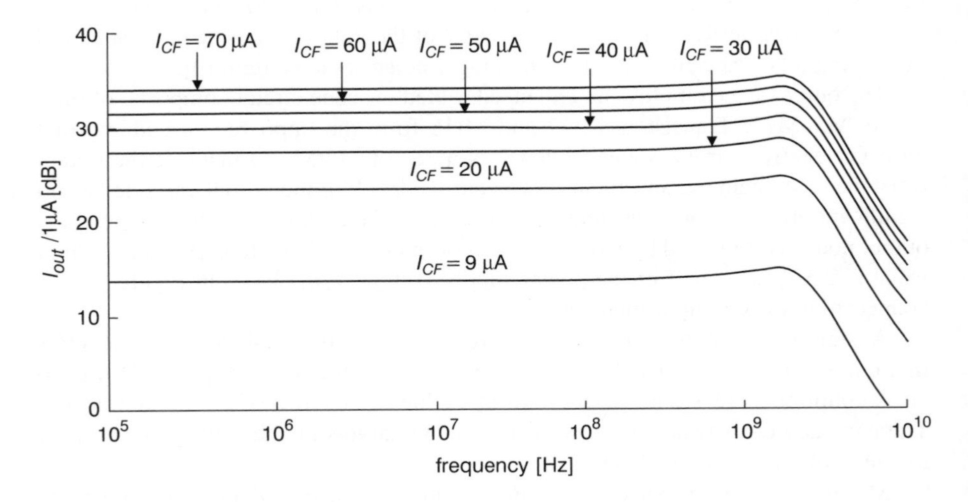

Figure 1.14 Simulated frequency response of the OTA (normalised to 1 μA*).*

simulated assuming equal 1 per cent mismatches in all transconductance parameters k_n and k_p, as well as in all threshold voltages V_{Tn} and V_{Tp} of MOS devices.

Figure 1.14 shows the OTA short-circuit frequency response [9]. The 3 dB frequency is in the GHz region. Therefore the OTA can be used for high frequency signal processing. The very high DC gain is another advantage, which makes the amplifier very useful for systems without Q-tuning.

1.6.2 Third-order elliptic g_m-C filter based on ladder simulation

A tunable 40 MHz third-order elliptic filter was designed and simulated using the OTA in Figure 1.13. The normalised passive prototype and the corresponding active implementation are given in Figures 1.15 and 1.16, respectively [9]. The input OTA

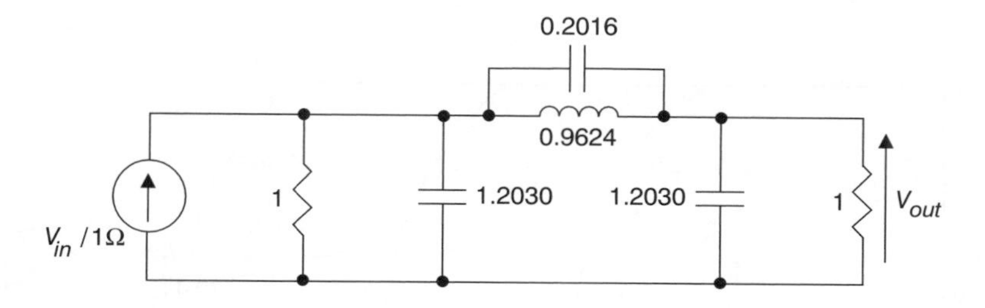

Figure 1.15 RLC prototype of third-order elliptic filter with current input

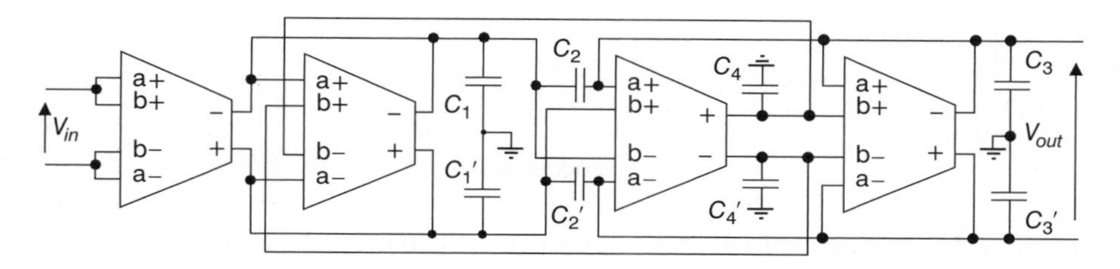

Figure 1.16 Balanced g_m-C implementation of third-order elliptic filter

has twice the transconductance to compensate for the 6 dB loss in the passband. It can be seen that at every node two or four OTA stages share a capacitor. Realisation based on two input pair OTAs results in half the number of active elements.

OTA parasitic capacitances must be taken into account when designing a filter for a high frequency range. Fortunately, in the filter structure of Figure 1.16, all OTA input and output capacitances and designed capacitances are in parallel. Subtracting the parasitic capacitances from the designed capacitances gives the real values of the capacitors:

$$C_1 = C_1' = 638\,\text{fF} - 2(2C_i + 2C_o) = 314\,\text{fF},$$
$$C_3 = C_3' = 638\,\text{fF} - 2(2C_i + C_o) = 380\,\text{fF},$$
$$C_4 = C_4' = 511\,\text{fF} - 2(2C_i + C_o) = 253\,\text{fF},\ C_2 = C_2' = 115\,\text{fF}$$

The simulation results are shown in Figure 1.17 [9]. Although the filter is designed to a particular frequency, it is possible to tune the cut-off frequency by varying I_{CF}, in a wide range from 4 MHz up to 40 MHz. High attenuation at transmission zeros is achieved due to the excellent OTA output resistance. The attenuation of the RLC prototype in the stopband equal to 28 dB is also achieved in the active implementation.

1.6.3 Seventh-order all-pole multiple loop feedback g_m-C filters

A fully balanced seventh-order lowpass IFLF g_m-C filter is presented in Figure 1.18 [9]. The OTA in Figure 1.13 is used. Symbols 2× and 4× mean 2 or

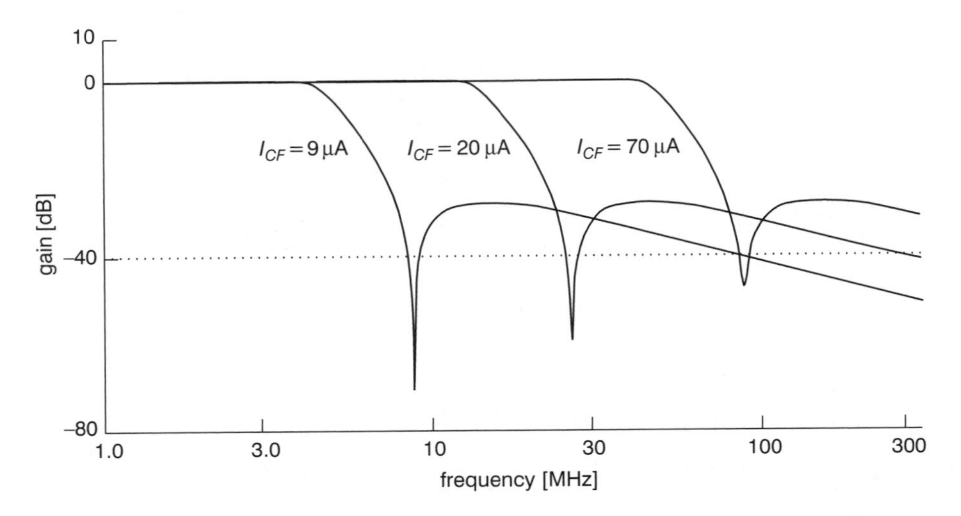

Figure 1.17 Simulated frequency response of g_m-C third-order elliptic filter

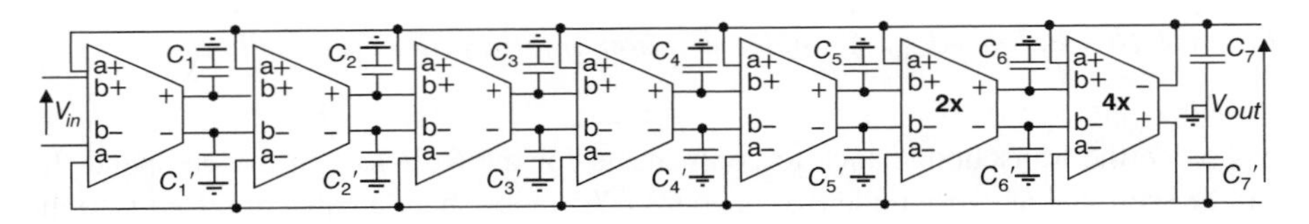

Figure 1.18 Fully balanced seventh-order lowpass IFLF g_m-C filter

4 unit OTAs connected in parallel. The cut-off frequency is chosen to be 4 MHz. With the transconductance of value equal to 50 µS at $I_{CF} = 7$ µA, the following values of capacitors can be obtained:

$$C_1 = C_1' = 12.64\text{ pF} - 2(C_o + C_i) = 12.48\text{ pF}$$
$$C_2 = C_2' = 5.86\text{ pF} - 2(C_o + C_i) = 5.7\text{ pF}$$
$$C_3 = C_3' = 3.62\text{ pF} - 2(C_o + C_i) = 3.46\text{ pF}$$
$$C_4 = C_4' = 2.39\text{ pF} - 2(C_o + C_i) = 2.23\text{ pF}$$
$$C_5 = C_5' = 1.71\text{ pF} - 2(C_o + 2C_i) = 1.45\text{ pF}$$
$$C_6 = C_6' = 2.12\text{ pF} - 2(2C_o + 4C_i) = 1.6\text{ pF}$$
$$C_7 = C_7' = 3.04\text{ pF} - 2(4C_o + 11C_i) = 1.72\text{ pF}$$

In the range of bias current I_{CF} of the OTA, the cut-off frequency of the filter varies by ten times, as can be seen from the simulated frequency response in Figure 1.19 [9].

In general, there is a simple way to introduce gain-boost real-axis zeros or any desired transmission zeros to the transfer function by adding an input or output OTA

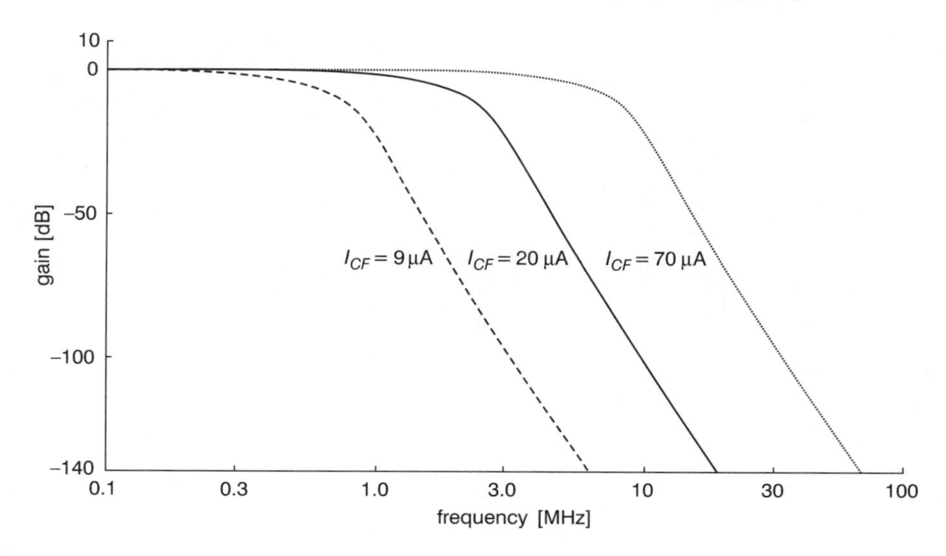

Figure 1.19 Simulated frequency response of the balanced IFLF g_m-C filter

network [3, 23–26] as discussed in Section 2.4.2. For example, a seventh-order filter with equiripple group delay and 3 dB gain-boost zeros and a tenth-order equaliser based on the IFLF structure with additional two input OTAs were implemented in [28, 29].

1.7 Automatic tuning of g_m-C filters

On-chip automatic tuning is crucial for the designed filter to operate as desired due to process parameter variations, thermal effects and mismatches [4]. Having an OTA of both tunable and programmable transconductance (g_m) and a programmable capacitor, it is possible to build g_m-C filters for a very wide frequency range [73, 74]. For most CMOS processes it is not possible to realise a capacitor, operating at high frequencies, with programmable capacitance over a wide range. Thus it is very important to have a widely programmable transconductance. On-chip automatic tuning of filters will be thoroughly dealt with in Chapter 7. In this section, we discuss some aspects of the issue including programmability and configurability in order to complete the discussion of the g_m-C technique and to provide some background for Chapters 2–6.

1.7.1 Tunable and programmable OTA

The simplified schematic diagram of an OTA based on two cross-coupled differential MOS pairs [10] and digitally programmable current mirrors is shown in Figure 1.20 [74]. Using the standard square-law model for MOS devices and assuming that the current gains of the programmable current mirrors are equal to A, the output current

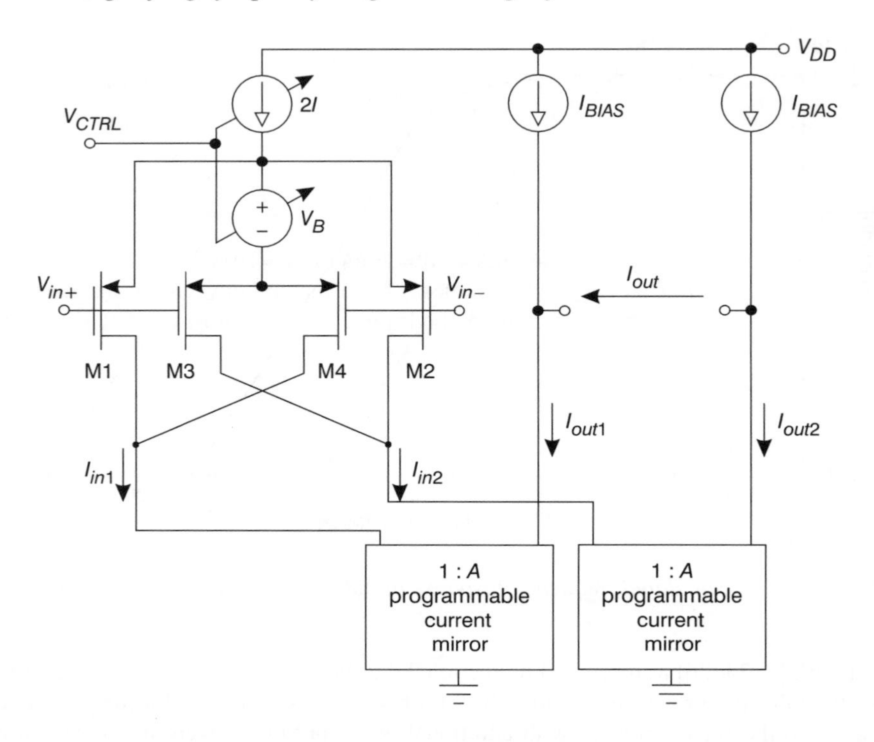

Figure 1.20 Simplified schematic diagram of programmable CMOS OTA

of the OTA can be derived as

$$I_{out} = kV_B A(V_{in+} - V_{in-}) = kV_B AV_{id} = g_m V_{id} \tag{1.36}$$

where $k = 0.5\,\mu_o C_{ox} W/L$ is the transconductance parameter of transistors M1–M4, having the same W/L ratios, V_B is the voltage of the floating DC source, A is the programmable current ratio and $V_{id} = V_{in+} - V_{in-}$ is the differential input voltage. The overall transconductance g_m of the OTA is then obtained from Eqn. (1.36), and is given by

$$g_m = kV_B A \tag{1.37}$$

From Eqn. (1.37) it can be seen that two methods can be used to change the transconductance g_m. One method is to tune the floating bias voltage V_B by an analogue voltage V_{CTRL}. The other is to make the gain A of the output current mirrors programmable in a digital way. The total tuning range will be the product of $V_B A$. For example, the transconductance parameter g_m of the OTA can be adjusted over 600 times: 31 times by the digitally programmable current mirror using a 5-bit control word and more than 20 times by analogue voltage V_{CTRL}, which changes the transconductance of the input cross-coupled MOS pairs [74].

In filter tuning, the transconductance parameters of all of the OTAs are controlled by the external analogue voltage V_{CTRL} in the input differential stage, adjusted by the automatic tuning circuitry, and by digital switching of the output current mirrors. While voltage V_{CTRL} is common to all amplifiers in the filter, each OTA can have its own dividing factor for the output stage. To obtain an accurate and wide range g_m/C ratio, a programmable capacitor array can also be used.

1.7.2 An analogue tuning scheme

The most popular method for tuning a continuous-time OTA-C filter is the master-slave structure [4]. This approach assumes a good match between the master OTA and the slave OTAs. Adaptive tuning methods have also been utilised [59] and, more recently, some attractive tuning methods for high-Q bandpass IF/RF filters have been proposed [45–48].

Figure 1.21 shows an automatic tuning circuit proposed in [9]. This is a typical phase locked loop (PLL) with an on-chip loop filter, thus requiring no external components. The open-loop transfer function of the PLL can be written as

$$A(s) = K_{PD} \cdot K_F(s) \cdot \frac{K_{VCO}}{s} \tag{1.38}$$

where K_{PD} is the phase comparator gain, $K_F(s)$ represents the lowpass filter transfer function and K_{VCO} is the oscillator gain. K_{VCO} is divided by s, because the frequency of the VCO output is converted to the phase at the input of the phase comparator.

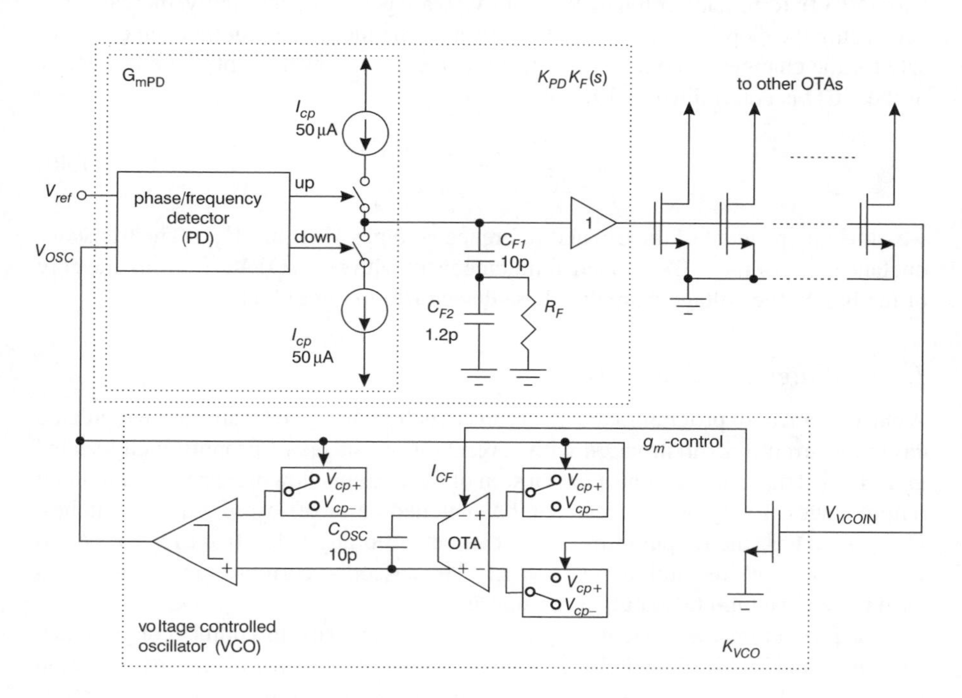

Figure 1.21 Automatic filter tuning circuit structure [9]

The phase detector circuit is in fact a phase-frequency one with two edge-triggered D-type flip-flops and a 3-state output stage. By using current sources as the charge pump the dead zone can be reduced to zero. The PLL loop filter is of the lag-lead type, where the charge pump works as an active part. The transfer function of this block can be expressed as [9]

$$K_{PD} \cdot K_F(s) = \frac{1 + sR_F(C_{F1} + C_{F2})}{s(C_{F1}/G_{mPD})(1 + sR_F C_{F2})} = \frac{1 + s\tau_2}{s\tau_1(1 + s\tau_3)} \tag{1.39}$$

G_{mPD} is the transconductance of the charge pump, which is equal to $I_{CP}/2\pi$, i.e. 8 nS. A simple RC filter can result in a very small phase margin. If a zero associated with time constant τ_2 is introduced to the filter, the phase margin of the PLL can be increased to a desired value, for example, 60 degrees, avoiding instability. Adding C_{F2} improves the attenuation of the current spikes from the phase detector. Since the typical ratio of C_{F2}/C_{F1} is about 0.1, C_{F1} can be realised as the floating capacitor between poly1 and poly2, and C_{F2} as the parasitic one to the substrate. R_F represents the resistance equivalent to the connection of several transistors biased by a certain current. It should be noted that the PLL lock range for this type of phase detector is equal to the capture range and is independent of the lowpass filter.

The connection of the tunable g_m-C integrator based on the same OTA as those inside the filter and the comparator with hysteresis forms the voltage-controlled oscillator (VCO). Reference voltages V_{cp+} and V_{cp-} are generated internally. Increasing or decreasing the slope of the triangle waveform at the integrator's output can compensate for the changes of comparator's trip voltages with power supply. The frequency of the VCO is determined by [9]

$$f = \frac{g_m(I_{CF})}{2C_{OSC}} = \frac{g_m(V_{VCOIN})}{2C_{OSC}} \tag{1.40}$$

which is independent of the trip and reference voltages V_{cp+} and V_{cp-}. The transconductance, g_m, of the OTA is tuned simultaneously with slave OTAs (forming the core of the IC) by the voltage from the phase detector/loop filter block.

1.7.3 A digital tuning method

A current mirror of programmable gain can be realised using the well known regulated cascode current mirror structure with several output stages of proportionally scaled gains [74]. The output current I_{out} is a sum of the currents flowing through individual output stages. Every output stage can be switched on or off by MOSFET switches. Using n stages, the output current I_{out} can be set to any value from $I_{out}^{min} = I_{in}$ to $I_{out}^{max} = nI_{in}$ with resolution $\Delta I_{out} = I_{in}$. The cascoded current mirror structure is used to achieve high OTA output resistance.

The programmable capacitor array can be constructed using a number of grounded capacitors and MOSFET switches in a binary format [9]. The switch transistors should have a width large enough to achieve a phase of the capacitor impedance in the range of $-90^\circ \pm 1^\circ$ for a wide frequency range.

1.7.4 *Configurable and FPAA-based g_m-C filters*

The analogue and digital tuning schemes keep the filter structure unchanged during the tuning. Similar to the field programmable analogue array (FPAA) concept, a universal filter circuit can be built which may be configured/programmed to realise different filter functions. Here configuration of the circuit by programming changes the filter structure. Configurable g_m-C filter design has been investigated in the literature [73, 74]. A single processor of different functions is constructed with OTAs, capacitors and MOSFET switches. There are two types of switches, that is, those for programming the output mirrors in the OTA and capacitors in digital tuning and those outside the OTAs and capacitors for configuring the processor to achieve a certain filter function.

Two types of signal occur in such filtering systems: analogue and digital. Analogue signals include the working signals and analogue tuning signals, whilst digital signals are the control signals embracing those for digital tuning and those for configuring. Thus, in programmable and configurable filter systems, there are analogue bus and digital busses, which serve as paths for analogue and digital control signals [73].

1.8 Conclusions

This chapter has addressed some topics in g_m-C filter design. Two integrator loop, LC ladder-based and multiple loop feedback OTA-C filter structures have been studied. Design examples using a wideband high output resistance OTA have been presented. Current-mode g_m-C filters have been discussed and compared with voltage-mode counterparts. Tuning issues including both analogue and digital schemes have been discussed.

In recent years low voltage/power OTAs and OTA-C filters have attracted great interest [12, 13, 38, 49, 50, 70]. Low voltage/power design is important for portable equipment. To push the working frequency of CMOS OTA-C filters up to the GHz range remains a challenge. Integrated OTA-C filter design for both low voltage and high frequency operation will continue to be an active research topic. Dynamic range needs to be further enhanced for applications such as communication front-ends. High-Q RF/IF bandpass OTA-C filter design is a specific challenge for wireless communication applications [45–48]. On the other hand, linear phase OTA-C lowpass filters/equalisers in the hundred MHz range for computer hard disk drive systems [51–58] are commercially available.

Finally, we mention that active filters using a single OTA, resistors and capacitors have been presented in Reference 3. Single OTA active filters have advantages such as low power consumption, noise, parasitic effects and cost [76, 77]. The resistors can be implemented using MOSFETs, which results in g_m-MOSFET-C filters which will be discussed in Chapter 2. Very wide tuning ranges can be achieved [78] using this technique. Although single OTA active filters have not attracted wide attention, their proven advantages may encourage researchers to look further at this method. Of course, replacing resistors by OTA simulated equivalents, OTA-C filters can be obtained from single-OTA-RC counterparts [3].

1.9 References

1 SUN, Y. (Guest Editor): Special Issue on 'High-frequency Integrated Analogue Filters', *IEE Proceedings: Circuits, Devices and Systems*, Vol. 147, No. 1, February 2000.
2 TSIVIDIS, Y. and VOORMAN, J. O. (Editors): 'Integrated Continuous-time Filters', IEEE Press, 1993.
3 DELIYANNIS, T., SUN, Y. and FIDLER, J. K.: 'Continuous-time Active Filter Design', CRC Press, Florida, USA, January 1999, ISBN: 0-8493-7893-1.
4 SCHAUMANN, R., LAKER, K. R. and GHAUSI, M. S.: 'Active Filter Design: Passive, Active and Switched-capacitor', Prentice Hall, 1990.
5 JOHNS, D. A. and MARTIN, K.: 'Analog Integrated Circuit Design', John Wiley & Sons, 1997.
6 DUPUIE, S. T. and ISMAIL, M.: 'High Frequency CMOS Transconductors', Chapter 5, in TOUMAZOU, C., LIDGEY, F. J. and HAIGH, D. G. (Eds.): 'Analogue IC Design', Peter Peregrinus Ltd, 1990.
7 SANCHEZ-SINENCIO, E. and SILVA-MARTINEZ, J.: 'CMOS Transconductance Amplifiers, Architectures and Active Filters: a Tutorial', in [1], pp. 3–12.
8 HILL, C., SUN, Y. and SZCZEPANSKI, S.: 'Low Noise Low Distortion High Frequency Fully Differential CMOS Transconductor', Proc. IEEE Asia and Pacific Conference on Circuits and Systems, Tianjin, China, December 2000.
9 GLINIANOWICZ, J., JAKUSZ, J., SZCZEPANSKI, S. and SUN, Y.: 'A High-frequency Two-input CMOS OTA for Continuous-time Filter Applications', in [1], pp. 13–18.
10 SZCZEPANSKI, S., JAKUSZ, J. and SCHAUMANN, R.: 'A Linear Fully Balanced CMOS OTA for VHF Filtering Applications', *IEEE Trans. Circuits and Systems, II*, Vol. 44, No. 3, pp. 174–187, 1997.
11 GATTI, U., MALOBERTI, F., PALMISANO, G. and TORELLI, G.: 'CMOS Triode-transistor Transconductor for High-frequency Continuous-time Filters', *IEE Proceedings: Circuits, Devices and Systems*, Vol. 141, No. 6, pp. 462–468, December 1994.
12 BLALOCK, B. J. and ALLEN, P. E.: 'A One-volt, 120-uW, 1-MHz OTA for Standard CMOS Technology', Proc. IEEE Int. Symp. Circuits and Systems, pp. 305–307, 1996.
13 LI, M. F., DASGUPTA, U., ZHANG, X. W. and LIM, Y. C.: 'A Low-voltage CMOS OTA with Rail-to-rail Differential Input Range', *IEEE Trans. Circuits and Systems, I*, Vol. 47, No. 1, pp. 1–8, January 2000.
14 HAYAHARA, E. and ENOMOTO, M.: 'Deriving Technique of Equivalent Forms from General Filter Circuits and its Application', *IEEE Trans. Circuits and Systems, I*, Vol. 46, No. 9, pp. 1037–1041, September 1999.
15 GEIGER, R. L. and SANCHEZ-SINENCIO, E.: 'Active Filter Design Using Operational Transconductance Amplifiers: a Tutorial', *IEEE Circuits and Devices Magazine*, Vol. 1, pp. 20–32, March 1985.

16 SANCHEZ-SINENCIO, E., GEIGER, R. L. and NEVAREZ-LOZANO, H.: 'Generation of Continuous-time Two Integrator Loop OTA Filter Structures', *IEEE Trans. Circuits and Systems*, Vol. 35, No. 8, pp. 936–946, 1988.
17 SUN, Y. and FIDLER, J. K.: 'Resonator-based Universal OTA-grounded Capacitor Filters', *International Journal of Circuit Theory and Applications*, Vol. 23, pp. 261–265, 1995.
18 SUN, Y.: 'Note on Two Integrator Loop OTA-C Configurations', *Electronics Letters*, Vol. 34, No. 16, pp. 1533–1534, 1998.
19 SUN, Y.: 'Second-order OTA-C Filters Derived from Nawrocki-Klein Biquad', *Electronics Letters*, Vol. 34, No. 15, pp. 1449–1450, 1998.
20 SUN, Y.: 'OTA-C Filter Design Using Inductor Substitution and Bruton Transformation Methods', *Electronics Letters*, Vol. 34, No. 22, pp. 2082–2083, 1998.
21 DE QUEIROZ, A. C. M., CALOBA, L. P. and SANCHEZ-SINENCIO, E.: 'Signal Flow Graph OTA-C Integrated Filters', Proc. IEEE Int. Symp. Circuits Systems, pp. 2165–2168, 1988.
22 SCHAUMANN, R.: 'Simulating Lossless Ladders with Transconductance-C Circuits', *IEEE Trans. Circuits Syst., II*, Vol. 45, No. 3, pp. 407–410, 1998.
23 SUN, Y. and FIDLER, J. K.: 'OTA-C Realization of General High-order Transfer Functions', *Electronics Letters*, Vol. 29, No. 12, pp. 1057–1058, 1993.
24 SUN, Y. and FIDLER, J. K.: 'Synthesis and Performance Analysis of Universal Minimum Component Integrator-based IFLF OTA-grounded Capacitor Filter', *IEE Proceedings: Circuits, Devices and Systems*, Vol. 143, pp. 107–114, 1996.
25 SUN, Y. and FIDLER, J. K.: 'Structure Generation and Design of Multiple Loop Feedback OTA-grounded Capacitor Filters', *IEEE Trans. Circuits and Systems, I: Fundamental Theory and Applications*, Vol. 44, No. 1, pp. 1–11, 1997.
26 SUN, Y. and FIDLER, J. K.: 'Fully-balanced Structures of Continuous-time MLF OTA-C Filters', Proc. IEEE Int. Conf. Electronics, Circuits and Systems, pp. 157–160, Portugal, 1998.
27 NEDUNGADI, A. P. and GEIGER, R. L.: 'High-frequency Voltage-controlled Continuous-time Low-pass Filter Using Linearised CMOS Integrators', *Electronics Letters*, Vol. 22, pp. 729–731, 1986.
28 CHIANG, D. H. and SCHAUMANN, R.: 'A CMOS Fully-balanced Continuous-time IFLF Filter Design for Read/write Channels', Proc. IEEE Int. Symp. Circuits and Systems, Vol. 1, pp. 167–170, 1996.
29 CHIANG, D. H. and SCHAUMANN, R.: 'Design of a CMOS Fully-differential Continuous-time Tenth-order Filter Based on IFLF Topology', Proc. IEEE Int. Symposium on Circuits and Systems, Vol. 1, pp. 123–126, 1998.
30 CHIANG, D. H. and SCHAUMANN, R.: 'Performance Comparison of High-order IFLF and Cascade Analogue Integrated Lowpass Filters', in [1], pp. 19–27.
31 SUN, Y., JEFFERIES, B. and TENG, J.: 'Universal Third-order OTA-C Filters', *International Journal of Electronics*, Vol. 85, No. 5, pp. 597–609, 1998.
32 SU, H. and SUN, Y.: 'Performance Comparison of Continuous-time IFLF and LF Lowpass All-pole OTA-C Filters', Proc. IEEE Asia and Pacific Conference on Circuits and Systems, Tianjin, China, December 2000.

33 SU, H., SUN, Y. and GORDON, R.: 'Performance Analysis and Comparison of High-frequency CMOS OTA-C Filters', Proc. IEE Symposium on Analogue Signal Processing, Oxford, November 2000.

34 GROENEWOLD, G.: 'The Design of High Dynamic Range Continuous-time Integratable Band-pass Filters', *IEEE Trans. Circuits and Systems*, Vol. 38, No. 8, pp. 838–852, August 1991.

35 EFTHIVOULIDIS, G., TÓTH, L. and TSIVIDIS, Y.: 'Noise in G_m-C Filters', *IEEE Trans. Circuits and Systems*, Vol. 45, No. 3, pp. 295–302, March 1998.

36 MAHATTANAKUL, J. and TOUMAZOU, C.: 'Current-mode Versus Voltage-mode G_m-C Biquad Filters: What the Theory Says', *IEEE Trans. Circuits and Systems*, Part II, Vol. 45, No. 2, pp. 173–186, February 1998.

37 ALINI, R., BASCHIROTTO, A. and CASTELLO, R.: 'Tunable BiCMOS Continuous-time Filter for High-frequency Applications', *IEEE J. Solid-State Circuits*, Vol. 27, No. 12, pp. 1905–1915, December 1992.

38 YANG, F. and ENZ, C. C.: 'A Low-distortion BiCMOS Seventh-order Bessel Filter Operating at 2.5V Supply', *IEEE J. Solid-State Circuits*, Vol. 31, No. 3, pp. 321–330, March 1996.

39 WANG, Y. T. and ABIDI, A. A.: 'CMOS Active Filter Design at Very High Frequencies', *IEEE J. Solid-State Circuits*, Vol. 25, No. 6, pp. 1562–1574, December 1990.

40 GOPINATHAN, V., TSIVIDIS, Y., TAN, K. S. and HESTER, R. K.: 'Design Considerations for High-frequency Continuous-time Filters and Implementation of an Antialiasing Filter for Digital Video', *IEEE J. Solid-State Circuits*, Vol. 25, No. 25, pp. 1368–1378, 1990.

41 CHANG, Z. Y., HASPESLAGH, D. and VERFAILLIE, J.: 'A Highly Linear CMOS G_m-C Bandpass Filter with On-chip Frequency Tuning', *IEEE J. Solid-State Circuits*, Vol. 32, No. 3, pp. 388–397, March 1997.

42 SILVA-MARTINEZ, J., STEYAERT, M. S. J. and SANSEN, W.: 'A 10.7-MHz 68-dB CMOS Continuous-time Filter with On-chip Automating Tuning', *IEEE J. Solid-State Circuits*, Vol. 27, No. 12, pp. 1843–1853, December 1992.

43 MARTIN SNELGROVE, W. and SHOVAL, A.: 'A Balanced 0.9 μm CMOS Transconductance-C Filter Tunable over the VHF Range', *IEEE Trans. Circuits and Systems, II*, Vol. 39, No. 6, March 1992.

44 NAUTA, B.: 'A CMOS Transconductance-C Filter Technique for Very High Frequencies', *IEEE Journal of Solid-State Circuits*, Vol. 27, No. 2, pp. 142–153, February 1992.

45 STEVENSON, J. M. and SANCHEZ-SINENCIO, E.: 'An Accurate Quality Factor Tuning Scheme for IF and High-Q Continuous-time Filters', *IEEE Journal of Solid-State Circuits*, Vol. 33, No. 12, pp. 1970–1978, 1998.

46 KARSILAYAN, A. I. and SCHAUMANN, R.: 'Mixed-mode Automatic Tuning Scheme for High-Q Continuous-time Filters', pp. 57–64 in [1].

47 MANETAKIS, K. and TOUMAZOU, C.: 'A 50-MHz High-Q Bandpass CMOS Filter', Proc. IEEE Int. Symp. Circuits and Systems, pp. 309–312, Hong Kong, June 1997.

48 CHOI, Y. W. and LUONG, H. C.: 'A High-Q Wide-dynamic Range CMOS IF Bandpass Filter for GSM Receivers', Proc. IEEE Int. Symp. Circuits and Systems, Switzerland, 2000.

49 DE LIMA, J. A. and DUALIBE, C.: 'On Designing Linearly-tunable Ultra-low Voltage CMOS gm-C Filters', Proc. IEEE Int. Symp. Circuits and Systems., Switzerland, 2000.

50 TAJALLI, A., ATARODI, M. and ABIDI, A. A.: 'A 1.5-V Supply, Video Range Frequency Gm-C Filter', Proc. IEEE Int. Symp. Circuits and Systems, Switzerland, 2000.

51 KHOURY, J. M.: 'Design of a 15-MHz CMOS Continuous-time Filter with On-chip Tuning', *IEEE J. Solid-State Circuits*, Vol. SC-26, No. 12, pp. 1988–1996, December 1991.

52 RAO, N., BALAN, V. and CONTRERAS, R.: 'A 3V 10-100MHz Continuous-time Seventh-order 0.05° Euiripple Linear-phase Filter', Proc. IEEE Int. Solid-State Circuits Conference, pp. 44–45, February 1999.

53 DE VEIRMAN, G. A. and YAMASAKI, R. G.: 'Design of a Bipolar 10-MHz Programmable Continuous-time 0.05° Equiripple Linear Phase Filter', *IEEE J. Solid-State Circuits*, Vol. 27, No. 3, pp. 324–331, March 1992.

54 MEHR, I. and WELLAND, D. R.: 'A CMOS Continuous-time Gm-C Filter for PRML Read Channel Applications at 150 Mb/s and Beyond', *IEEE J. Solid-State Circuits*, Vol. 32, No. 4, pp. 499–513, April 1997.

55 DEHAENE, W., STEYAERT, M. S. J. and SANSEN, W.: 'A 50-MHz Standard CMOS Pulse Equalizer for Hard Disk Read Channels', *IEEE J. Solid-State Circuits*, Vol. 32, No. 7, pp. 977–988, July 1997.

56 PAI, P. K. D., BREWSTER, A. D. and ABIDI, A. A.: 'A 160-MHz Analog Front-end IC for EPR-IV PRML Magnetic Storage Read Channels', *IEEE J. Solid-State Circuits*, Vol. 31, No. 11, pp. 1803–1816, November 1996.

57 REZZI, F., BIETTI, I., CAZZANIGA, M. and CASTELLO, R.: 'A 70 mW Seventh-order Filter with 7–50 MHz Cutoff Frequency and Programmable Boost and Group Delay Equalization', *IEEE Journal of Solid-State Circuits*, Vol. 32, No. 12, pp. 1987–1999, December 1997.

58 ALINI, R., BETTI, G., BIETTI, I., BOLLATI, G., BRIANTI, F., DATI, A. *et al.*: 'A 200-Msample/s Trellis-coded PRML Read/write Channel with Analog Adaptive Equalizer and Digital Servo', *IEEE J. Solid-State Circuits*, Vol. 32, No. 11, pp. 1824–1838, November 1997.

59 CARUSONE, A. and JOHNS, D. A.: 'Analogue Adaptive Filters: Past and Present', in [1], pp. 82–90.

60 LU, Y., GREER, N. P. J. and SEWELL, J. I.: 'Efficient Design of Ladder-based Transconductor-capacitor Filters and Equalisers', *IEE Proc. Circuits Devices and Systems*, Vol. 142, No. 4, pp. 263–272, August 1995.

61 GOPINATHAN, V. and TSIVIDIS, Y.: 'Design Considerations for Integrated Continuous-time Video Filters', Proc. IEEE Int. Symp. on Circuits and Systems, pp. 1177–1180, 1990.

62 DE FIGUEIREDO GARRIDO, N., FRANCA, J. E. and KENNEY, J. G.: 'A Comparative Study of Two Adaptive Continuous-time Filters for Decision Feedback

Equalization Read Channels', Proc. IEEE Int. Symp. on Circuits and Systems, pp. 89–92, Hong Kong, June 1997.

63 AL-HASHIMI, B. M., DUDEK, F. and SUN, Y.: 'Current-mode CMOS Delay Equaliser Design Using Multiple Output OTAs', *International Journal of Analog Integrated Circuits and Signal Processing*, Vol. 24, pp. 163–169, August 2000.

64 SUN, Y. and FIDLER, J. K.: 'Structure Generation of Current-mode Two Integrator Loop Dual Output-OTA Grounded capacitor Filters', *IEEE Transactions on Circuits and Systems, II: Analog and Digital Signal Processing*, Vol. 43, No. 9, pp. 659–663, 1996.

65 SUN, Y. and FIDLER, J. K.: 'Current-mode OTA-C realization of Arbitrary Filter Characteristics', *Electronics Letters*, Vol. 32, No. 13, pp. 1181–1182, 1996.

66 SUN, Y. and FIDLER, J. K.: 'Current-mode Multiple Loop Feedback Filters Using Dual Output OTAs and Grounded Capacitors', *International Journal of Circuit Theory and Applications*, Vol. 25, No. 2, pp. 69–80, 1997.

67 SUN, Y. and FIDLER, J. K.: 'General Current-mode MLF MO-OTA-C Filters', Proc. European Conference on Circuit Theory and Design, Stresa, Italy, 1999.

68 RAMIREZ-ANGULO, J., ROBINSON, M. and SANCHEZ-SINENCIO, E.: 'Current-mode Continuous-time Filters: Two Design Approaches', *IEEE Trans. Circuits and Systems, II*, Vol. 39, pp. 337–341, 1992.

69 RAMIREZ-ANGULO, J. and SANCHEZ-SINENCIO, E.: 'High Frequency Compensated Current-mode Ladder Filters Using Multiple Output OTAs', *IEEE Trans. Circuits and Systems, II*, Vol. 41, pp. 581–586, 1994.

70 SANCHEZ-SINENCIO, E. and SMITH, S. L.: 'Continuous-time Low-voltage Current-mode Filters', Chapter 11, in SANCHEZ-SINENCIO, E. and ANDREOU, A. G. (Eds.): 'Low-voltage/Low-power Integrated Circuits and Systems', IEEE Press, 1999.

71 SUN, Y. and FIDLER, J. K.: 'Some Design Methods of OTA-C and CCII-RC Filters', Proc. IEE Saraga Colloquium on Electronic Filters, pp. 7/1–7/8, London, UK, 1993.

72 ROBERTS, G. W. and SEDRA, A. S.: 'Adjoint Networks Revisited', Proc. IEEE Int. Symp. on Circuits and Systems, pp. 540–544, 1990.

73 LOH, K. H., HISER, D. L., ADAMS, W. J. and GEIGER, R. L.: 'A Versatile Digitally Controlled Continuous-time Filter Structure With Wide Range and Fine Resolution Capability', *IEEE Trans. Circuits and Systems, II*, Vol. 39, pp. 265–276, 1992.

74 PANKIEWICZ, B., WÓJCIKOWSKI, M., SZCZEPAŃSKI, S. and SUN, Y.: 'A CMOS Field Programmable Analog Array and its Application in Continuous-time OTA-C Filter Design', Proc. IEEE Int. Symp. on Circuits and Systems, Sydney, May 2001.

75 LI, D. and TSIVIDIS, Y.: 'Active LC Filters on Silicon', in [1], pp. 49–56.

76 HILL, C., SUN, Y. and SU, H.: 'Design of a Low Power HF CMOS Filter', IEE Colloquium on Low Power IC Design, London, January 2001.

77 SCHMID, H. and MOSCHYTZ, G. S.: 'Active-MOSFET-C Single-amplifier Biquadratic Filters for Video Frequencies', in [1], pp. 35–41.

78 GROENEWOLD, G.: 'Low-power MOSFET-C 120 MHz Bessel Allpass Filter With Extended Tuning Range', in [1], pp. 28–34.

Chapter 2

The MOSFET-C technique: designing power efficient, high frequency filters

Mihai Banu and Yannis Tsividis

2.1 Introduction

The MOSFET-C design technique [1–11] was introduced in 1983 [1] as a simple and effective method for integrating high complexity continuous-time filters on a single CMOS chip with large dynamic range and accurate frequency response. Prior to this invention, precision fully integrated analogue filters for communications, mass storage, consumer electronics and other applications, were realised either with sampled-data circuits such as switched capacitors [12] or with open-loop active blocks operating in continuous time such as g_m-C circuits (see Chapter 1 or References 8, 9 and 12 and the many references therein). The switched-capacitor technique required no filter tuning, ensured excellent precision in the frequency response, and yielded high dynamic range, but required the use of additional continuous-time input anti-aliasing filters and output smoothing filters. In addition, although switched-capacitor filters have been extremely successful in numerous low frequency applications, they are difficult to apply at high frequencies due to signal sampling. The original open-loop active-block approach was proposed for easier high frequency designs based on continuous-time operation but required on-chip tuning and suffered from excessive nonlinear effects and noise. Despite advances, these filters still have a substantially lower dynamic range than necessary in many applications [10]. MOSFET-C filters demonstrated continuous-time operation coupled with very high dynamic range. This prompted their introduction in IC products such as disk drive and consumer electronics chips.

The unique feature of MOSFET-C circuits is their fundamental similarity and first-order equivalence to certain important classical active-RC topologies. These define the class of filters using as active elements only operational amplifiers (OA) or operational transconductance amplifiers (OTA) connected in feedback configurations.

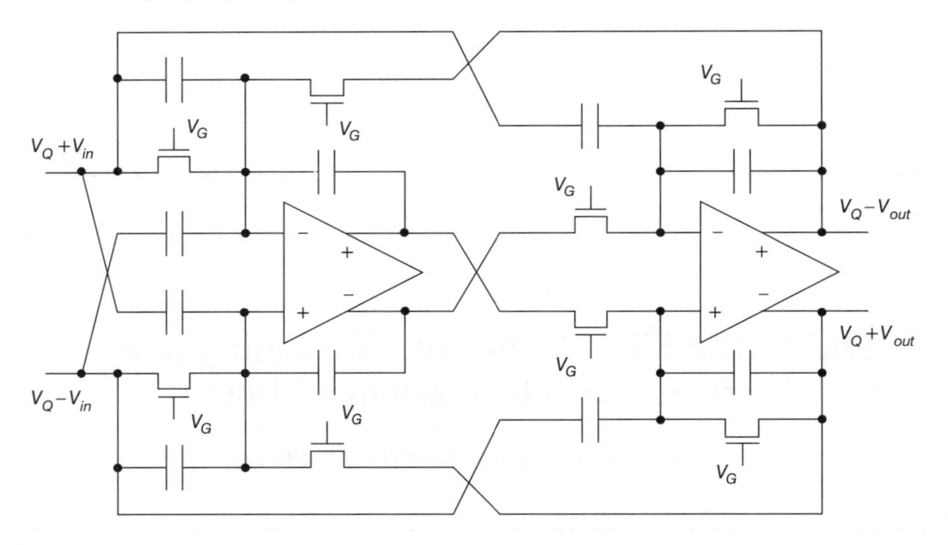

Figure 2.1 A typical MOSFET-C filter

A typical example is shown in Figure 2.1 [13], with input and output signals defined around a signal-ground level V_Q. Just as in the classical case, in MOSFET-C filters, the operational amplifier nonlinear effects are drastically reduced by high-gain negative feedback. The difference between MOSFET-C filters and classical active-RC filters is in the type of passive dissipating components utilised. Instead of resistors, MOSFET-C filters use MOS transistors operating in the triode region. The tuning capability introduced through the gate voltage is for correcting the frequency response drifts due to process and temperature variations. For small-signal operation at low enough frequencies such that the transistor parasitic capacitances can be neglected, MOSFET-C filters are indistinguishable topologically from balanced versions of their classical active-RC counterparts. For this reason, for each MOSFET-C filter there is a classical active-RC prototype.

For large signals and low frequencies as described above, the transistors in MOSFET-C filters behave like highly nonlinear tunable resistors. The strength of the MOSFET-C design technique comes from the fact that, despite the use of nonlinear elements and despite having internal nonlinear behaviour, the overall input-output filter characteristics are excellent approximations of the classical linear active-RC prototype characteristics. Some small residual nonlinear terms are present [1, 2], but the achievable dynamic range is significantly higher than that of other types of integrated continuous-time filters. For frequencies where the MOS transistor parasitic capacitances are not negligible, the filter dynamic range does not degrade significantly but the frequency response may be affected. In low-Q applications, these errors are usually compensated through simple element-value perturbations.

Traditionally, it has been suggested that MOSFET-C filters have limited application at high frequencies due to the use of high gain amplifiers. In this chapter, it will be shown that the MOSFET-C technique is in fact uniquely suited for high frequency operation when all things such as linearity and power dissipation are considered.

2.2 The MOSFET-C concept

Figure 2.2 illustrates the simplest MOSFET-C configuration using one fully differential and balanced operational amplifier (in short, balanced OA), two transistors and two linear passive feedback elements. If these elements are capacitors, the circuit in Figure 2.2 becomes an integrator. The gate and body (substrate) voltages V_G and V_B are DC quantities and all nodes, not including those inside the OA, rest at the quiescent signal-ground level V_Q when $V_{in} = 0$ (absence of signal). Naturally, this assumes that the balanced OA has a very high open-loop DC gain. Later, it will be demonstrated that, unlike the classical active-RC case, the voltage V_x, common to both OA inputs, does not remain zero in the presence of input signals.

The essential design objectives of the circuit in Figure 2.2 are the realisation of structural symmetry and the enforcement of accurate input and output voltage balancing. The first condition is accomplished by using on-chip device matching techniques such that the two transistors and the two linear feedback elements are practically identical. By definition, the balanced operational amplifier has a symmetrical functional structure and it enforces output voltage balancing. A common design technique for balancing a fully differential OA is by detecting and accurately cancelling the output common-mode signal component [14]. It is emphasised that balancing, i.e. having a negligible common-mode output component, is an extra which is not automatically met in simple fully differential topologies; rather, it must be imposed by proper amplifier design [14]. The input balancing in Figure 2.2 may be provided by the outputs of a similar MOSFET-C block driving the circuit within a filter (for example, see Figure 2.1). Next, it will be shown that structural symmetry and input/output balancing are sufficient conditions for obtaining practically linear input-output characteristics.

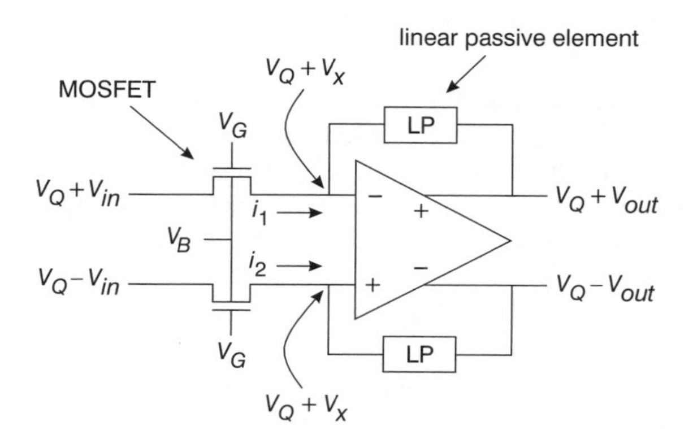

Figure 2.2 Simple MOSFET-C building block

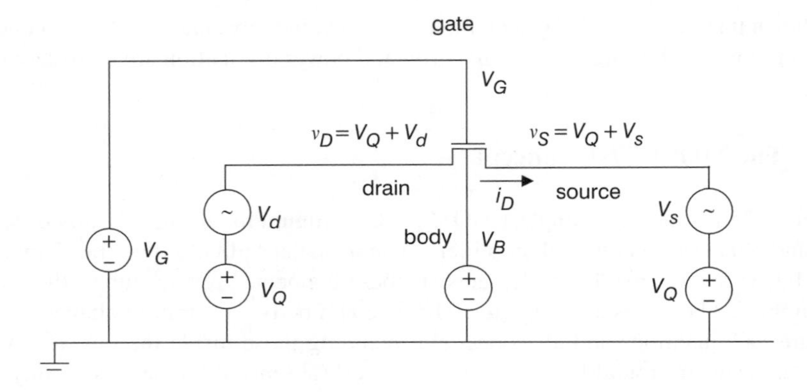

Figure 2.3 Biased MOSFET

The MOSFET low-frequency drain current in the triode region is accurately modelled by the following relationship [2]:

$$i_D = K[f(V_G, V_B, V_Q, V_d) - f(V_G, V_B, V_Q, V_s)] \tag{2.1}$$

where K is a constant proportional to the transistor width-to-length ratio, V_G and V_B are gate and body DC voltages, V_Q is the signal-ground level, and V_d and V_s are drain and source voltage signals. These quantities are shown in Figure 2.3. Notice from Eqn. (2.1) that V_d and V_s are equivalent and decoupled variables. This is meant in the sense that (a) the drain current magnitude is preserved when V_d and V_s values are interchanged and (b) the current dependence on V_d and V_s occurs only through independent terms. Furthermore, from device physics considerations, the function $f(V_G, V_B, V_Q, V)$ can be approximated with a high degree of accuracy as

$$f(V_G, V_B, V_Q, V) \cong aV + bV^2 \tag{2.2}$$

where a is a strong function of $(V_G - V_Q)$ and $(V_B - V_Q)$, and b depends practically only on $(V_B - V_Q)$ [2]. Equation (2.2) emphasises that the nonlinear characteristic of a MOS transistor in the triode region is mainly of the second order. The product Ka is the small-signal conductance of the transistor and can be tuned through the voltages V_G and/or V_B (typically only V_G) assuming that V_Q is a fixed DC bias. In Reference 2 it is shown that

$$Ka = 2K(V_G - V_Q - V_T) \tag{2.3}$$

where V_T is the transistor threshold voltage, depending on $(V_B - V_Q)$.

Using Eqns. (2.1) and (2.2), and Figure 2.2, one can easily show how the MOSFET-C concept works. The balanced OA connected in negative feedback forces equal voltages at its inputs and balanced voltages at its outputs. If balanced signals

V_{in} and $-V_{in}$ are applied to the MOSFET-C block, we have:

$$i_1 = K[f(V_G, V_B, V_Q, V_{in}) - f(V_G, V_B, V_Q, V_x)] \quad (2.4)$$

$$i_2 = K[f(V_G, V_B, V_Q, -V_{in}) - f(V_G, V_B, V_Q, V_x)] \quad (2.5)$$

$$V_{out} = L\{i_1\} + V_x \quad (2.6)$$

$$-V_{out} = L\{i_2\} + V_x \quad (2.7)$$

where $L\{\ \}$ is a linear operator describing the current-to-voltage relationship for the linear feedback element. In the case of purely capacitive feedback (MOSFET-C integrator), this operator is a time integral. From Eqns. (2.6) and (2.7) we eliminate V_x and calculate the output voltage:

$$V_{out} = \tfrac{1}{2}[L\{i_1\} - L\{i_2\}] = \tfrac{1}{2}L\{i_1 - i_2\} \quad (2.8)$$

The last equality in Eqn. (2.8) follows from the basic properties of linear operators. Next, Eqns. (2.4) and (2.5) are subtracted to calculate the difference between the two currents and we notice that the dependence on V_x is eliminated again. Substituting the result into (2.8) we obtain:

$$V_{out} = \tfrac{1}{2}KL\{f(V_G, V_B, V_Q, V_{in}) - f(V_G, V_B, V_Q, -V_{in})\} \quad (2.9)$$

Finally, Eqn. (2.2) is applied and we obtain

$$V_{out} = KaL\{V_{in}\} \quad (2.10)$$

The factor a can be taken outside the linear operator only when it remains constant, such as during normal filter operation. When the circuit is being tuned and the voltage V_G is changing, Eqn. (2.10) does not hold; however, this transient condition does not affect the regular operation of the filter. The important relation in Eqn. (2.10) shows that the circuit in Figure 2.2 exhibits a linear large-signal input/output characteristic and its effective linear input conductance is equal to the transistor (nonlinear device!) small-signal conductance Ka. For this reason, within MOSFET-C filters, it is appropriate to call the transistors MOSFET "resistors".

It is emphasised that the voltage V_x is NOT zero and varies in a strongly nonlinear fashion with respect to the input voltage, as proven by the following argument. If V_x were zero, Eqn. (2.6) would represent a linear dependence between i_1 and V_{out} and Eqn. (2.4) would represent a nonlinear (second order) dependence between i_1 and V_{in} in accordance with Eqn. (2.2). Therefore, a zero V_x would imply a strongly nonlinear dependence between V_{out} and V_{in}, which is in contradiction with the mathematically correct result Eqn. (2.10). We conclude that V_x must be varying in a nonlinear fashion with V_{in}. This is confirmed by simulations and measurements. The values of V_x, i_1 and i_2 automatically adjust in nonlinear relationship to V_{in} such as to produce output voltages linearly dependent on the input voltage. In general, MOSFET-C circuits are internally nonlinear and externally linear. If resistors are used in addition to MOSFETs and capacitors [15], it is possible to design MOSFET-R-C circuits, which are externally and internally linear [16]. However, these circuits lack some of the important properties of MOSFET-C networks to be discussed later, such as insensitivity to parasitic source/drain capacitances.

A further insight into the MOSFET-C concept can be attained by the following observation. We may regard the topologically symmetrical nodes as pairs and analyse the differential-mode and common-mode components of the respective voltage–signal pairs with respect to the quiescent voltage V_Q. For example, in Figure 2.2 we have an input node pair with signals V_{in} and $-V_{in}$, a middle node pair with signals V_x and V_x (OA inputs), and an output node pair with signals V_{out} and $-V_{out}$. Notice that for any of these signal pairs either the common-mode or the differential-mode component is absent. The use of the balanced operational amplifier ensures this condition throughout the circuit. The same is true for any MOSFET-C filter, as will be obvious after the discussion of general MOSFET-C topologies in the next section. Therefore, in MOSFET-C filters, any node voltage outside the active elements belongs either to a pure differential signal pair or to a pure common signal pair but never to both.

2.3 Practical considerations

2.3.1 *Linearity in presence of nonidealities*

The discussion in the previous section assumed perfect matching of devices and ideal amplifiers. In reality, small transistor and capacitor mismatches exist and the OAs have finite gain, finite input offset voltages, etc. In addition, the exact transistor characteristics contain more nonlinear terms than are shown in Eqn. (2.2), which are small but still present. Since the basic linearity estimates of MOSFET-C circuits, as done in the previous section, give perfect performance, the higher order effects are responsible for the practical limitation of this technique. A detailed analysis of these effects was given in Reference 2 and is beyond the scope of this discussion. Here, it is only pointed out that the MOSFET-C technique has excellent linearity performance even in the presence of nonidealities, as attested by the actual practical performance.

To help the understanding of the following DC biasing discussion, it is necessary to mention the following qualitative fact. The detrimental second-order effects due to device mismatch and due to the presence of higher order nonlinear terms are minimised by making $(V_G - V_Q)$ and $(V_B - V_Q)$ as large as possible (in absolute value). This is simply because the single transistor nonlinear behaviour is reduced. This optimisation condition is more efficient for $(V_G - V_Q)$ than for $(V_B - V_Q)$ since the gate terminal controls the transistor current more strongly than the body terminal.

2.3.2 *DC biasing and tuning*

For the proper application of the MOSFET-C technique, the signal ground V_Q and the maximum signal level must be selected such that the MOSFETs operate in the triode region. If these devices are allowed to enter saturation, the relations in the previous section cease to be valid and the circuit input/output characteristic degrades rapidly toward nonlinear behaviour. Naturally, the DC design strategy should be chosen to maximise the circuit linear range.

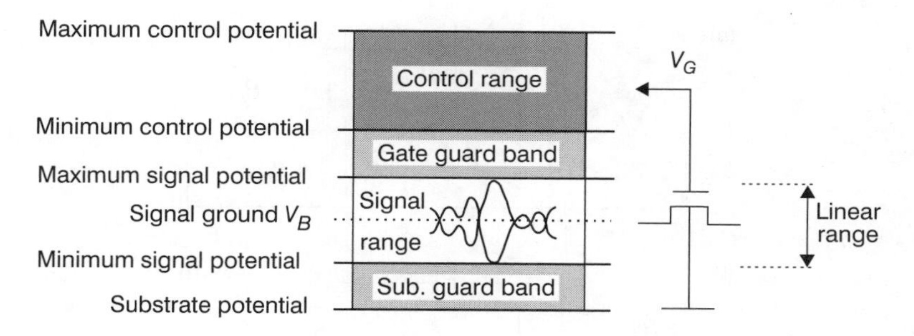

Figure 2.4 Control and signal ranges for n-channel transistor in MOSFET-C circuits

To discuss the optimum DC biasing, the highest voltage on chip is divided into four adjacent bands, as illustrated in Figure 2.4 for n-channel transistors (for p-channel devices a corresponding picture and arguments apply). Based on the observation in the previous subsection, positioning the signal ground closer to the chip substrate potential than the control voltage enlarges the linear region, typically, although the negative peaks of the signal should remain somewhat above the substrate potential. By taking into account the control signal range and the mismatches, a precise optimum can be identified with simple computer simulations. In general, a gate guard band and a substrate guard band are necessary to minimise the residual distortion due to nonidealities and to cover the device threshold voltage. Naturally, the linearity of the MOSFET-C technique suffers as the maximum available on-chip DC voltages decrease.

The tuning requirements determine the control range. In most designs the maximum voltage on chip is the positive power supply voltage but larger values can be used, e.g. through charge pumping, as long as the transistor breakdown conditions are not reached [17–19]. Typically, when no additional techniques are used, the tuning requirements are satisfied through relative resistance changes by a factor of two to three with respect to the smallest actual values obtained after fabrication. Most of these changes are necessary to compensate the process and temperature variations of the MOSFET resistors themselves. A considerably lower tuning range is sufficient if parallel MOSFET resistors are switched in and out of the network for coarse tuning [11]. Naturally, since the second approach minimises the size of the tuning band, it allows more room for the signal swing, accomplishing a larger linear range.

2.3.3 Use of OTAs

Since the matter to be considered in this section [7–9, 13] is no different for MOSFET-C networks than for single-ended active-RC networks, the latter will be discussed here for simplicity. Figure 2.5 shows two integrators driving similar circuits

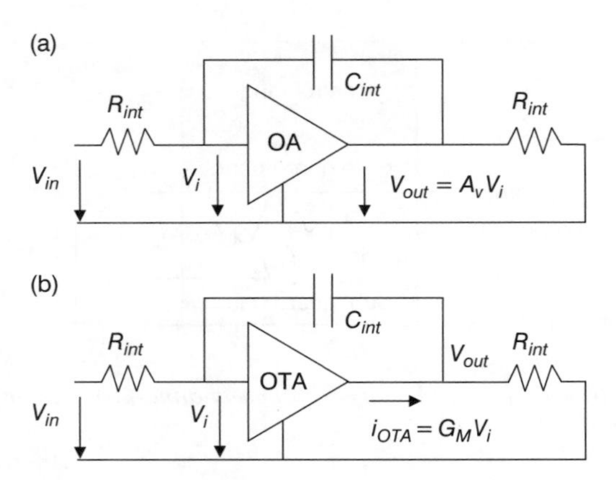

Figure 2.5 Active integrators: (a) OA-based; (b) OTA-based

(only input effective resistors shown), one using an OA and the other one using an OTA (operational transconductance amplifier). It is often argued incorrectly that only the OA-based integrator operates properly because the OTA "cannot drive" resistor loads. A simple calculation shows that the two circuits are equivalent if the following condition holds [9]:

$$\beta G_M R_{int} = A_V \gg 1 \tag{2.11}$$

where G_M is the OTA open-loop transconductance, A_V is the OA open-loop voltage gain and β is a general load factor. The value of the latter is 1 for the case of Figure 2.5, and would be 0.5 for a load of two parallel equal resistors, etc. Equation (2.11) clearly demonstrates that OTAs can drive resistors just as naturally and easily as OAs if G_M is large enough. In other words, all that is required is a sufficiently high DC gain in both cases. An essential observation is that, as long as the OA and the OTA provide the same current, they produce the same network response. Important practical implications of this fact will be discussed later regarding power dissipation considerations in MOSFET-C/active-RC filters.

The use of OTAs in active-RC blocks has certain similarities with those in the g_m-C technique. Indeed, comparing Figure 2.6 showing a g_m-C integrator with Figure 2.5(b), we notice that both circuits contain a voltage-to-current conversion active element and a capacitor. The essential difference between them is that in the OTA-RC integrator the g_m element is connected in a closed loop and has a large value, while in the g_m-C integrator, the g_m element is connected in an open loop and has a substantially smaller value. For the same integrating time constant, the latter is related to the integrating resistor in Figure 2.5(a) by

$$g_m = \frac{1}{R_{int}} \tag{2.12}$$

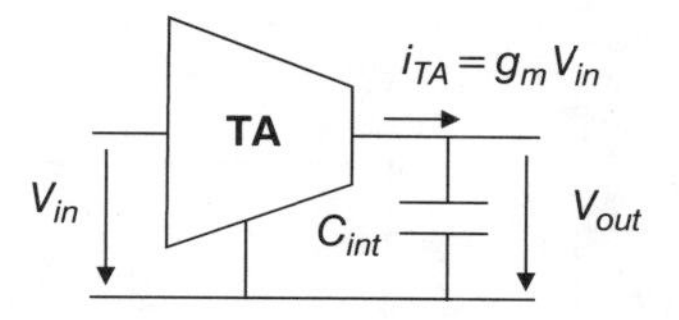

Figure 2.6 g_m*-C integrator*

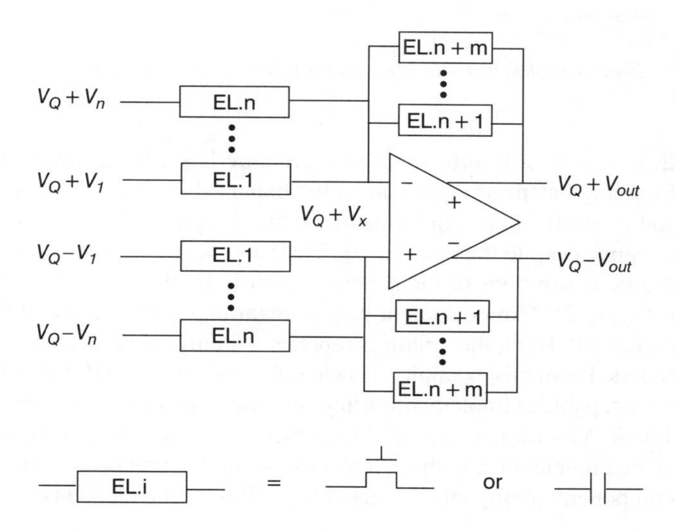

Figure 2.7 General MOSFET-C building block

One of the most important consequences of the previous observation is that the voltage-to-current function in the open-loop g_m-C integrator must be linear by definition and requires complex linearisation schemes. In contrast, the same function in the closed-loop OTA may be nonlinear as long as high voltage-to-current gain is present, and no linearisation circuit is needed, just as in the OA case. While in the g_m-C circuit the integrating current is generated by the g_m element, in the MOSFET-C circuit the integrating current is generated by the input resistor. This has important further consequences for the noise performance, as will be discussed later.

2.3.4 *General filter building blocks*

The circuit of Figure 2.2 is the simplest MOSFET-C configuration possible, useful mainly for the demonstration of the principle. However, a straightforward generalisation yields building blocks sufficient for the realisation of any filter transfer function, theoretically. Figure 2.7 shows a general MOSFET-C configuration containing *n* input matched device pairs and *m* feedback matched device pairs. Each device pair consists of either MOSFETs or capacitors. Using similar calculations to the ones in

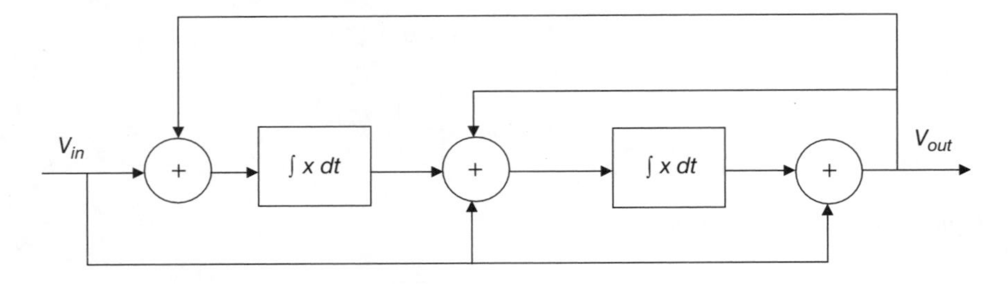

Figure 2.8 Functional schematic diagram of filter in Figure 2.1

Section 2.2, one can easily prove that this structure has a linear input/output characteristic from any balanced input pair to the balanced output pair. In addition, the superposition property applies in terms of balanced input and output signals just as for linear systems, despite the internal nonlinear character of this network.

Two specific realisations of the general circuit in Figure 2.7 can be identified in the filter of Figure 2.1. An equivalent way of regarding the structure of this filter is shown in Figure 2.8. Here, the circuit is represented only in terms of integrators and weighted adders. From this example, it is clear that the general MOSFET-C structure in Figure 2.7 is capable of implementing any combination of integrators and adders, as a building block. According to classical filter theory, any rational polynomial transfer function can be implemented with such blocks. Naturally, practical limitations apply, related to component quality factors, matching and parasitic elements.

2.3.5 Frequency response

Ideal active-RC integrators, implicit in classical filter topologies, have infinite DC gain and a single-pole roll-off at all frequencies. In practice the finite DC gain of OA/OTAs and their own frequency response roll-off can modify substantially the integrator AC characteristics. This effect is increasingly troublesome at high frequencies, limiting the AC performance of these filters. MOSFET-C circuits obviously suffer from the same fundamental limitation. Nevertheless, recently published circuits accomplished excellent AC performance at frequencies in excess of 100 MHz [11].

Figure 2.9 shows an OTA-based active-RC integrator, in which the OTA has a one-pole current roll-off, with ω_T the unity current-gain frequency. This could be a first-order AC model for a single-stage OTA. The integrator drives a resistive load, assumed to be β times R_{int}. Typically, β is between 0.2 and 2, assuming the filter integrators have similar resistance values. A more accurate model would include a G_M frequency response roll-off and extra parasitic capacitances and resistances. For practical purposes, it can be shown that the integrator nonideal frequency response behaviour is due mostly to a parasitic high frequency pole and a parasitic right half-plane zero. The latter can be compensated by adding either a parallel capacitor to the integrating resistor or a series resistor to the integrating capacitor [7, 9]. Nevertheless, unavoidable integrator magnitude and phase errors occur and increase with frequency.

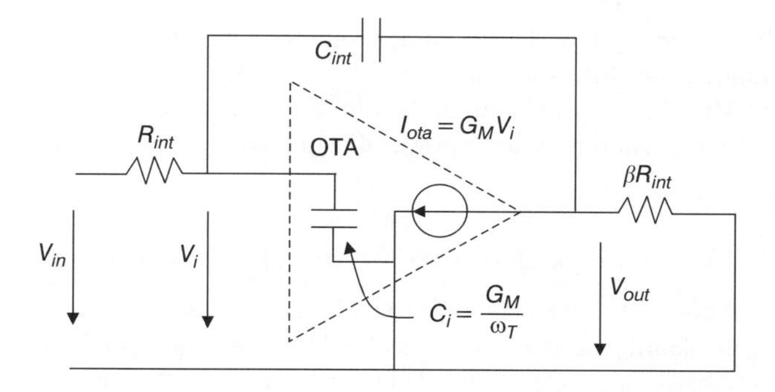

Figure 2.9 First-order AC model for active-RC integrator

Incidentally, similar limitations apply to the g_m-C integrators, as attested by the fact that their practical high frequency performance is not substantially better that that of MOSFET-C filters.

The realisation of precision high frequency filters requires very small integrator errors, especially within the passband. As a consequence, the ratio between the OTA and the integrator unity-gain frequencies, equal to $\omega_T R_{int} C_{int}$, must be larger than 10, typically, or even larger than 100 in critical cases. For example, in [11] this ratio is approximately 16. In the past, this may have precluded MOSFET-C filters for application at very high frequencies, but the situation is changing rapidly with the advent of extremely high speed IC technologies. For example, modern, inexpensive SiGe BiCMOS technologies provide transistors with a unity-gain frequency in excess of 75 GHz [20], and the next generation devices will attain over 100 GHz performance. A simple extrapolation of the result in [11], where only 15 GHz transistors were used, shows clearly that above 500 MHz MOSFET-C precision filters are well within reach. g_m-C filters will also benefit from these IC technology developments.

2.3.6 Noise and dynamic range

The noise of MOSFET-C filters equals that of their active-RC prototypes and has been analysed in detail in References 21 and 22. The two major sources are the thermal noise in MOSFETs and the OTA noise. The first mechanism produces a least filter degradation condition in the sense that no other methods are possible for linearly generating an integrating current with a lower noise than applying a voltage on a resistor. Any other IC integrator, including the g_m-C circuit in Figure 2.6, has significantly more noise than this. The OTA input devices dominate the amplifier noise.

The dynamic range DR is calculated based on the total output noise and the maximum output signal acceptable before harmonic and/or inter-modulation distortion occur. The excellent linearity of MOSFET-C filters may allow the use of signals, which are a large percentage of the power supply voltage available on chip. The g_m-C and other continuous-time filtering techniques are typically restricted to substantially smaller signals [10]. A comprehensive study of dynamic range in integrated continuous-time filters was done in Reference 23, and clear evidence was given showing that MOSFET-C filters attain the highest dynamic range for a given power dissipation. A discussion regarding power dissipation will be presented in the next section.

2.3.7 IC technology considerations for high speed filter design

Two important concepts have been discussed in the previous subsections with direct implications in identifying the best IC technology for high speed continuous-time filters. MOS transistors are clearly needed in order to apply the MOSFET-C method, shown to produce the highest dynamic range. However, bipolar transistors produce the highest OTA gain and unity-gain frequency needed for maximum-speed MOSFET-C operation. We conclude that BiCMOS is the optimum technology for high frequency high dynamic range filters. This was realised early in Reference 7. A BiCMOS MOSFET-C filter makes the best use of each transistor type. As mentioned before, in the future, advanced BiCMOS technologies will enable the application of this technique close to and even in the microwave region. Although not as high in gain-bandwidth product as BiCMOS, the CMOS capability for high speed MOSFET-C filters will also continue to improve with transistor scaling. CMOS MOSFET-C filters operating above 100 MHz have already been used in certain industrial applications.

2.3.8 "Double balanced" MOSFET-C

An extension of the MOSFET-C technique [24] is shown in Figure 2.10 (the four-transistor combination used in this integrator was proposed independently in Reference 25 for passive mixers). This configuration has important properties that go beyond the capabilities of the original technique. The residual third-order and higher-order transistor nonlinear terms are cancelled and the integrating current becomes independent of the MOSFET threshold voltage. However, practical device mismatches prevent this technique from improving the linearity of regular MOSFET-C circuits by a substantial amount. Also, additional resistor thermal noise is generated due to current subtraction. Nonetheless, this topology has inherent advantages in terms of high frequency MOSFET resistor limitations due to channel-to-gate and channel-to-substrate parasitic capacitances [5].

2.3.9 Other practical aspects

An important topological property of MOSFET-C filters is the fact that all passive components are connected only between low impedance nodes: inputs and outputs of OTAs (OAs). This gives these IC filters a very low sensitivity to parasitic capacitances, which typically are present at the component terminals. Furthermore, the intrinsic balanced nature of these circuits not only realises differential signal processing for doubling the signal level but also gives improved power supply rejection and substrate noise immunity. Next, the fundamental strength of the MOSFET-C technique in terms of power dissipation properties is discussed in detail.

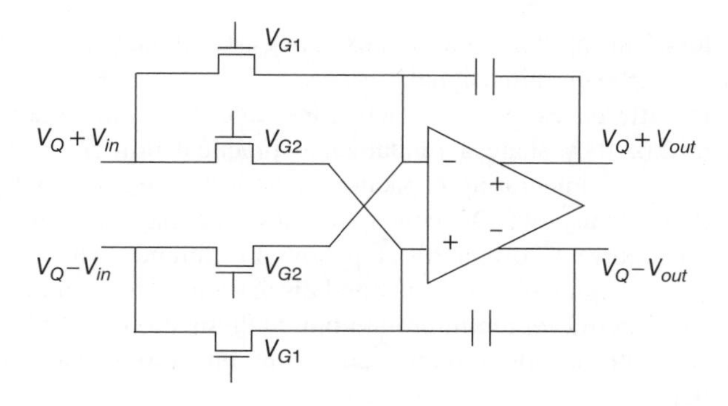

Figure 2.10 Double balanced MOSFET-C (Czarnul-Song) integrator

2.4 Power efficiency in MOSFET-C designs

The matter of power dissipation in MOSFET-C filters in comparison to other techniques and in relation to fundamental signal-to-noise properties has already been covered in several publications [10, 23]. A fundamental capability of this filter design method for power efficient implementations has been pointed out. However, due to the importance of this topic in practice, we will complement the previous arguments with additional support based on topological observations. As in previous treatments, the discussion here on active-RC filters applies almost identically to the MOSFET-C case.

2.4.1 Basic power dissipation considerations in IC filters

In principle, properly synthesised classical RLC passive filters dissipate no power in addition to that required for driving the resistive termination. Since the filter core consists of reactive components, the generator and the load resistance are the only real-impedance elements in the network. For this reason, passive filters have a fundamental capability for optimum power dissipation performance. Furthermore, as a related property, the only noise degradation in addition to that produced by the generator itself comes from the load. In other words, for ideal components, the generator signal-to-noise ratio is degraded only outside the filter core. However, the synthesis of such filters requires the use of both capacitors and inductors.

Active filters have been introduced for the practical situation in which only one type of lossless component is available. For example, in the important case of integrated circuits it is practically impossible to manufacture useful on-chip inductors but at very high frequencies. A common and efficient design methodology is to synthesise active filters using only integrators and summers, including positive or negative weights in general. Typically, the integrators are connected in a ring structure. The active-RC prototype of the MOSFET-C circuit of Figure 2.1 is such an example with the functional diagram shown in Figure 2.8. From Figure 2.1 it is clear that in MOSFET-C filters the summers can be implemented using the same OTA/OAs used for the integrators. This property does not exist in g_m-C filters, where extra g_m blocks are necessary for each summing signal.

Despite the differences between active-RC and g_m-C integrators shown in Figures 2.5 and 2.6, they share a fundamental practical limitation. The charging and discharging of the integrating capacitors cannot be done without energy loss. All active IC elements use two DC power supplies (one may be ground) and power is dissipated as net current flows across a potential difference. This is equivalent to charging a capacitor from a power supply and discharging it to ground, as is the case in digital circuits. Therefore, the minimum power dissipation possible for practical active integrators is proportional to the value of the integrating capacitor and to the operating frequency.

Figure 2.11 shows the simplest practical fully differential capacitor-charging/discharging scheme. For generality, a resistor is connected in series with the integrating capacitor. The controlled (control not shown) current source pairs J1A/J2B and J1B/J2A are assumed on and off alternately according to class-B operation. It is important to observe and remember that the amount of energy loss through this process is independent of the series resistor. Depending on how the integrating current is generated, from this simple circuit one can derive all three integrator schemes of Figures 2.5 and 2.6. The current sources J1A, J1B, J2A and J2B are inside the active elements. Next, we show that the active-RC filter sub-network made of interconnected integrators is a combination of charging/discharging circuits as in Figure 2.11 with added high-gain voltage feedback.

2.4.2 *Topological advantages of active-RC integrators for power efficiency*

Figure 2.12(a) illustrates two active-RC differential integrators interconnected in a typical filter and Figure 2.12(b) shows the same circuit in which a high level model for the OTA/OA is included. The "input circuit" block could be just a differential transistor combination for a one-stage OTA/OA design (in which case J2A and J2B

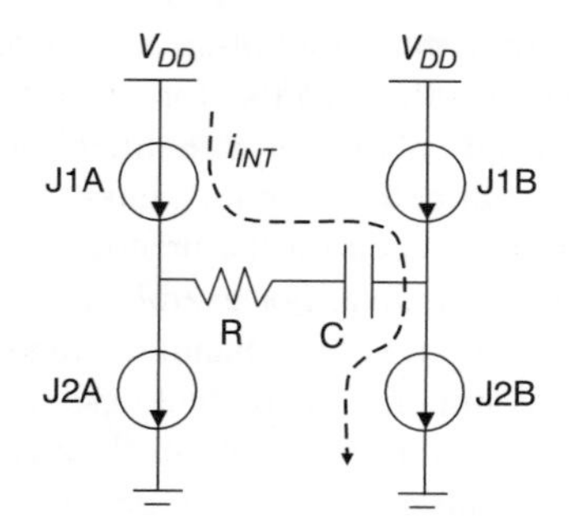

Figure 2.11 Capacitor charging/discharging scheme. The current sources represent the output stages of active elements

may be connected slightly differently while performing the same operation) or a whole input stage for a typical two-stage design. Since the OTA/OA output drivers are assumed to operate in class-B, as far as the charging/discharging of the integrating capacitors is concerned, it makes no difference if these drivers are current sources as illustrated in Figure 2.12(b) or voltage sources (not shown). This confirms the earlier observation that OTAs and OAs may be interchanged in this topology. The important thing is to have high-gain voltage feedback such that the common capacitor-resistor terminal is at a virtual signal ground. In this way the integrating current is inversely proportional to the respective series resistor. Inspecting the diagram, it is clear that, in general, an active-RC integrator chain consists of charging/discharging circuits as in Figure 2.11, implemented over consecutive integrators. This observation has an important implication regarding the common claim that active-RC integrators require more current than other active integrators due to the need to drive resistors. *The "extra" current each integrator delivers into adjacent integrators is not wasted but rather used to charge/discharge other integrating capacitors in the filter.* Therefore, active-RC integrator chains or rings dissipate the minimum possible power for any such active circuits if class-B operation is assumed.

Furthermore, from Figure 2.12(b) it is important to notice that the active-RC topology allows high-gain voltage feedback only weakly dependent on the integrating current. In other words, as the integrating current (provided by the OTA/OA drivers) varies, including the condition of having very low values, high-gain voltage feedback can still be guaranteed through appropriate design. This, in fact, is the fundamental reason class-B stages are possible in such filters.

It is instructive to analyse a practical g_m-C integrator from the same perspective. Figure 2.13(a) shows a familiar realisation using resistive degeneration and Figure 2.13(b) shows the same circuit with transistors replaced by first-order DC models. The latter are simply two transconductors: TA and TB. Assuming large values for GM in TA and TB, the input voltages are "copied" on the resistor terminals: $V_{in+} = V_{i+}$ and $V_{in-} = V_{i-}$ (source/emitter followers). The currents obtained on the resistors pass unaltered through TA and TB and get integrated on the capacitors, assuming that J1A, J1B, J2A and J2B are fixed current sources. It is clear that this circuit is a simple modification of the scheme in Figure 2.11 and it can be regarded as a pseudo active-RC integrator because applying a voltage on a resistor generates the integrating current. Furthermore, the operation of this circuit can be interpreted through a high-gain feedback explanation. For example, when V_{in+} moves up instantaneously, a large current is generated by the large g_m of TA, which in turn raises the voltage V_{i+}. This decreases the TA input voltage ($V_{in+} - V_{i+}$), effectively acting as high-gain negative feedback. Therefore, the g_m-C integrator in Figure 2.13 contains the same basic "ingredients" of the active-RC integrator: current derived by applying a voltage on a resistor, and high-gain feedback. However, the topological details are fundamentally different. While in the active-RC case the integrating capacitors are inside the feedback loops, in the g_m-C case both the resistors and the capacitors are outside any feedback. For this reason, we can legitimately refer to the active-RC integrator as a closed-loop approach and to the g_m-C integrator as an open-loop approach.

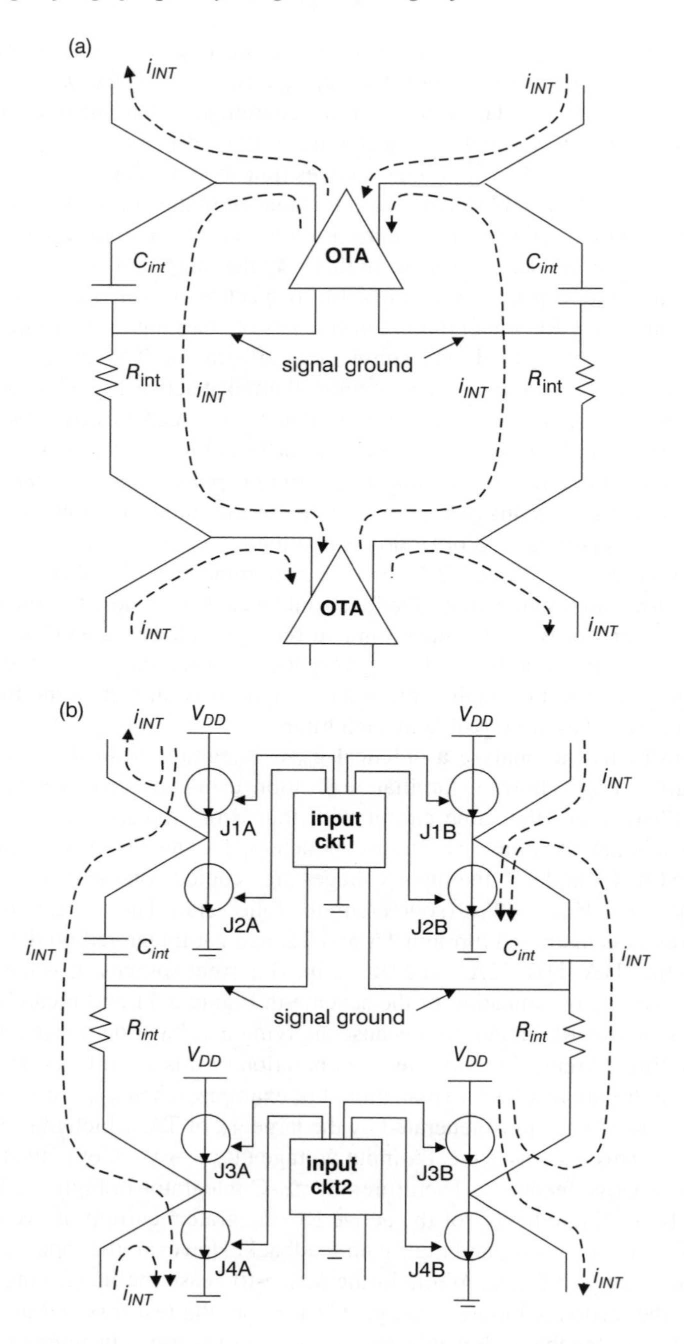

Figure 2.12 (a) Capacitor charging/discharging in active-RC filters; (b) circuit in (a) with each OTA represented by an input stage and a four current source output stage

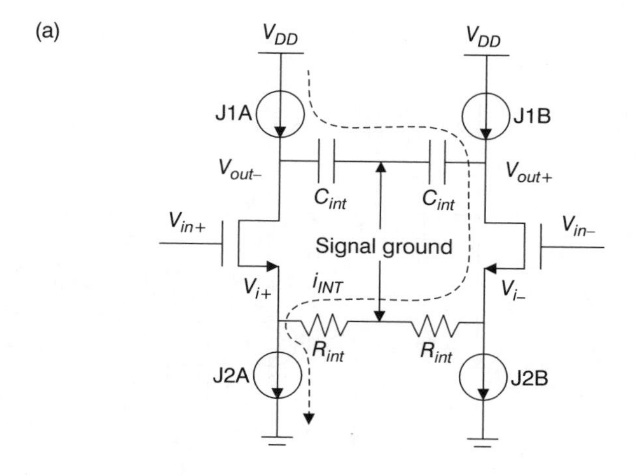

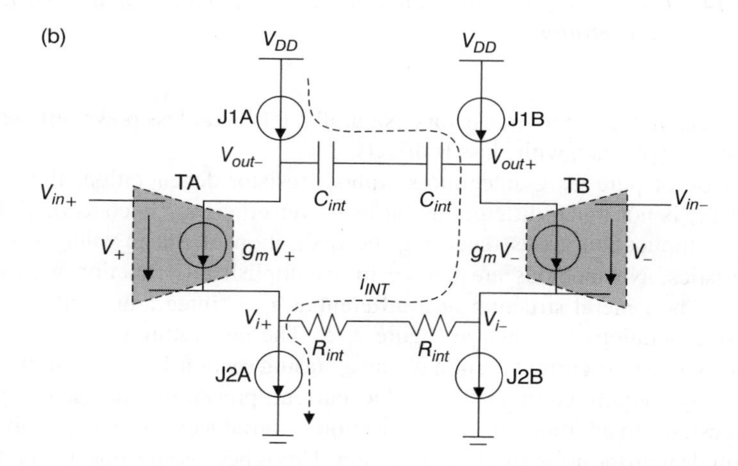

Figure 2.13 (a) Capacitor charging/discharging in g_m-C integrator with resistive degeneration; (b) circuit in (a) with each transistor represented by a first-order circuit model

Unlike the active-RC circuit, the g_m-C open-loop topology in Figure 2.13 imposes further limitations with serious repercussions regarding the power dissipation. Notice that the current sources cannot be turned off to accomplish class-B operation because the transistor g_m would drop to zero and no input signals could be applied to the resistors. An alternative way of explaining this limitation is by observing that the effective negative feedback described above for the g_m element in the g_m-C integrator is strongly dependent on the integrating current. For this fundamental reason, the g_m-C integrators with resistive degeneration use class-A or class-AB operation, requiring

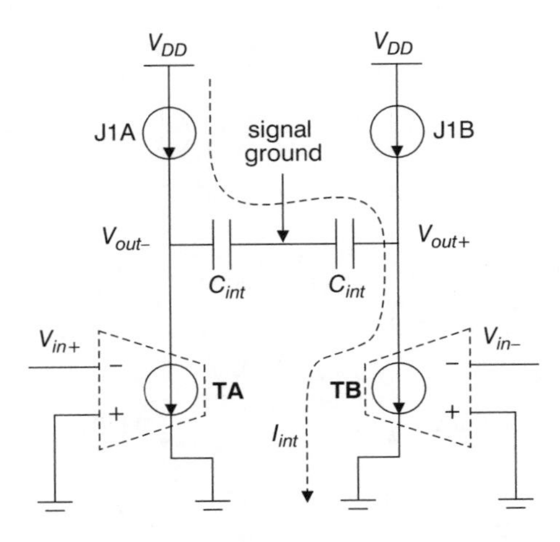

Figure 2.14 Capacitor charging/discharging in g_m-C integrator without resistive degeneration

DC current at all time in all transistors. Naturally, this is far less power efficient than the active-RC approach with class B drivers.

The case of pure g_m-C integrators without resistor degeneration, illustrated in Figure 2.14, is not much different as far as power efficiency is concerned. In fact, an extra complication arises regarding the realisation of linear voltage-to-current characteristics. No methods are known to accomplish this function with class-B operation. The general structure of a differential g_m-C integrator (with or without resistor degeneration) is shown in Figure 2.15. The integrating capacitor is placed between a voltage-to-current section X and a biasing section Y. Practical implementations always require constant flow of DC current, preventing the use of optimum class-B designs. In addition, the biasing section Y introduces noise in the same order of magnitude with the noise from the X section. This is because the transistor transconductances in Y will not be that different from those in X since all devices carry the same DC biasing current. Typically, an "excess noise factor" of 2 to 3 results in g_m-C topologies, and as a consequence, these designs require proportionally larger power dissipation than MOSFET-C filters [10, 23].

2.4.3 Power dissipation in summers

Finally, the power dissipation due to summers will be discussed. As mentioned earlier, the integrator-based filter structures require such functions. For the active-RC approach, the signal summing is implemented with conventional resistor/OTA or capacitor/OTA circuits. Having identical topology with the integrators, these networks can reuse the OTAs of the latter. For example, comparing Figures 2.1 and 2.8, we notice that the addition of the input signal into the final summer on the right hand side of Figure 2.8 is implemented with capacitors (feed-forward from the input).

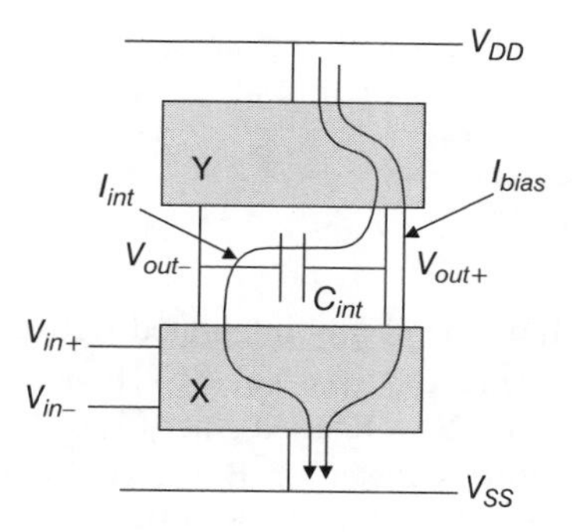

Figure 2.15 Integrating and biasing currents in g_m-C integrators

Despite the convenience of hardware reuse, the presence of summers increases the filter power dissipation. The resistive/OTA networks carry currents at all frequencies while the capacitor/OTA networks carry currents mostly at high frequencies. However, the class-B OA/OTA operation, as required by optimum integrating capacitor charging/discharging, ensures that the active-RC filters are globally power efficient. All currents drawn from the power supply are used either for charging the integrator capacitors or for realising signal summation – both necessary functions.

The realisation of summers in g_m-C filters is not as convenient and power efficient as in the previous case. For summing before integration, extra voltage-to-current converters (block X in Figure 2.15) must be connected in parallel. This increases the DC bias currents by a straight linear scaling, thus lowering the power efficiency. The situation is worse than in the active-RC case since class-B operation is not possible, as discussed previously.

2.5 Conclusions

The MOSFET-C method for fully integrated filters is a mature and proven design technique. Over the last 14 years, it has been used in high dynamic-range, continuous-time, large-volume applications at low and medium frequencies, and recently at frequencies above 100 MHz. The traditional strength of this method has been its capabilities for best power efficiency and best dynamic-range realisations among all integrated continuous-time filters. Conventional and new theoretical support explaining this performance has been presented in detail. The next generation MOSFET-C filters will cover increasingly higher frequencies. Clear experimental and theoretical evidence shows that an extension of this technique to frequencies approaching 1 GHz is within reach. The crucial IC technology advancement enabling this development is the advent of very high speed, inexpensive BiCMOS. The MOSFET-C method is ideally suited for this technology. It uses MOS transistors for obtaining tuned analogue cells with the largest possible on-chip dynamic range for a given power dissipation, and bipolar transistors for obtaining amplifiers with the largest possible on-chip gain-bandwidth product.

2.6 References

1 BANU, M. and TSIVIDIS, Y.: 'Fully Integrated Active RC Filters in MOS Technology', *IEEE J. Solid-State Circuits*, Vol. SC-18, pp. 644–651, December 1983.
2 BANU, M. and TSIVIDIS, Y.: 'Detailed Analysis of Nonidealities in MOS Fully Integrated Active RC Filters Based on Balanced Networks', *IEE Proceedings*, Vol. 131, Pt.G, No. 5, pp. 190–196, October 1984.
3 BANU, M. and TSIVIDIS, Y.: 'An Elliptic Continuous-Time CMOS Filter with On-Chip Automatic Tuning', *IEEE J. Solid-State Circuits*, Vol. SC-20, No. 6, pp. 1114–1121, December 1985.
4 TSIVIDIS, Y., BANU, M. and KHOURY, L.: 'Continuous-Time MOSFET-C Filters in VLSI', *IEEE J. Solid-State Circuits*, Vol. SC-21, No. 1, pp. 15–30, February 1986.
5 ISMAIL, M. and RUBIN, D.: 'Improved Circuits for the Realization of MOSFET-Capacitor Filters', Proceedings, 1986 IEEE Int. Symp. Circuits and Systems, pp. 1186–1189, San Jose, 1986.
6 KHOURY, J. H. and TSIVIDIS, Y. P.: 'Analysis and Compensation of High-Frequency Effects in Integrated MOSFET-C Continuous-Time Filters', *IEEE Trans. Circuits and Systems, II: Analog and Digital Signal Processing*, Vol. CAS-34, No. 8, pp. 862–875, August 1987.
7 VAN BEZOOIJEN, A., RAMALHO, N. and VOORMAN, J. O.: 'Balanced Integrator Filters at Video Frequencies', Digest ESSCIRC'91, pp. 1–4, 1991.
8 TSIVIDIS, Y. P. and VOORMAN, J. O. (Eds): *Integrated Continuous-Time Filters*, New York, IEEE Press, 1993.
9 TSIVIDIS, Y. P.: 'Integrated Continuous-Time Filter Design – An Overview', *IEEE J. Solid-State Circuits*, Vol. SC-29, No. 3, pp. 166–176, March 1994.
10 GROENEWOLD, G., MONNA, B. and NAUTA, B.: 'Micro-Power Analog-Filter Design', in VAN DE PLASSCHE, R. J., SANSEN, and HUIJSING, J. H. (Eds.): 'Analog Circuit Design Low-Power Low-Voltage Integrated Filters and Smart Power', Kluwer Academic Publishers, 1995.
11 GROENEWOLD, G.: 'Low-power MOSFET-C 120 MHz Bessel Allpass Filter with Extended Tuning Range', *IEE Proceedings – Circuits, Devices and Systems*, Vol. 147, No. 1, pp. 28–34, February 2000.
12 SCHAUMANN, R., GHAUSI, M. S. and LAKER, K. R.: *Design of Analog Filters: Passive, Active RC and Switched Capacitor*, Prentice-Hall, Englewood Cliffs, 1990.
13 VOORMAN, J. O., VAN BEZOOIJEN, A. and TAMAHLO, N.: 'On Balanced Integrator Filters', in [8].
14 BANU, M., KHOURY, J. and TSIVIDIS, Y.: 'Fully Differential Operational Amplifiers with Accurate Output Balancing', *IEEE J. Solid-State Circuits*, Vol. SC-23, pp.1410–1414, December 1988.

15 MOON, U. and SONG, B.: 'A Low-Distortion 22 kHz 5th-Order Bessel Filter', *IEEE J. Solid-State Circuits*, Vol. SC-28, pp. 1254–1264, December 1993.

16 BANU, M. and LIFSHITZ, N.: 'A MOSFET-R-C Filtering Technique with Improved Linearity', Proceedings of the European Conference on Circuit Theory and Design ECCTD'99, Vol. 1, pp.1–4, September 1999.

17 MONNA, G. L. E., SANDEE, J. C., VERHOEVEN, C. J. M., GROENEWOLD, G. and VAN ROERMUND, A. H. M.: 'Charge Pump for Optimal Dynamic Range Filters', Proceedings 1994 IEEE International Symposium on Circuits and Systems, Vol. 5, pp. 747–750, 1994.

18 SCHMID, H. and MOSCHYTZ, G. S.: 'A Charge-pump-controlled MOSFET-C Single-Amplifier Biquad', Proceedings, ISCAS, Geneva, Switzerland, May 28–31, Vol. 2, pp. 677–680, February 2000.

19 YOSHIZAWA, A. and TSIVIDIS, Y.: 'An Anti-blocker Structure MOSFET-C Filter for a Direct Conversion Receiver', *Digest of Technical Papers, 2001 Custom Integrated Circuits Conference*, pp. 5–8.

20 KING, C.A. *et al.*: 'Very Low Cost Graded SiGe Base Bipolar Transistors for a High Performance Modular BiCMOS Process', 1999 IEDM Tech. Dig., pp. 565–568, 1999.

21 TOTH, L., EFTHIVOULIDIS, G., GOPINATHAN, V. and TSIVIDIS, Y. P.: 'General Results for Resistive Noise in Active RC and MOSFET-C Filters', *IEEE Trans. Circuits and Systems, II: Analog and Digital Signal Processing*, Vol. 42, pp. 785–793, December 1995.

22 EFTHIVOULIDIS, G., TOTH, L. and TSIVIDIS, Y. P.: 'Further Results for Noise in Active RC and MOSFET-C Filters', *IEEE Trans. Circuits and Systems, II: Analog and Digital Signal Processing*, Vol. 45, No. 9, pp. 1311–1315, September 1998.

23 GROENEWOLD, G.: 'Optimal Dynamic Range Integrators', *IEEE Trans. Circuits and Systems, I: Fundamental Theory and Applications*, Vol. 39, No. 8, pp. 614–627, August 1992.

24 CZARNUL, Z.: 'Modification of the Banu-Tsividis Continuous-Time Integrator Structure,' *IEEE Trans. Circuits and Systems, II: Analog and Digital Signal Processing*, Vol. CAS-33, pp. 714–716, July 1986.

25 SONG, B. S.: 'CMOS RF Circuits for Data Communications Applications', *IEEE J. Solid-State Circuits*, Vol. SC-21, pp. 310–317, April 1986.

Chapter 3

Active filters using integrated inductors[1]

Dandan Li and Yannis Tsividis

3.1 Introduction

The recent research interest in integrated active LC filters on silicon [1–6] can be largely attributed to:

- the advent of highly integrated wireless communication transceivers. This provides potential applications for integrated active LC filters, since on-chip spiral inductors can have usable Q and inductance in the GHz frequency range. While at present the noise of active LC filters is not sufficiently low to allow their use in receivers, it may be sufficiently low for certain applications in transmitters, e.g. to clean up the signal following a mixing operation.
- persistent efforts to improve the quality of on-chip spiral inductors. Not only can the physical layout of the inductors be optimised with the aid of computer simulators [7–9], but some processes have been tailored to produce high-Q spiral inductors [10–12]. As a result, spiral inductors with values of several nH and Q of more than 6 in the 1–5 GHz frequency range have become practically available on silicon chips.
- the achievable superior dynamic range[2] (DR) performance of active LC filters compared to that of other integrated continuous-time filters not using inductors

[1] This chapter is based on "Active LC filters on silicon", *IEE Proceedings on Circuits, Devices and Systems*, Vol. 147, No. 1, pp. 49–56, February 2000.

[2] DR is generally defined as the ratio of the maximum input power, above which the signal will be unacceptably distorted by the circuit, to the minimum input power, at which the circuit provides a reasonable signal-to-noise ratio at the output [13]. Depending on how the maximum input level is specified, DR can be further classified into several types. For example, 1 dB compression dynamic range is defined with the maximum input signal level taken as that which produces 1 dB compression of the fundamental component at the output. Spurious-free dynamic range (SFDR) is defined with the maximum input signal level taken as that which produces third-order intermodulation distortion (IM3) products power equal to the noise power.

[2, 14]. The DR of active LC filters can be further increased with the improvement of the quality of on-chip reactive components.

Integrated active LC filters are not simply copies of their discrete counterparts. The difficulty of integration mostly results from two problems:

1. Reactive components integrated on silicon are more nonideal than the corresponding discrete parts. Such integrated components not only are much more lossy but also have significant parasitic capacitances associated with the silicon substrate. Both the losses and the parasitic capacitances are not easily absorbed into the filter's transfer function, a situation which makes it extremely challenging to design an integrated active LC filter with the desired exact frequency response.
2. Because the inductance of an integrated spiral inductor is mainly determined by its lateral physical layout, the inductance can be predicted with decent accuracy (e.g. within 5%) with the aid of computer simulators [9, 15–17], and it stays almost stable when the temperature changes [18]. But this is not the case with capacitors, resistive losses and parasitics which have high initial tolerance and vary with environmental changes. It is thus necessary to make active LC filters automatically tunable. As will be seen later, the automatic tuning of an active LC filter in the GHz range is a design challenge.

3.2 Principles of Q enhancement

Q enhancement has long been used in discrete LC circuits for signal amplification, signal selection and generating stable oscillation [19–24]. In this section, our discussion is focused on the use of Q-enhancement in integrated LC filters.

It can be shown that for a passive LC filter, the losses associated with the reactive components shift the pole positions towards the left in the s-plane. The frequency response is thus distorted if those lossy components are used in a design intended for lossless reactances. The losses in the reactive components can be approximately modelled to first order as lumped positive resistors. If we have at hand ideal negative resistors (the exact opposite of positive resistors), we can use them to undo the harm caused by the losses of the reactive components. Once a lossy reactive component is combined with a negative resistor, the real part of the total impedance can be reduced, eliminated or even made negative depending on the value of the negative resistance that is used. Hence, we can claim that the Q of the original lossy component can be *enhanced* by negative resistors, and the resulting Q-enhanced component can replace the lossy component used in the LC filter. Hopefully, by replacing lossy components with Q-enhanced components, the poles of the filter would be at the positions intended for a filter using lossless reactances; practically, though, this often cannot be achieved. One reason is that the total impedance of an integrated reactive component may be a complicated function of frequency; another reason is that ideal negative resistors do not exist in the real world, and the impedance of a negative resistor realised on-chip generally has frequency-dependent real and imaginary parts.

Both problems complicate the transfer function of the Q-enhanced LC filter. Let us first look into the characteristics of the total impedance of Q-enhanced components.

3.2.1 *Compensation using negative resistances*

We assume for now that a lossy inductor and a lossy capacitor can be simplistically modelled as shown in Figure 3.1(a), in which R_L and R_C represent the losses in the inductor and the capacitor, respectively. We can compensate the losses by connecting negative resistors in series with the inductor and in parallel with the capacitor as shown in Figure 3.1(b), where R_1 and R_2 denote the absolute values of those negative resistors. By making $R_1 = R_L$ and $R_2 = R_C$, the total impedances become ideal. We can also connect negative resistors in parallel with the inductor and in series with the capacitor as shown in Figure 3.2(a); the equivalent circuits are derived in Figure 3.2(b) [25]. In this case, we observe that not only L' and C' become frequency-dependent

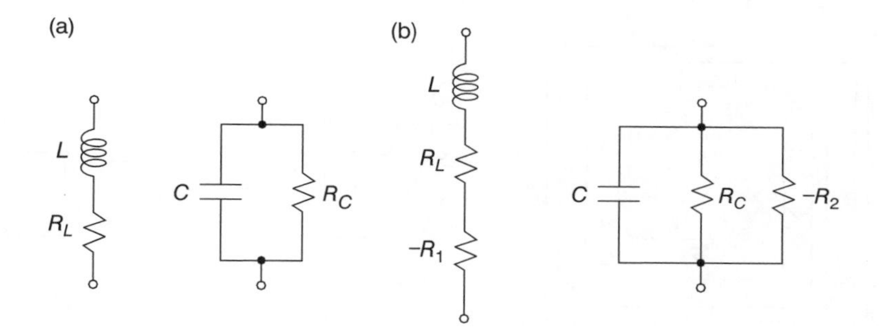

Figure 3.1 (a) Lossy reactive components. (b) Q-enhanced components

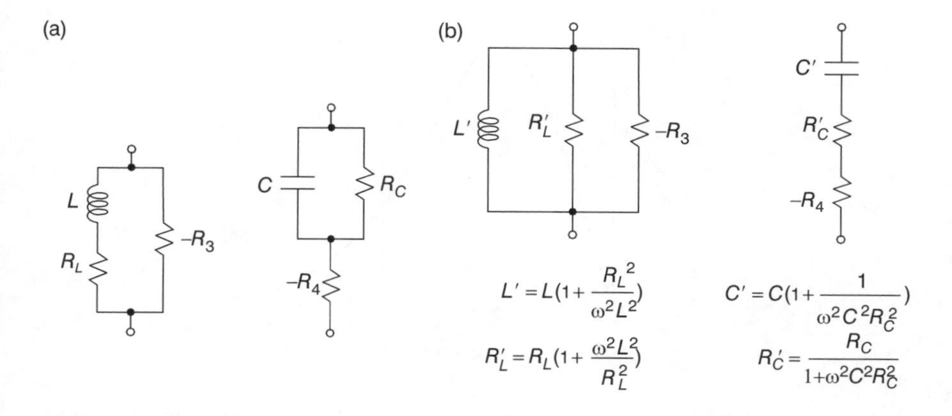

Figure 3.2 (a) Q-enhanced components. (b) Equivalent circuit for (a)

but also the equivalent losses R'_L and R'_C, and thus the losses cannot be perfectly compensated by fixed negative resistors at all frequencies. Hence, to compensate the loss which can be modelled as a series resistor, connecting a negative resistor in series with the lossy component may be a better choice [1]; on the other hand, if the loss can be modelled as a parallel resistor, connecting a negative resistor in parallel with the lossy component may be a better choice [2–6]. The next question is then how to compensate the losses of integrated reactive components, realistic models for which are much more complicated than those in Figure 3.1(a). Since on-chip spiral inductors are usually much more lossy than capacitors, we concentrate the discussion on spiral inductors.

3.2.2 *Integrated inductors*

Figure 3.3(a) shows a typical top view of an integrated spiral inductor, in which the top metal layer, which has lower resistivity and smaller capacitance to the substrate,

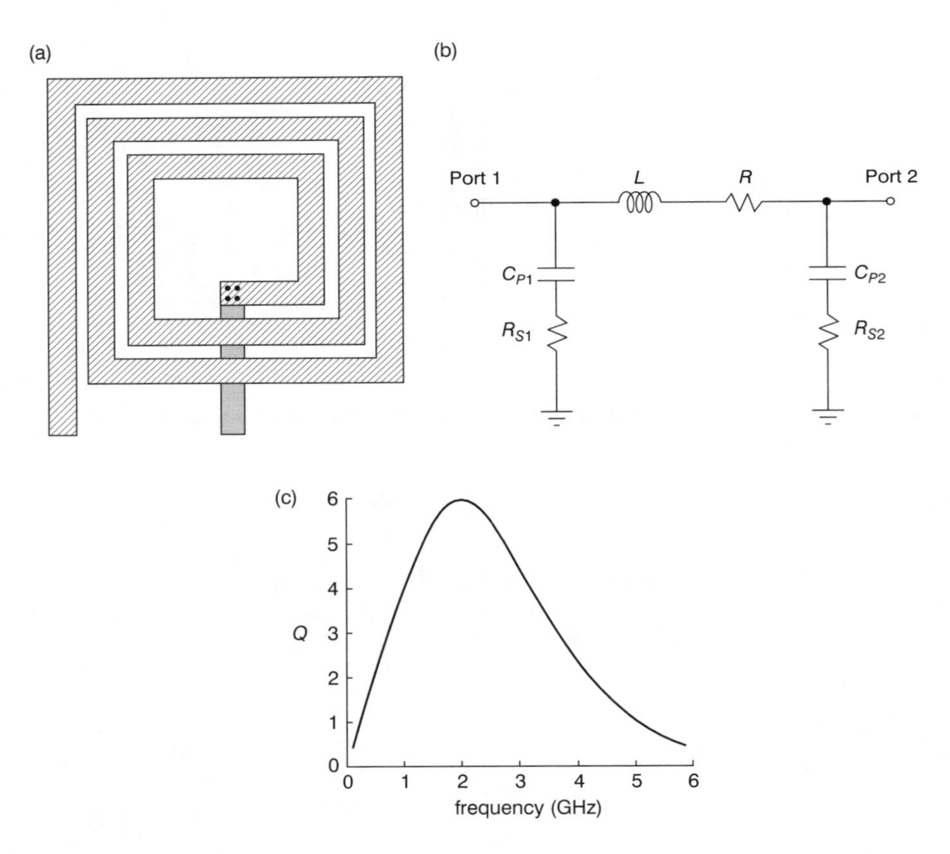

Figure 3.3 *(a) Top view of an on-chip spiral inductor. (b) Model for a spiral inductor (more sophisticated models can also be used); example values are* $L = 4\,\text{nH}$, $R = 6\,\Omega$, $C_{P1} = C_{P2} = 200\,\text{fF}$ *and* $R_{S1} = R_{S2} = 90\,\Omega$. *With these values, the relation between* Q *and frequency is as shown in (c)*

is used for the spiral turns and a lower metal layer is used for connection to the spiral's centre. It is known that the losses of on-chip inductors come from two sources: the ohmic loss in the metal trace and contacts, and the loss in the silicon substrate due to the bulk resistance and eddy currents. An often-used model, called the π model, is shown in Figure 3.3(b); the ohmic resistance of the metal and contacts and the loss due to eddy currents in the substrate are modelled by R, the metal-to-substrate parasitic capacitance is modelled as C_{P1} and C_{P2}, and the loss due to the substrate resistance is modelled as R_{S1} and R_{S2}. Example values for the model parameters are given in the figure caption, and will later be used in the simulation of high order filters. The Q of the inductor connected as one port (i.e. with Port 2 in Figure 3.3(b) shorted to ground) is used here to evaluate the quality of the inductor. This Q can be defined as the ratio of the imaginary part of the one-port impedance to the real part of the one-port impedance.[3] Figure 3.3(c) shows how the Q of the inductor modelled by Figure 3.3(b), with the element values given in the caption, varies with frequency. Before Q rises to its peak, R is the dominant loss, and, to cancel it, it is preferable to connect a negative resistor in series with the inductor. But often, after the Q of the inductor is maximised within the signal band to achieve better noise performance, the Q is also found to reach its peak inside/close to the signal band. In these cases not only R but R_{S1} and R_{S2} count, and no matter whether we connect a negative resistor in series or in parallel with the inductor, the Q-enhanced inductor impedance will be a complicated function of frequency. Because of this, Q-enhanced integrated LC filters are more attractive for narrow-band applications, in which the frequency dependence of the Q-enhanced impedances can have little effect within the narrow signal band.

3.3 Negative resistor implementations

The first extensive discussion of negative resistors dates back to 1918, when Hull presented a type of vacuum tube called dynatron, which exhibited negative resistance over a limited range [29]. Solid-state diodes exhibiting negative resistance were described by O. Losev in 1924 [30]. The shape of the $I-V$ curve of the above devices is similar to that of tunnel diodes. Later, it was discovered that negative resistance can also be obtained by using feedback [19, 31, 32]. Active components can be used to build voltage-controlled or current-controlled circuits whose terminal impedances have negative real parts. Components like tunnel diodes are not available on ICs, and we thus limit our discussion to the type of negative resistors which are built by using feedback.

3.3.1 Single-ended schemes

Integrated negative resistors can be single-ended or differential. Figure 3.4(a) shows an example [33] of a single-ended circuit which has been simplified for AC analysis

[3] The Q of an inductor can be defined in several ways [7, 11, 26–28]. The most fundamental definition is based on energy loss and storage [26, 27]; impedance/admittance of an inductor connected as one port has also been used to define its Q [7, 11]. The impedance-based Q definition is used here due to its popularity among circuit designers.

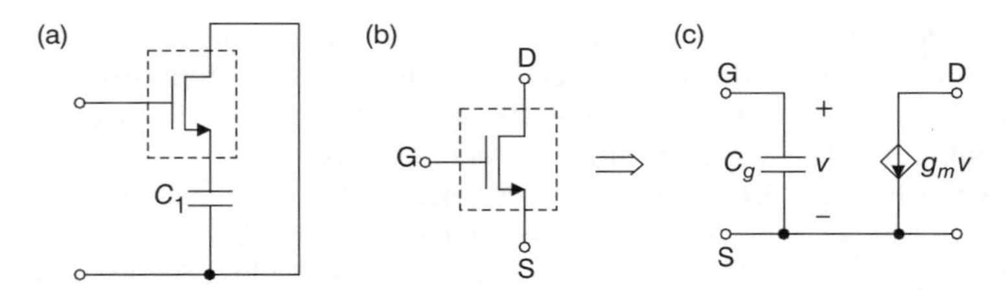

Figure 3.4 *(a) A negative resistor. The circuit in the dashed box is meant to indicate the transistor's small signal equivalent circuit. (b) Simplified model for the transistor*

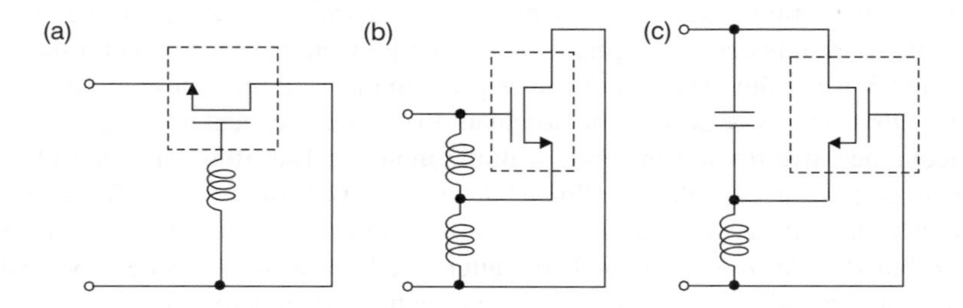

Figure 3.5 *Single-ended negative resistors. The circuit in the dashed box is meant to indicate the transistor's small signal equivalent circuit*

(the bias is omitted), and a simplified high frequency transistor model is shown in Figure 3.4(b). It can be easily derived that the terminal impedance of the circuit in Figure 3.4(a) is

$$Z(j\omega) = -\frac{g_m}{\omega^2 C_1 C_g} + \frac{C_1 + C_g}{j\omega C_1 C_g} \tag{3.1}$$

This equation shows that $Z(j\omega)$ has a negative real part, which is frequency-dependent and inversely proportional to ω^2. We also see that the impedance has a nonzero imaginary part; this is usually the case for on-chip realisations of negative resistors.

Figure 3.5 shows some other single-ended designs that can be found in the literature [34–36]. They all share the two properties discussed above: that the negative real part is strongly frequency-dependent, and that the impedance has a nonzero imaginary part which should be taken into account in circuit synthesis. When using these negative resistors in a filter, one should make sure that the above two properties will not unacceptably distort the filter's frequency response.

A major problem with the single-ended designs is that they are susceptible to noise and interference coupled through supply lines and the substrate. Hence, although the single-ended designs can often be found in circuits using GaAs technologies, where the substrate is semi-insulating, and where digital circuits do not usually co-exist with analogue circuits, they are not as attractive for use in silicon technologies. Another problem with many single-ended designs is that they have rather high even-order distortion.

3.3.2 Balanced schemes

It is possible to combine two single-ended negative resistors and construct a balanced structure, but this is not a common practice. A differential negative resistor is often designed by connecting a differential transconductor in a way shown in Figure 3.6(a). Many kinds of differential transconductors from the literature on g_m-C filters [37–41] can be used to implement negative resistors, and Figure 3.6(b)–(d) shows a few examples. Compared with single-ended designs, the differential structures have the following advantages:

- They are insensitive to noise and interference coupled through supply lines and the substrate.
- They have smaller even-order distortion.
- Many linearisation methods used for balanced transconductor stages can also be used for negative resistors.
- For the single-ended examples given in Figures 3.4(a) and 3.5, we have seen that the real parts of their impedances are strongly frequency-dependent, and

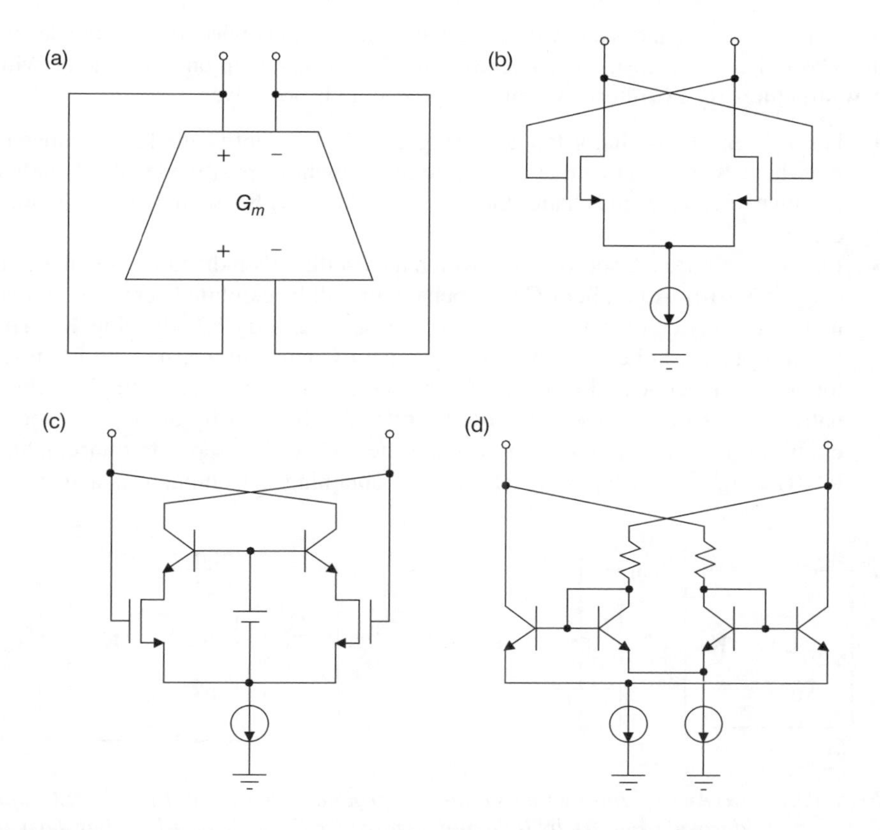

Figure 3.6 (a) Negative resistor built by a differential transconductor. (b)–(d) Examples for (a)

thus these single-ended negative resistors are often used for narrow-band applications. However, in the case of negative resistors using the reported balanced transconductors [37–41], the real part of the impedance is usually not strongly dependent on frequency, until a certain "cutoff" frequency is reached. If that cut-off frequency is far above the signal band, the negative resistor can be used for wide-band applications.

When designing negative resistors for GHz-range active LC filters, one should aim at low noise, good linearity and low power consumption. Simple circuits usually work better than complicated ones for GHz operation. The negative resistors should also be made tunable since the losses in the integrated components have high initial tolerance and vary with environmental changes.

3.4 Second-order sections

3.4.1 A synthesis procedure

Since negative resistances are widely used in oscillators, knowledge about the design of LC oscillators (at least, those operating in class A) can help one to come up with new structures for LC filters. A synthesis procedure is now given.

- For a given LC oscillator topology (Figure 3.7(a)), identify the key parameters **K** (where **K** is a vector) that determine the amount of positive feedback and/or negative resistance, and change them to new values, say **K**′, so that the oscillations cease.
- Identify the nodes A and B across which the oscillator's inductance is connected (Figure 3.7(b)). A parallel LC tank behaviour will be exhibited across these two nodes, the losses of which can be compensated exactly by adjusting **K**′. The resulting tank can be used in lieu of passive LC tanks in well-known LC filter topologies, like coupled-resonator filters (see Section 3.5 for an example). Alternatively, a self-contained second-order filter can be built by driving the above circuit at a suitable point, or by driving the tank with a separate transconductor (G_m) fed by an input signal v_{in}, and obtaining the output v_{out} across the

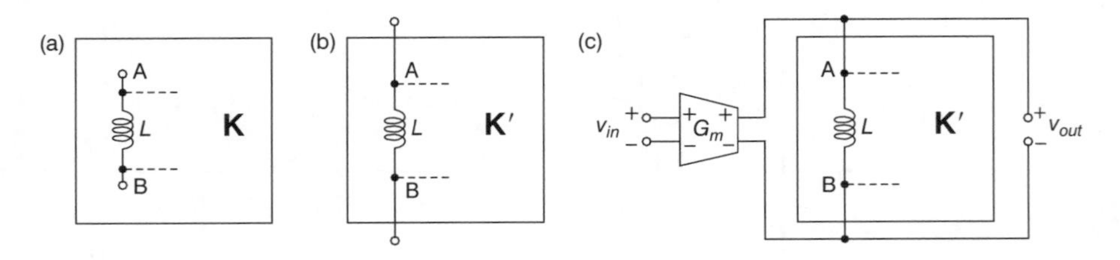

Figure 3.7 *(a) An oscillator with its inductor terminals identified. (b) Active LC tank derived from (a), by reducing positive feedback. (c) g_m-LC filter derived from (b)*

same two nodes (Figure 3.7(c)). For most oscillator topologies, the resulting transfer function V_{out}/V_{in} will be a bandpass one, of essentially second order. The value $\mathbf{K}'$ can be used to adjust the Q of the poles for a desired frequency response.

- If necessary, replace the active device(s) with ones of higher linearity (this may make necessary the use of composite devices).
- (Recommended:) convert to balanced topology.

Variations of the above procedure can be used; for example, one can use nodes other than A and B above to feed the transconductor current, or take the output voltage across nodes other than A and B, depending on the LC oscillator prototype.

Using the above procedure, a large number of topologies for active LC filters can be generated and compared, notably in terms of noise and dynamic range. For example, Figure 3.8(a) shows a Colpitts oscillator; the nodes across the inductor are denoted as A (AC ground) and B. We now reduce the biasing current I_1 to a value I_2 to stop the oscillation. The input signal (plus bias) can then be fed to the transistor gate as shown in Figure 3.8(b). Alternatively, we can drive node B with a transconductor as shown in Figure 3.8(c) (modifications may have to be made, in order to make sure that the output of the transconductor is biased at an appropriate DC point; for example, V_{DD} can be lowered to a different value, V'_{DD}, through a low-impedance level shifter). The above single-ended active LC biquads can be converted to balanced ones as shown in Figure 3.8(d) and (e). The Q of the biquads can be tuned by changing I_2.

3.4.2 Noise, dynamic range and power

An active LC biquad can have larger DR than a g_m-C biquad with the same frequency response and power consumption.[4] This is pointed out in Reference 14, where an active LC biquad, as shown in Figure 3.9(b), is compared against a g_m-C biquad shown in Figure 3.9(a), which has an optimal dynamic range [42]. According to [42], the optimum DR which can be achieved by a high-Q g_m-C biquad is

$$DR_{opt} = \frac{V_{max}^2 C_{total}}{4kT\xi Q} \tag{3.2}$$

in which k is Boltzmann's constant, T is the absolute temperature, ξ is the noise factor used to account for excess noise contributed by the transconductors, V_{max} is the allowed maximum RMS signal voltage for specified linearity performance and C_{total} is the total capacitance in the biquad.

One can also calculate the DR of the active LC biquad shown in Figure 3.9(b), under the assumption that the original lossy LC tank can be modelled as an inductor, a capacitor and a resistor (R_L) all in parallel, and those values are not frequency-dependent. Although this assumption may not completely describe a real situation, it

[4] In this chapter, we do not consider at all the so-called "active inductors", i.e. active circuits not containing actual inductors, but which exhibit inductive behaviour. Some of these circuits are quite noisy and nonlinear, and the resulting DR is poor. Unfortunately, it is common in the literature on active inductors to ignore noise and DR issues.

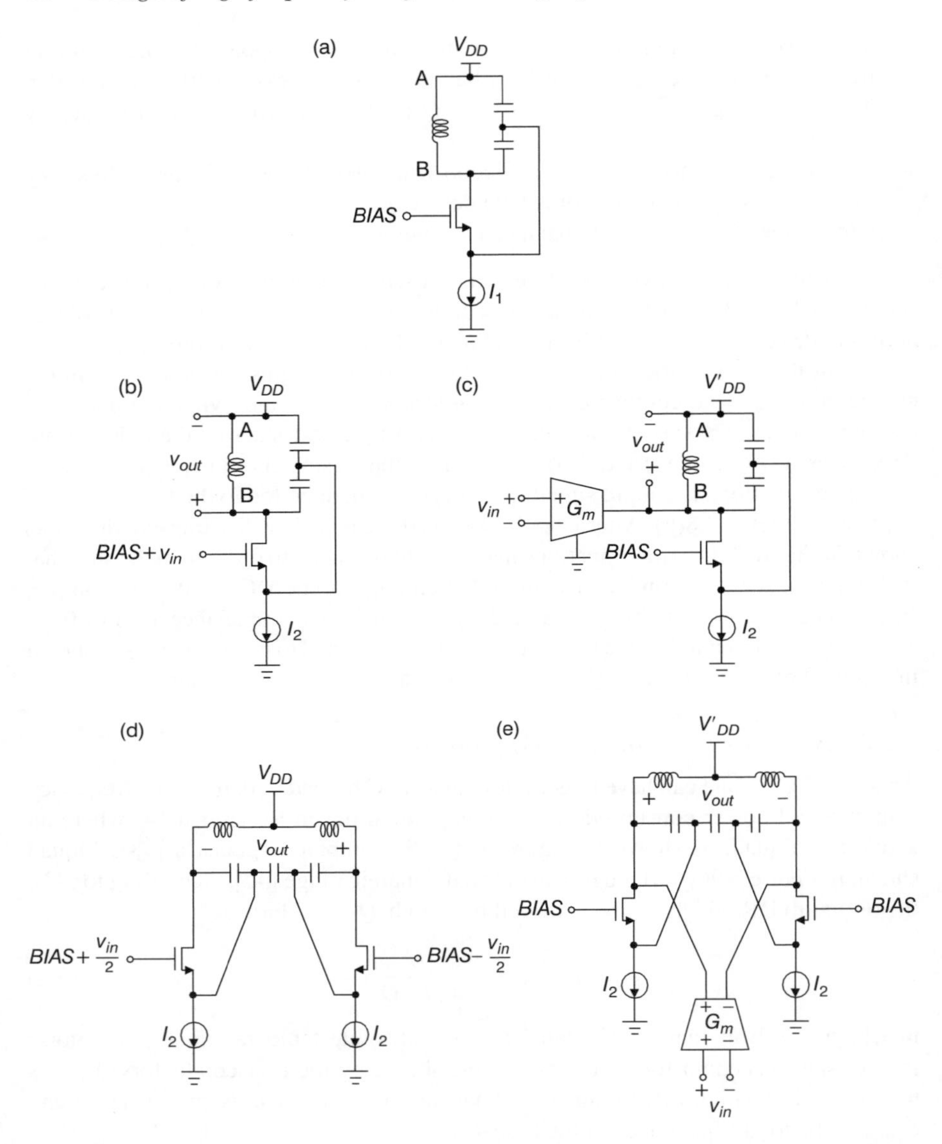

Figure 3.8 (a) A Colpitts oscillator. (b,c) Single-ended active LC biquads drived from (a). (d,e) Balanced LC biquads converted from (b) and (c), respectively

simplifies analysis and gives usable results. For the filter in Figure 3.9(b), if the Q of the filter is high, but the Q of the original LC tank is low (which is the case that we are interested in), G_{m1} can be much smaller than G_{m2}, and thus its noise contribution can be neglected. For such a filter, a straightforward calculation shows that the output

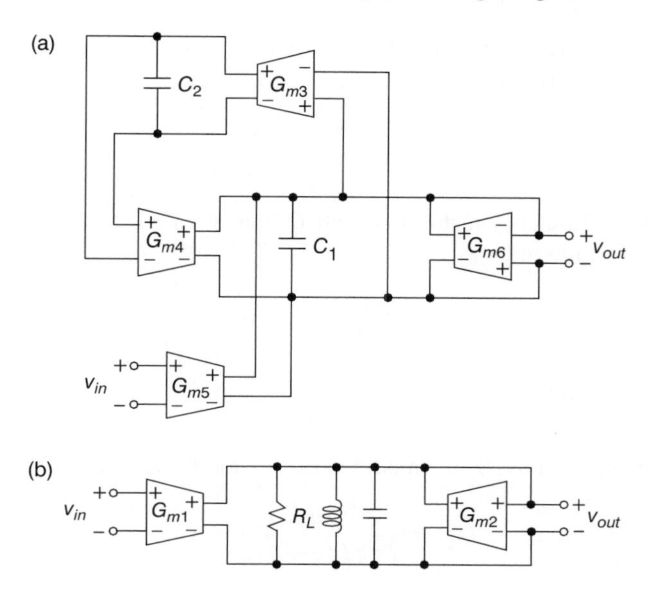

Figure 3.9 (a) A g_m-C biquad. (b) An active LC biquad

noise power can be approximated as

$$\overline{v_{nout}^2} = \frac{kT}{C}(\xi + 1)\frac{Q}{Q_0} \tag{3.3}$$

where Q_0 is the quality factor of the original lossy LC tank. The DR of this filter is thus

$$DR = \frac{V_{max}^2}{\overline{v_{nout}^2}} = \frac{V_{max}^2 C}{kT(\xi + 1)} \frac{Q_0}{Q} \tag{3.4}$$

Comparing Eqns. (3.2) and (3.4), we can see that if V_{max} is identical for both cases, the possible DR improvement due to the use of inductors is proportional to Q_0.[5]

Eqns. (3.2) and (3.4) can be extended to show the dependence of DR on power dissipation P_{diss} and -3 dB bandwidth B [14]. For the active LC biquad shown in Figure 3.9(b), when the Q of the filter is much larger than the original Q of the LC tank, G_{m2} can be much larger than G_{m1}, and thus for similar linearity and circuit structure, G_{m2} consumes much more power. The total power dissipation of the biquad (P_{diss}) is approximately equal to the power dissipation of G_{m2}. The maximum output signal power of G_{m2} is

$$P_{sm2} = \frac{V_{max}^2}{R_N} \tag{3.5}$$

[5] The dynamic range calculated by Eqn. (3.4) is a little different from the one given in [14], where the noise generated by the loss was exaggerated to be equal to the noise generated by the negative resistor, and thus 2ξ was used in lieu of $\xi + 1$ in Eqns. (3.3) and (3.4).

where $R_N = 1/G_{m2}$ is the absolute value of the negative resistor. The power dissipation can be written as

$$P_{diss} = \frac{P_{sm2}}{\eta} = \frac{V_{max}^2}{\eta R_N} \tag{3.6}$$

in which η is an efficiency factor. For high-Q biquads, we have $R_N \approx R_L$, and thus Eqn. (3.6) can be rewritten as

$$\eta P_{diss} \approx V_{max}^2/R_L = V_{max}^2 \frac{\omega_0 C}{Q_0} \tag{3.7}$$

Using Eqn. (3.7) and

$$\omega_0 = 2\pi BQ \tag{3.8}$$

with B the bandwidth in Hz, the modified expression for Eqn. (3.4) becomes

$$DR = \frac{\eta P_{diss}}{2\pi kT(\xi + 1)BQ^2} Q_0^2 \tag{3.9}$$

For the g_m-C filter shown in Figure 3.9(a), we follow a derivation similar to that in Reference 43. When the Q is high, $G_{m3,4}$ can be much larger than $G_{m5,6}$, and thus for similar linearity and circuit structure, $G_{m3,4}$ consume much more power. The total power dissipation of the biquad (P_{diss}) is approximately equal to the power dissipation of G_{m3} and G_{m4}. At the centre frequency ω_0, the quantity corresponding to P_{sm2} in Eqn. (3.5) is

$$P_{sm3} = V_{max}^2 \omega_0 C_2 \tag{3.10}$$

for G_{m3}, and is

$$P_{sm4} \approx V_{max}^2 \omega_0 C_1 \tag{3.11}$$

for G_{m4} since G_{m6} is negligible compared to $\omega_0 C_1$ for high-Q cases. Using Eqns. (3.10) and (3.11), the power dissipation can be written as

$$P_{diss} = \frac{P_{sm3} + P_{sm4}}{\eta} = \frac{V_{max}^2 \omega_0 C_{total}}{\eta} \tag{3.12}$$

in which η is, again, an efficiency factor corresponding to the η in Eqn. (3.6). Using Eqns. (3.12) and (3.8), the modified expression for Eqn. (3.2) becomes

$$DR_{opt} = \frac{\eta P_{diss}}{8\pi kT\xi BQ^2} \tag{3.13}$$

Comparing Eqns. (3.9) and (3.13), we can see that for the same bandwidth and power dissipation,[6] the use of inductors can improve the DR of a biquad by at least Q_0^2

[6] If G_{m2}, G_{m3} and G_{m4} share similar circuit structures, η and ξ in Eqn. (3.9) should be similar to those in Eqn. (3.13).

times[7] (assuming that $\xi \geq 1$). For a Q_0 of 6, ideally, DR can increase more than 36 times, or 15.6 dB. This advantage is clear motivation for research on Q-enhanced LC filters. As a performance example, we mention that a 1.9 GHz fourth-order bandpass filter with a Q of 12.5 has been implemented [6], which dissipates 54 mW, and has a 1 dB compression DR of 63 dB and an SFDR of 49 dB.

3.5 High-order filters

3.5.1 Design issues and examples

It is common practice in low frequency design to construct a high-order filter by cascading biquads. However, the required isolation between two biquad stages usually cannot be achieved at GHz frequencies. Hence, active LC filters have to use other structures, and classic LC filter synthesis comes in handy [44]. With the desired frequency response in mind, we can design a lossless LC filter as a prototype. Once losses are introduced into the reactive components, negative resistors can be added to enhance Q, but whether the desired frequency response can still be preserved is questionable, as Sections 3.2 and 3.3 have shown that using negative resistors may complicate the transfer function of the filter. One needs to consider where in the network it is appropriate to connect the negative resistors, and whether we can pre-distort the lossless LC filter, so that the filter's frequency response can be least distorted after introducing losses and negative resistors. We discuss these questions by using a simple example.

Assume that we want a filter having a fourth-order bandpass Butterworth frequency response with a centre frequency of 2 GHz and −3 dB bandwidth of 200 MHz. Figure 3.10 shows a possible lossless LC prototype and its frequency response. We now replace the lossless inductors with lossy inductors whose model is shown in Figure 3.3(b) (it can be shown that the impedance of the resulting grounded inductor is inductive at the frequencies of interest, and the equivalent parallel inductance is 4.62 nH at 2 GHz). Figure 3.3(c) shows that the peak of the inductor's Q is around 2 GHz, which is the centre frequency of the desired filter. Hence, in this filter design neither the series loss nor the substrate loss of the inductor can be neglected, and thus as far as frequency response is concerned, no obvious advantage is gained whether we connect a negative resistor in parallel or in series with the lossy inductor. In what follows, we choose to connect negative resistors in parallel with the inductors.

One can compensate the losses of inductors individually by connecting a negative resistor in parallel with each inductor. One can also try to use only one negative resistor and connect it with either the inductor close to the source or the one close to the load (simulation shows that the frequency response of the two cases is identical);

[7] The derivation of Eqn. (3.9) implies that, if Q_0 increases n times but the power dissipation does not change, for similar linearity performance, V_{max} will increase $\sqrt{n}$ times (see Eqn. (3.7)). This may not be true when V_{max} is limited by factors such as supply voltage, other than power dissipation. Hence, the comparison between Eqns. (3.13) and (3.9) should be used with caution.

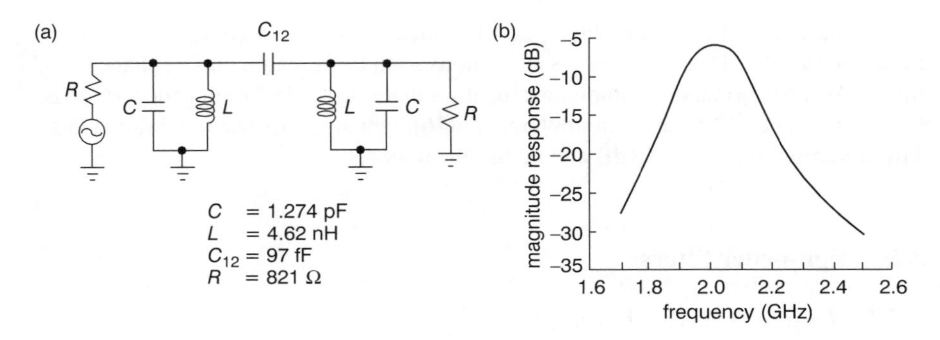

Figure 3.10 A fourth-order lossless LC filter and its frequency response

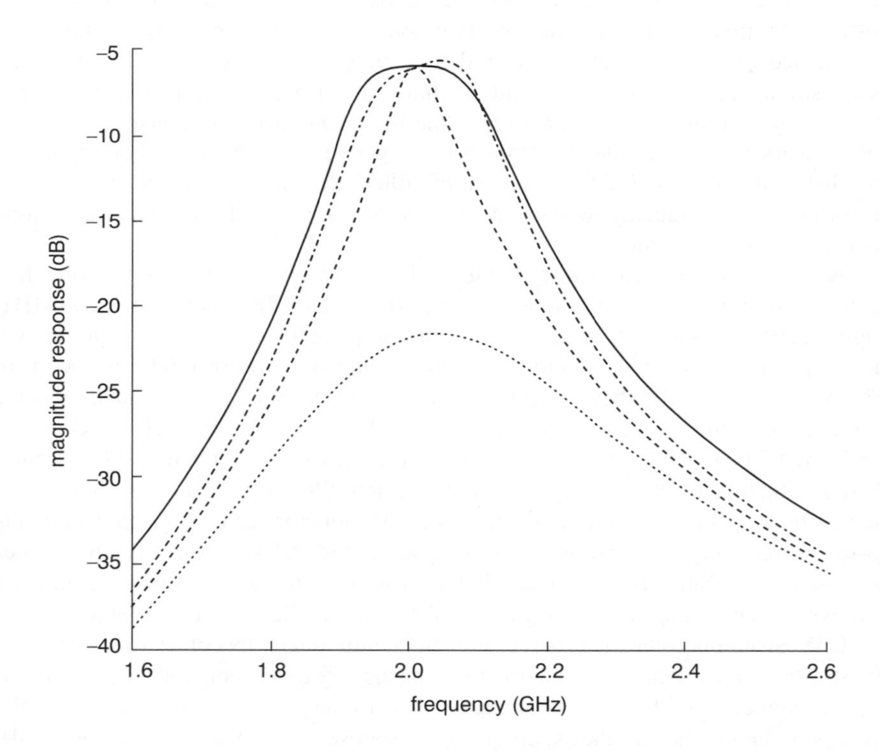

Figure 3.11 Magnitude response of the lossless LC filter (solid line), the lossy LC filter not using negative resistors (dotted line), the LC filter using only one negative resistor (dashed line) and the LC filter using two negative resistors (dotted-dashed line)

by choosing an appropriate value for the negative resistor, all the losses can be compensated at least at one frequency. The frequency response of the filter using only one negative resistor is compared to that of the filter using two negative resistors in Figure 3.11; also shown are the frequency responses of the lossless filter used as prototype and of the lossy filter not using any negative resistors. It can be observed that the negative resistors compensate the losses but introduce distortion in the frequency response compared with the lossless filter, and that the frequency response of the filter using only one negative resistor is more distorted than that of the filter using two negative resistors.

It should be noted that the frequency response of the filter using two negative resistors may still be unacceptably distorted if the starting inductor Q is low. A possible solution to this problem is to predistort the frequency response of the lossless LC prototype, a known practice in passive lossy LC filter synthesis [44–47]; however, predistortion does not work well when the Q of the reactive components is lower than the highest Q of the conjugate poles of the filter. The situation with active LC filters is different in that we use negative resistors to enhance the Q of the reactive components. Even for passive LC filters, predistortion is a difficult subject, and assumptions, such as lossless capacitors or uniform dissipation, have to be made in some of the techniques in the literature in order to come up with systematic solutions. For active LC filters, not only do we have negative resistors, but also their impedances may be complicated functions of frequency, and these make predistortion extremely difficult.

One challenge in high-order active LC filters is the analysis of their DR, which heavily depends on the structures of the filters. Systematic analysis of DR for high order active LC filters is lacking. Generally we do not have enough degrees of freedom to scale the peak of the voltage amplitude response at the output of each negative resistor, and from experience with active filters not using inductors, one expects that this will place a penalty on the DR of active LC filters.

3.5.2 Predistortion

For the fourth-order filters used as examples in the previous section, we found that if mutual inductive coupling, instead of capacitive coupling, is used, the passband frequency response will be distorted in the opposite direction from that shown in Figure 3.11. By using both types of coupling mechanism, we can reduce the passband distortion. The prototype in Figure 3.10(a) is then changed to the one in Figure 3.12(a), and the frequency response of the Q-enhanced LC filter using lossy inductors, two negative resistors and the new prototype is compared to the ideal frequency response in Figure 3.12(b).

As will be discussed in the next section, it is necessary to make on-chip active LC filters automatically tunable. This restricts the freedom of predistortion. When indirect tuning schemes are used, in order to achieve decent matching, it is desirable that all the inductors in the filter be identical. This desirable property should be preserved during the predistortion process.

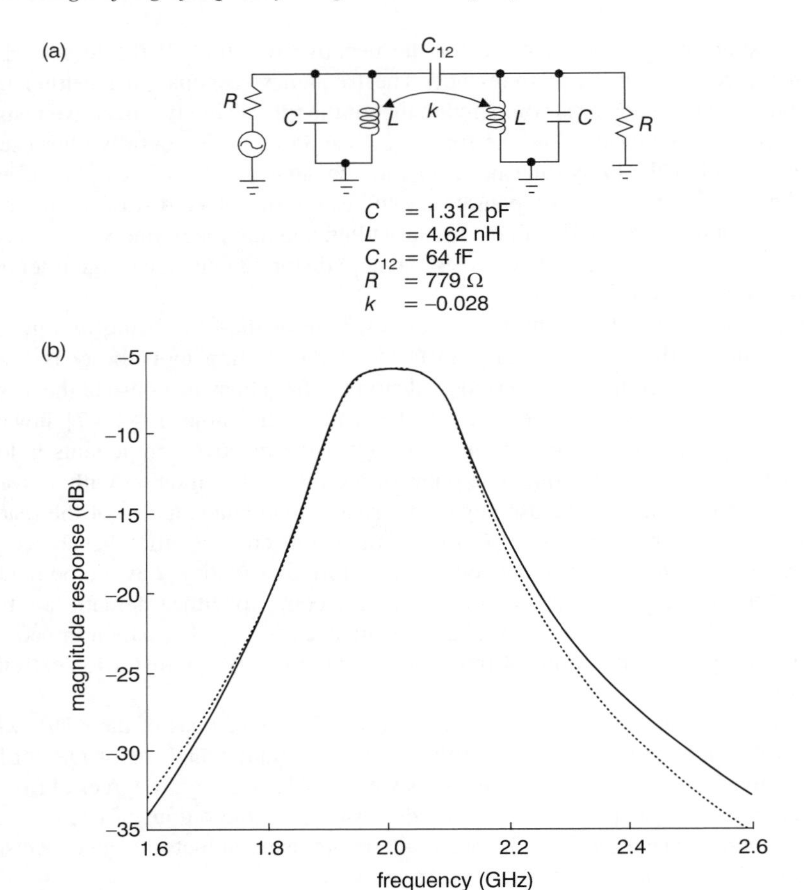

Figure 3.12 *(a) Predistorted lossless LC filter prototype. (b) Frequency response of the Q-enhanced LC filter using lossy inductors, two negative resistors and the prototype in (a) (dotted line), compared to the ideal frequency response (solid line)*

In order to reduce the complexity of the tuning system, mutual inductive coupling is more appealing than capacitive coupling,[8] because mutual inductive coupling coefficients, which are mostly determined by the coupled inductors' lateral physical layout, are stable in the presence of technology variations and temperature changes, and thus do not require tuning.

The example given in Figure 3.12 used both mutual inductive coupling and capacitive coupling, and this is not desirable, according to the above discussion. But for such a fourth-order filter having two LC tanks, this is the only apparent solution for

[8] Because spiral inductors on silicon are lossy and area-consuming, it is not wise to use self-inductive (nonmutual) coupling.

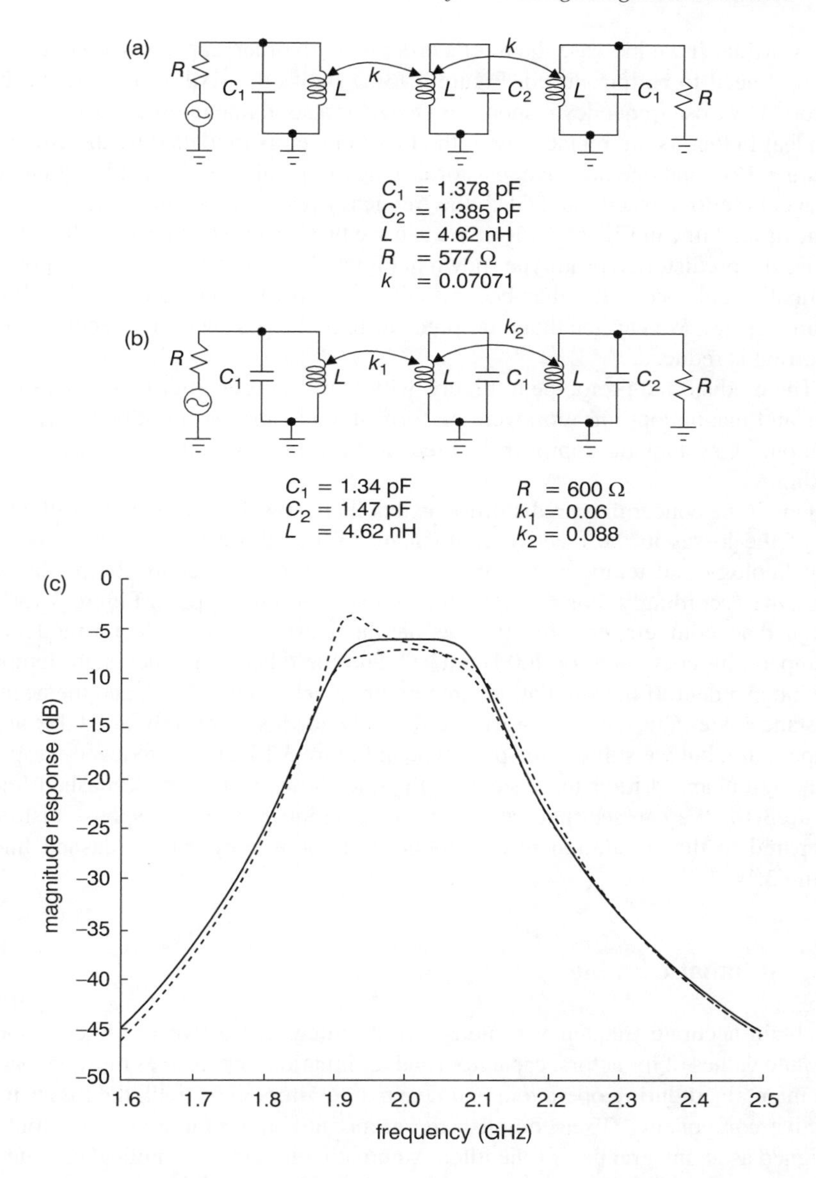

Figure 3.13 (a) Original filter prototype. (b) Predistorted filter prototype. (c) Frequency response for ideal lossless filter (solid line), Q-enhanced filter using the prototype in (a) (dashed line) and Q-enhanced filter using the prototype in (b) (dotted-dashed line)

predistortion. If we increase the filter's order to sixth order, capacitive coupling then may not need to be introduced. Figure 3.13(a) shows the sixth-order original filter prototype, whose frequency response is shown as a solid line in Figure 3.13(c). After the ideal inductors are replaced with the lossy inductors modelled by the circuits in Figure 3.3(b), and one negative resistor is added in parallel with each lossy inductor to cancel the loss exactly at 2 GHz, the frequency response becomes the one shown as the dashed line in Figure 3.13(c). We can see that the passband is now distorted. If we use the predistorted prototype shown in Figure 3.13(b), the frequency response of the final Q-enhanced filter then becomes the one shown as the dotted-dashed line in Figure 3.13(c). We can see that after predistortion, the passband frequency response distortion is reduced.

The predistorted prototype in Figure 3.13(b) was chosen after numerous simulations and manual optimisation were performed, but it may still not be the optimum solution. A systematic approach to predistorting on-chip active LC filters is still lacking.

One issue concerning predistortion is that its results depend on the absolute values of the losses in the reactive components. When those values change because of technology and temperature variations, we may need to adapt the predistorted prototype accordingly. For example, the predistorted prototype in Figure 3.13(b) is designed at room temperature. Assume that the resistors in the inductor model have a temperature coefficient of 4000 ppm/°C, and the other components are temperature independent. If the simulation temperature increases to 100°C, and the negative resistances are adjusted to compensate the inductor losses exactly at 2 GHz at this temperature, but we still use the prototype in Figure 3.13(b), the frequency response of the Q-enhanced filter then becomes the one shown as the dotted-dashed line in Figure 3.14. We can see that we now have different frequency response distortion compared to the simulation result produced at room temperature (dashed line in Figure 3.14).

3.6 Automatic tuning

To obtain accurate frequency response for an integrated active LC filter, accurate absolute values of inductors, capacitors and termination impedances must be realised and maintained during operation, and negative resistors must track the losses in the reactive components. To accomplish these, an automatic tuning system should be designed as an integral part of the filter. Automatic tuning is a complicated issue. We refer the reader to Chapter 7 of this book and the literature [48–51] for the general principles and schemes involved. In this section, we only discuss certain additional issues that surface in the automatic tuning of active LC filters.

The use of inductors poses great challenges to automatic tuning. Because the losses of the on-chip inductors generally do not scale exactly with inductances, and cannot be accurately predicted, it is desirable to have all the inductors inside the filter identical, to simplify the tuning of negative resistances. Setting all the inductors at the same value requires freedom in choosing inductor values during synthesis; this

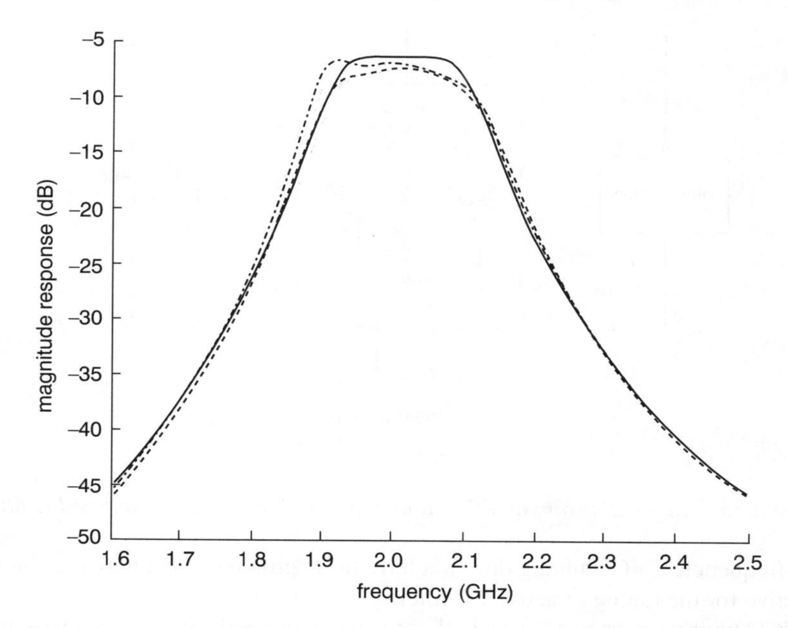

Figure 3.14 *Frequency response for ideal lossless filter (solid line), Q-enhanced filter using the prototype in Figure 3.13(b) at room temperature (dashed line) and Q-enhanced filter using the prototype in Figure 3.13(b) at* 100°C *(dotted-dashed line)*

freedom is provided for some filter structures such as coupled-resonator LC filters [45]. If the inductors in the filter are the same, identical inductors should also be used in the tuning system to improve the tuning accuracy if an indirect tuning system is used.

The equivalent losses of the on-chip spiral inductors are frequency-dependent. In the presence of all the nonidealities, in order for the inductors in an indirect tuning system to accurately track the ones in the filter, they should be tuned using a frequency inside the signal band. If the finite isolation between the filter and the tuning system cannot effectively prevent the feed-through of the in-band tuning signal to the filter, one cannot have the tuning system working at the same time as the filter. Signal processing has to be interrupted while the filter is being tuned. In some communication systems, like GSM, which use time-division multiple access, idle time slots between signal receiving and signal transmitting are available, in which the tuning system can be activated. Direct tuning [51] may be used in this case. At the end of the idle time slots, the tuning system can be turned off, and the tuning signals can be sampled and held or stored digitally, and used to control the filter in the following receiving or transmitting time slots.

Voltage-controlled filter (VCF) tuning [52–54] and voltage-controlled oscillator (VCO) tuning [6, 41, 55–60] are popular methods used for frequency tuning of continuous-time filters. Because VCF tuning needs a reference signal with low harmonic content and a phase detector having low offsets, it is very difficult to realise at

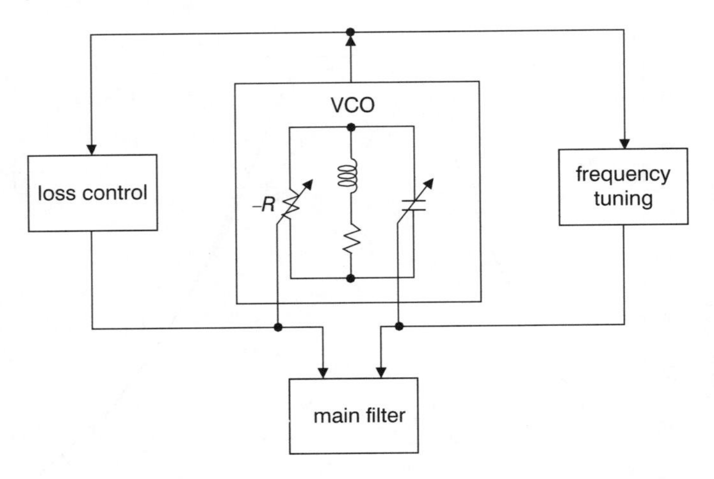

Figure 3.15 Block diagram of VCO tuning system for integrated active LC filters

GHz frequencies. VCO tuning does not have these problems, and may thus be more attractive for the tuning of active LC filters.

VCO tuning can be used to set both the capacitance and the negative resistance as shown in Figure 3.15 [6, 59, 60]. The capacitances are made variable, and the tuning of the capacitances is usually realised by using a phase-locked loop. There are several ways to make variable capacitors. In Reference 2 an impedance multiplier is used to obtain a large capacitance tuning range. However, because active components are used to build the impedance multiplier, this type of variable capacitor is noisy and nonlinear. Varactors have also been used for frequency tuning [6, 59]. They are less noisy than variable capacitors using impedance multipliers, but their limited capacitance tuning range makes the frequency tuning range of the filter smaller. Capacitance can also be varied in discrete steps by switching capacitors in and out of the circuit [5]. But the switches have parasitic capacitances which reduce the frequency tuning range, and parasitic resistances which decrease the Q of the capacitor.

Envelope information of the output of the VCO is often extracted for the use of loss control, and the control loop is expected to achieve a constant envelope when the loss becomes zero. A second-order, instead of high-order, LC VCO should be used in the loss control loop, so that multi-frequency oscillation, which can have varied envelope at zero loss, can be avoided. Utmost care should be taken when designing such a loop. In Reference 60 it is pointed out that if one directly uses the envelope as the control target, the loss control loop may be unstable for high frequency operation. To solve this problem, one can use a loss control method which extracts the information about Q of the VCO from the envelope and thus uses Q directly as the control target [60].

Using VCO tuning to set capacitances and negative resistances is still not enough to achieve accurate frequency response; the latter also requires precise damping resistances. Tuning the damping resistances will further increase the complexity of the tuning system. For the sake of simplicity, the tuning of the damping resistance is usually left out [6, 59].

3.7 Conclusions

The advent of highly integrated wireless transceivers is a platform for the research and development of active LC filters on silicon chips. The design of such filters presents great challenges to designers. The quality of on-chip reactive components is the main reason for this, since the losses and the parasitics in those components can result in frequency response distortion, and the need for active components, like negative resistors, limits dynamic range. Nevertheless, the performance already achievable may be sufficient for the implementation of such filters in certain parts of the transmitter, e.g. at the output of mixers, where a large signal must be bandpass-filtered. Many improvements are needed, and can be expected in the future, in the realisation of higher quality on-chip inductors and capacitors, in filter synthesis when the reactive component losses are large, in the accurate compensation of such losses using negative resistances, in the implementation of those resistances with low noise and high linearity, and in the automatic tuning of on-chip active LC filters.

3.8 References

1 DUNCAN, R. A., MARTIN, K. and SEDRA, A.: 'A Q-enhanced Active-RLC Filter', *IEEE Tran. Circuits and Systems, II*, Vol. 44, No. 5, pp. 341–347, May 1997.

2 PIPILOS, S., TSIVIDIS, Y., FENK, J. and PAPANANOS, Y.: 'A Si 1.8 GHz RLC Filter with Tunable Center Frequency and Quality Factor', *IEEE Journal of Solid-State Circuits*, Vol. 31, No. 10, pp. 1517–1524, October 1996.

3 KUHN, W. B., STEPHENSON, F. W. and ELSHABINI-RAID, A.: 'A 200 MHz CMOS *Q*-enhanced LC Bandpass Filter', *IEEE Journal of Solid-State Circuits*, Vol. 31, No. 8, pp. 1112–1121, August 1996.

4 GAO, W. and SNELGROVE, W. M.: 'A Linear Integrated LC Bandpass Filter with Q-enhancement', *IEEE Trans. Circuits and Systems, II*, Vol. 45, No. 5, pp. 635–639, May 1998.

5 KUHN, W. B., YANDURU, N. K. and WYSZYNSKI, A. S.: 'A High Dynamic Range, Digitally Tuned, Q-enhanced LC Bandpass Filter for Cellular/PCS Receivers', *IEEE Radio Frequency Integrated Circuits Symposium*, pp. 261–264, 1998.

6 LI, D. and TSIVIDIS, Y.: 'A 1.9 GHz Si Active LC Filter with On-chip Automatic Tuning', *Proc. 2001 IEEE International Solid-State Circuits Conference*, pp. 368–369, February 2001.

7 LONG, J. R. and COPELAND, M. A.: 'The Modeling, Characterization, and Design of Monolithic Inductors for Silicon RF ICs', *IEEE Journal of Solid-State Circuits*, Vol. 32, No. 3, pp. 357–369, March 1997.

8 CRANINCKX, J. and STEYAERT, M. S. J.: 'A 1.8-GHz Low-phase-noise CMOS VCO Using Optimized Hollow Spiral Inductors', *IEEE Journal of Solid-State Circuits*, Vol. 32, No. 5, pp. 736–744, May 1997.

9 KOUTSOYANNOPOULOS, Y., PAPANANOS, Y., ALEMANNI, C. and BANTAS, S.: 'A Generic CAD Model for Arbitrarily Shaped and Multi-layer

Integrated Inductors on Silicon Substrates', *Proc. 1997 European Solid-State Circuits Conf.*, pp. 320–323, Southampton, UK, September 1997.

10 CHANG, J. Y., ABIDI, A. A. and GAITAN, M.: 'Large Suspended Inductors on Silicon and Their Use in a 2-μ CMOS RF Amplifier', *IEEE Electron Device Lett.*, Vol. 14, pp. 246–248, May 1993.

11 ASHBY, K. B., FINLEY, W. C., BASTEK, J. J., MOINIAN, S. and KOULLIAS, I. A.: 'High Q Inductors for Wireless Applications in a Complementary Silicon Bipolar Process', *Proc. Bipolar and BiCMOS Circuits and Tech. Meeting*, pp. 179–182, 1994.

12 BURGHARTZ, J. N., EDELSTEIN, D. C., SOYUER, M., AINSPAN, H. A. and JENKINS, K. A.: 'RF Circuit Design Aspects of Spiral Inductors on Silicon', *IEEE Journal of Solid-State Circuits*, Vol. 33, No. 12, pp. 2028–2034, December 1998.

13 RAZAVI, B.: *RF Microelectronics*, Prentice Hall, 1998.

14 KUHN, W. B., STEPHENSON, F. W. and ELSHABINI-RAID, A.: 'Dynamic Range of High-Q OTA-C and Enhanced-Q LC RF Bandpass Filters', *Proc. Midwest Symp. on Circuits and Systems*, pp. 767–771, 1994.

15 FREEMAN, E. M.: 'MagNet 5 User Guide – Using the MagNet Version 5 Package from Infolytica', Infolytica, 1993.

16 NIKNEJAD, A. M. and MEYER, R. G.: 'Analysis, Design, and Optimization of Spiral Inductors and Transformers for Si RF ICs', *IEEE Journal of Solid-State Circuits*, Vol. 33, No. 10, pp. 1470–1481, October 1998.

17 ZHAO, J., DAI, W. W. M., KAPUR, S. and LONG, D. E.: 'Efficient Three-dimensional Extraction Based on Static and Full-wave Layered Green's Functions', *35th Design Automation Conference*, pp. 224–229, San Francisco, CA, June 1998.

18 GROVES, R., HARAME, D. L. and JADUS, D.: 'Temperature Dependence of Q and Inductance in Spiral Inductors Fabricated in a Silicon-Germanium/BiCMOS Technology', *IEEE Journal of Solid-State Circuits*, Vol. 32, No. 9, pp. 1455–1459, September 1997.

19 ARMSRONG, E. H.: 'Some recent developments in the Audion receiver', *Proc. IRE*, Vol. 3, no. 4, pp. 215–238, Sept. 1915.

20 CURTIS, L.: 'Selectivity Control for Radio', US Patent 2,033,330, 10 March 1936.

21 HARRIS, H.: 'Simplified Q Multiplier', *Electronics*, Vol. 24, pp. 130–134, May 1951.

22 MUEHLNER, J.: 'Transfer Properties of Single and Coupled Circuit Stages with and without Feedback', *Proc. IRE*, Vol. 39, pp. 939–945, August 1951.

23 BANGERT, J. T.: 'The Transistor as a Network Element', *Bell System Technical Journal*, Vol. 33, No. 2, pp. 329–352, March 1954.

24 KAWAKAMI, M., YANAGISAWA, T. and SHIBAYAMA, H.: 'Highly Selective Bandpass Filters Using Negative Resistances', *Proc. of Symp. on Active Networks and Feedback Systems*, pp. 369–378, New York, April 1960.

25 KRAUSS, H. L., BOSTIAN, C. W. and RAAB, F. H.: *Solid State Radio Engineering*, New York: Wiley, 1980.

26 BOOKER, H. G.: *Energy in Electromagnetism*, London/New York: Peter Peregrinus, 1982.
27 YUE, C. P. and WONG, S. S.: 'On-chip Spiral Inductors with Patterned Ground Shields for Si-based RF ICs', *IEEE Journal of Solid-State Circuits*, Vol. 33, No. 5, pp. 743–751, May 1998.
28 O, K.: 'Estimation Methods for Quality Factors of Inductors Fabricated in Silicon Integrated Circuit Process Technologies', *IEEE Journal of Solid-State Circuits*, Vol. 33, No. 8, pp. 1249–1252, August 1998.
29 HULL, A. W.: 'The Dynatron – a Vacuum Tube Possessing Negative Electric Resistance', *Proc. IRE*, Vol. 6, pp. 5–35, 1918.
30 LOSEV, O.: 'Oscillating Crystals', *The Wireless World and Radio Review*, pp. 93–96, 22 October, 1924.
31 CRISSON, G.: 'Negative Impedances and the Twin 21-type Repeater', *Bell System Technical Journal*, Vol. 10, pp. 485–513, July 1931.
32 HEROLD, E. W.: 'Negative Resistance and Devices for Obtaining It', *Proc. IRE*, Vol. 23, No. 10, pp. 1201–1223, October 1935.
33 KARACAOGLU, U. and ROBERTSON, I. D.: 'MMIC Active Bandpass Filter Using Negative Resistance Elements', *IEEE Microwave and Millimeter-Wave Monolithic Circuits Symposium*, pp. 171–174, 1995.
34 ADAMS, D. K. and HO, R. Y. C.: 'Active Filters for UHF and Microwave Frequencies', *IEEE Trans. Microwave Theory and Tech.*, Vol. MTT-17, No. 9, pp. 662–670, September 1969.
35 KATZIN, P., BEDARD, B. and AYASLI, Y.: 'Narrow-band MMIC Filters with Automatic Tuning and Q-factor Control', *IEEE Microwave and Millimeter-Wave Monolithic Circuits Symposium*, pp. 141–144, 1993.
36 HOPF, B., WOLFF, I. and GUGLIELMI, M.: 'Coplanar MMIC Active Bandpass Filter Using Negative Resistance Circuits', *IEEE Microwave and Millimeter-Wave Monolithic Circuits Symposium*, pp. 229–231, 1994.
37 TANIMOTO, H., KOYAMA, M. and YOSHIDA, Y.: 'Realization of a 1-V Active Filter Using a Linearization Technique Employing Plurality of Emitter-coupled Pairs', *IEEE Journal of Solid-State Circuits*, Vol. 26, No. 7, pp. 937–945, July 1991.
38 ALINI, R., BASCHIROTTO, A. and CASTELLO, R.: 'Tunable BiCMOS Continuous-time Filter for High-frequency Application', *IEEE Journal of Solid-State Circuits*, Vol. 27, No. 12, pp. 1905–1915, December 1992.
39 DE HEIJ, W. J. A., SEEVINCK, E. and HOEN, K.: 'Transconductor and Integrator Circuits for Integrated Bipolar Video Frequency Filters', *IEEE Proc. ISCAS*, pp. 114–117, 1989.
40 TSIVIDIS, Y., CZARNUL, Z. and FANG, S. C.: 'MOS Transconductors and Integrators with High Linearity', *Electronics Letters*, Vol. 22, pp. 245–246, 27 February, 1986; errata: *ibid.*, pp. 619, 22 May, 1986.
41 KRUMMENACHER, F. and JOEHL, N.: 'A 4-MHz CMOS Continuous-time Filter with On-chip Automatic Tuning', *IEEE Journal of Solid-State Circuits*, Vol. 23, No. 3, pp. 750–758, June 1988.

42 GROENEWOLD, G.: 'The Design of High Dynamic Range Continuous-time Integrable Bandpass Filters', *IEEE Trans. on Circuits and Systems*, Vol. 38, No. 8, pp. 838–852, August 1991.
43 BLOM, D. and VOORMAN, J. O.: 'Noise and Dissipation in Electronic Gyrators', *Philips Research Reports*, Vol. 26, pp. 103–113, 1971.
44 TEMES, G. C. and LAPATRA, J. W.: *Introduction to Circuit Synthesis and Design*, New York: McGraw-Hill, 1977.
45 ZVEREV, A. I.: *Handbook of Filter Synthesis*, New York: John Wiley & Sons, 1967.
46 CRAIG, J. W.: *Design of Lossy Filters*, The MIT Press, 1970.
47 WILLIAMS, A. B. and TAYLOR, F. J.: *Electronic Filter Design Handbook, Third Edition*, McGraw-Hill, Inc., 1995.
48 SCHAUMANN, R.: 'The Problem of Automatic Tuning in Continuous-time Integrated Filters', *Proc. 1989 ISCAS*, pp. 106–109.
49 TSIVIDIS, Y. and VOORMAN, J. O. (Eds.): *Integrated Continuous-Time Filters*, Englewood Cliffs: IEEE Press, 1993.
50 SCHAUMANN, R., GHAUSI, M. S. and LAKER, K. R.: *Design of Analog Filters*, Prentice Hall, Inc., 1990.
51 TSIVIDIS, Y.: 'Self-tuned Filters': *Electronics Letters*, Vol. 17, No. 12, pp. 406–407, June 1981.
52 CANNING, J. R. and WILSON, G. A.: 'Frequency Discriminator Circuit Arrangement', UK Patent, 1 421 093, 14 January, 1976.
53 KHORRANMABADI, H. and GRAY, P. R.: 'High Frequency CMOS Continuous-time Filters', *IEEE Journal of Solid-State Circuits*, Vol. 19, No. 12, pp. 939–948, December 1984.
54 GOPINATHAN, V., TSIVIDIS, Y., TAN, K. and HESTER, R. K.: 'Design Consideration for High-frequency Continuous-time Filters and Implementation of an Anti-aliasing Filter for Digital Video', *IEEE Journal of Solid-State Circuits*, Vol. 25, No. 6, pp. 1368–1378, December 1990.
55 TAN, K-S. and GRAY, P. R.: 'Fully Integrated Analog Filters using Bipolar-JFET Technology', *IEEE Journal of Solid-State Circuits*, Vol. 13, No. 6, pp. 814–821, December 1978.
56 BANU, M. and TSIVIDIS, Y.: 'An Elliptic Continuous-time CMOS Filter with On-chip Automatic Tuning', *IEEE Journal of Solid-State Circuits*, Vol. 20, pp. 1114–1121, December 1985.
57 SENDEROWICZ, D., HODGES, D. A. and GRAY, P. R.: 'An NMOS Integrated Vector-locked Loop', *IEEE Proc. ISCAS*, pp. 1164–1167, 1982.
58 WANG, Y. and ABIDI, A. A.: 'CMOS Active Filter Design at Very High Frequencies', *IEEE Journal Solid-State Circuits*, Vol. 25, pp. 1562–1574, December 1990.
59 APARIN, V. and KATZIN, P.: 'Active GaAs MMIC Band-pass Filters with Automatic Frequency Tuning and Insertion Loss Control', *IEEE Journal of Solid-State Circuits*, Vol. 30, pp. 1068–1073, October 1995.
60 LI, D. and TSIVIDIS, Y.: 'A Loss Control Feedback Loop for VCO Indirect Tuning of RF Integrated Filters', *IEEE Trans. on Circuits and Systems, II*, Vol. 47, No. 3, pp. 169–175, March 1998.

Chapter 4

Log domain filters

Douglas Frey

4.1 Introduction

Log domain filters are amongst the newest of ideas in the well developed field of electronic filter design. These filters offer reduced circuit complexity, wider bandwidth, wider dynamic range and lower power consumption in certain applications than their counterparts. For example, with typical signal swings of 100 mV, it is possible to achieve 60–80 dB of dynamic range in circuits operating in the MHz range using power supplies of 2.7 V or less. Moreover, these filters are widely tunable electronically, making them a very interesting choice in such applications as magnetic read channels. We will explore these possibilities and the limitations of log domain filters in this chapter, along with ways of designing and understanding these filters. Beyond the possible performance improvements which may be achievable, they offer a view into an entirely new way of looking at linear filters. Specifically, log domain filters are a type of externally linear, internally nonlinear (ELIN) filters, which we may call a special case of externally linear filters, which includes all circuits that provide a linear transfer function from the input to the output. This broader perspective has not been used historically because it has been tacitly assumed that input-output linear filters had to be made with linear components, or at least linearised components.

Linear filtering is a concept that has existed since the earliest stages of electronic design. Classically the 'linear' part of this terminology is a natural consequence of the linearity of the equations used to derive the design equations, coupled with the assumed linearity of the electronic components used to realise the design equations. Because the first filters developed were passive RLC filters, this assumption of linearity was quite reasonable. With the introduction of active electronics – that is, vacuum tubes, transistors and ultimately integrated electronics – the assumption of linearity became suspect. This is because active electronic elements, such as transistors, are inherently nonlinear. Consequently, active filters were developed under the technically invalid hypothesis that the embedded active elements were linear. Of course, passive RLC components have nonlinearity that was ignored in passive synthesis, but the degree of nonlinearity was almost always so slight as to be insignificant.

The linearity assumption in active filters has classically been justified via a 'small signal' argument. This small signal argument is predicated upon the fundamental idea that any reasonably well behaved nonlinear function looks locally linear, a fact which underlies many useful mathematical techniques from Newton's method in most practical equation solvers to the use of tangent spaces in differential geometry. Basically, we say that a 'small signal' is applied to a nonlinear element, such as a transistor, if the transfer characteristics of the device are well approximated by linear relations over the range of the net signal – that is, DC bias plus the 'small signal' – present. Using this assumption, the nonlinear characteristics of the active devices in a circuit are replaced by linear characteristics which then justify the assumption of overall linearity.

To understand this idea better and lay the groundwork for the understanding of log domain filters, let us consider a special active filter which requires the small signal assumption in deriving a linear filter relation between the input and the output. The circuit of Figure 4.1 is perhaps the simplest possible active filter. A voltage, V_{in}, is applied to the base of the transistor which drives a current into the capacitor. Due to the nonzero dynamic impedance seen looking back into the emitter, a lowpass filter is created from the input voltage to the output voltage, V_{out}, measured across the capacitor. The DC current source is added so that the transistor remains active at all times. This is crucial, since the emitter current must always be positive for this circuit to work properly. Specifically, signal swings across the capacitor require positive and negative current in the capacitor. Hence, the DC bias current must be large enough so that the total emitter current, which equals the sum of the DC bias and the capacitor current, is greater than zero at all times. However, in order for us to make the small signal assumption in this circuit, the DC bias current must not only ensure that the net emitter current remains positive, but it must also ensure that the net emitter current varies only slightly from its quiescent value over the complete range of signals applied at the input. Otherwise, the strong nonlinearity of the base-emitter junction of the bipolar transistor will impact the circuit linearity. The following equations highlight the basic ideas:

$$
\begin{aligned}
I_E = I_{DC} + I_{cap} &= I_S\, e^{V_{BE}/V_t} = I_S\, e^{(V_{in}-V_{out})/V_t} \\
\Rightarrow \quad V_{in} - V_{out} &= V_t \ln\left(\frac{I_E}{I_S}\right) = V_t \ln\left(\frac{I_{DC}+I_{cap}}{I_S}\right) \\
&= V_t \ln\left(\frac{I_{DC}}{I_S}\right) + V_t \ln\left(\frac{I_{DC}+I_{cap}}{I_{DC}}\right) \\
&\approx V_{BE-DC} + V_t \frac{I_{cap}}{I_{DC}}
\end{aligned}
\tag{4.1}
$$

where

$$
V_{BE-DC} = V_t \ln\left(\frac{I_{DC}}{I_S}\right); \quad I_{cap} \ll I_{DC}
$$

Notice that we have assumed that the emitter current is exponentially related to the base-emitter voltage. Technically, it is the collector current, but we have assumed negligible base current. Using the above results, it is a simple matter to verify that the peak to peak base-emitter voltage of the transistor must be significantly less than the thermal voltage, $V_t \approx 26\,\mathrm{mV}$, at room temperature, of the transistor if the linear approximation is to be valid. This translates to a constraint that the net emitter current must not change by more than, say, 10 per cent over the range of input voltages.

Restricting current swings to less than 10 per cent of the DC bias is an extreme restriction when considered from the perspective of dynamic range. This is analogous to restricting voltage swings to 1 volt peak in an op amp circuit powered from + and −12 V supplies. Such a restriction throws away about 20 dB of headroom, reducing the dynamic range by that much. Of course in the case of the op amp no significant distortion increase results from using the top 20 dB of headroom, unlike the case above. Not only does the signal restriction appear as a headroom reduction, but it also amounts to a decrease in power efficiency. Power dissipation in an amplifier is directly related to the bias current levels. In class A amplifiers the bias levels must exceed the largest signal excursions. Hence, restricting signal swings to 10 per cent of the bias level leads to about ten times the power dissipation that one would expect for a given circuit.

Clearly, if a way can be found to allow signal swings to approach the DC bias level in the transistor of Figure 4.1, then dynamic range and power efficiency will be greatly improved. This is exactly what log domain filters do, which explains a key reason for their potential value. Another reason for the interest in these filters is due to their potential for wide bandwidth operation. To understand this, suppose that we choose to improve the performance of the filter of Figure 4.1 by replacing the transistor with an active linearised buffer in series with a resistor as shown in Figure 4.2. We assume that the buffer output impedance is zero and r equals the quiescent dynamic impedance looking into the emitter of the transistor in Figure 4.1. Note that the buffer can be omitted without affecting the apparent transfer function, but we shall assume that the buffering operation is central to the benefit of this active filter to maintain the

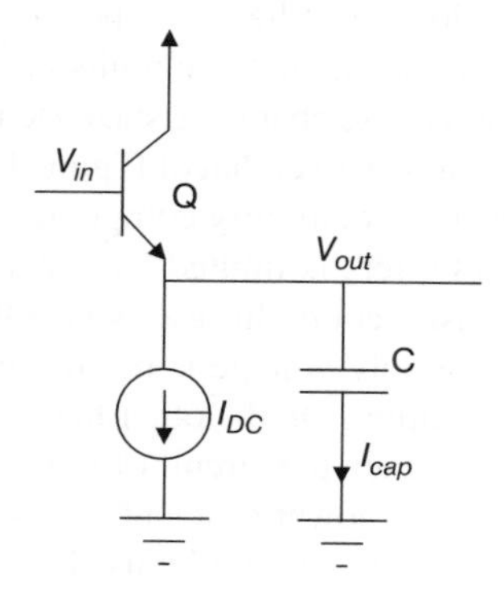

Figure 4.1 Simple active filter

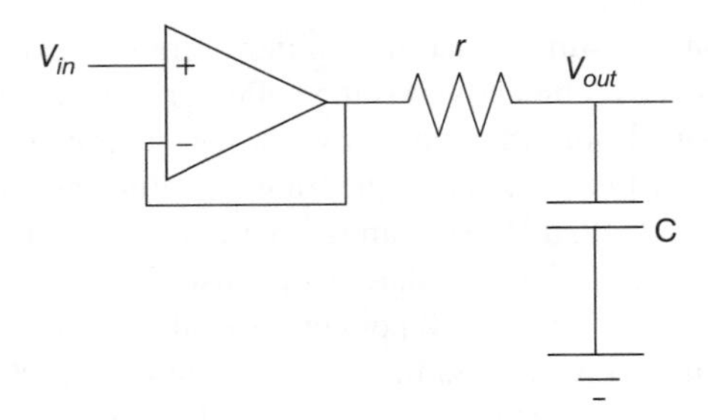

Figure 4.2 RC filter equivalent of Figure 4.1

appropriate correspondence between the examples. The linearised buffer could be an operational amplifier, for example.

Now let us consider the relative performance of the circuits of Figures 4.1 and 4.2. The use of the operational amplifier in Figure 4.2 allows a much wider signal swing in this circuit without significantly compromising linearity. Thus, the dynamic range can be expected to be considerably enhanced. Note, however, that the operational amplifier is likely to be noisier than the single transistor so the dynamic range will be increased by a factor less than the increase in signal swing due to the increased noise floor. The power efficiency may or may not be better depending on the power requirements of the operational amplifier. Nevertheless, in a typical scenario one would expect the power efficiency to improve as well. These facts explain the popularity of operational amplifiers.

Despite these advantages, however, operational amplifiers do present limitations. For example, if the power supply voltage is very low then the input range and output swing of the op amp will limit the usable upper limit of circuit operation and, at the very least, will dramatically reduce the dynamic range advantage described above. Perhaps more importantly, the circuit with the op amp is guaranteed to have less bandwidth capability than the single transistor circuit, making reasonable normalising assumptions about the bandwidth of the op amp relative to the f_T of the transistor. Specifically, the cutoff of the lowpass filter must be significantly less than the gain bandwidth product of the active device in the circuits in order to ensure precision in the lowpass filter amplitude and phase characteristics. Hence, the circuit of Figure 4.1 has an inherent bandwidth advantage over that of Figure 4.2. This will be the case any time the linearised buffer consists of circuitry comprising several transistors. Indeed, the price paid for linearity in buffers is limited bandwidth and an increase in noise and, of course, increased transistor count. In cases where the bandwidth requirements are not severe and the power supply voltage is not too low, however, the circuit of Figure 4.1 is not an optimal solution. On the other hand, if bandwidth is critical and the power supply voltage is low, then the circuit of Figure 4.1 may be preferable. Of course its low dynamic range and power efficiency make the choice difficult. Only in the most demanding cases will this circuit be used. Using a log domain approach as described later will allow this circuit to be considered an attractive alternative in a much wider variety of cases.

4.2 Log domain filtering

As we have seen from the discussion above, the circuit of Figure 4.1 suffers from the fact that its small signal range is significantly limited, in that the circuit loses 20 dB of potential dynamic range. This limitation is due essentially to the nonlinearity of the transistor. The log domain filtering idea allows us to recover this lost dynamic range by pre- and post-distorting the signals in the filter. Specifically, consider the circuit of Figure 4.3, under the assumption that all base currents may be neglected. Please note that we will continue to make this assumption in all of the discussion to follow, until later when we specifically address the impact of nonideal transistor performance on log domain filters. Q_1 pre-processes the input current, I_{in}, producing a voltage, V_0, which is logarithmically related to it. This predistorted input is applied to the base of Q_3 with a level shift (Q_2 accomplishes this), V_{LS}. Q_3 and the capacitor and current source are the 'core filter' of Figure 4.1. Finally, the output, V, of this core filter is post-processed via Q_4 to produce the final output, I_{out}, which is exponentially related to the capacitor voltage. To the uninitiated, it is not clear that the additional nonlinear processing in this circuit does not simply further corrupt the already distorted signal. However, as the following analysis shows, the additional nonlinear processing exactly cancels the nonlinearity of the 'core filter' (Figure 4.1).

The following analysis fully describes the operation of this circuit. We begin by writing the KVL equation corresponding to the 'translinear loop' (we will discuss translinear loops later):

$$V_0 + V_{LS} = V_{BE-Q3} + V \tag{4.2}$$

Now, using the diode law and some basic properties of logarithms, we have

$$V_t \ln\left(\frac{I_{in}}{I_S}\right) + V_t \ln\left(\frac{I_{DC}}{I_S}\right) = V_t \ln\left(\frac{I_{E-Q3}}{I_S}\right) + V_t \ln\left(\frac{I_{in}}{I_S}\right)$$

$$\Rightarrow \quad I_{in} I_{DC} = I_{E-Q3} I_{out} \tag{4.3}$$

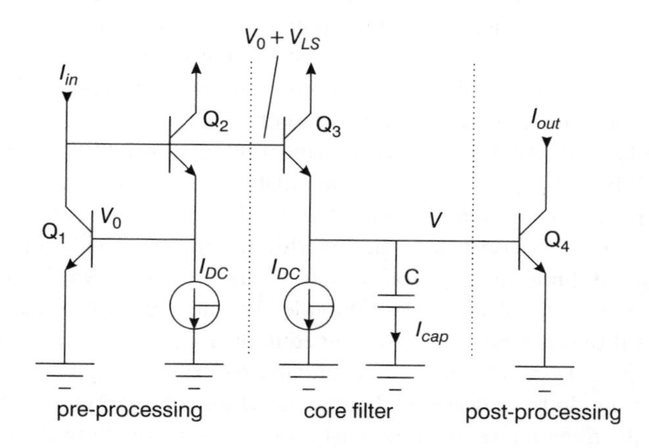

Figure 4.3 Basic log domain filter

Writing a KCL equation at the node at the top of the capacitor and rearranging, we may derive an equation relating the input and the output. We have

$$\begin{aligned} I_{E-Q3} &= I_{DC} + I_{cap}; \quad I_{cap} = C\dot{V} \\ \therefore I_{in}\, I_{DC} &= (I_{DC} + I_{cap})\, I_{out} = I_{DC}\, I_{out} + C\dot{V}\, I_{out} \end{aligned} \tag{4.4}$$

We may eliminate the derivative of the capacitor voltage by exploiting our knowledge of the relationship between the capacitor voltage, V, and the output current, I_{out}:

$$\begin{aligned} I_{out} = I_S\, e^{V/V_t} \Rightarrow \dot{I}_{out} &= \frac{\dot{V}}{V_t} I_S\, e^{V/V_t} = \frac{\dot{V}}{V_t} I_{out} \\ \therefore C\dot{V}\, I_{out} &= C V_t\, \dot{I}_{out} \end{aligned} \tag{4.5}$$

Finally, by combining the above results, we arrive at the differential equation relating the output to the input in the filter. Specifically,

$$I_{in} I_{DC} = C V_t \dot{I}_{out} + I_{DC} I_{out} \Rightarrow \dot{I}_{out} = -\omega_0 I_{out} + \omega_0 I_{in} \tag{4.6}$$

where

$$\omega_0 = \frac{I_{DC}}{C V_t}$$

This first-order differential equation describes a first-order lowpass filter whose cutoff frequency is given by ω_0. The remarkable aspect of this result is that the input and output are related by a *linear* differential equation. This, of course, guarantees that the input–output behaviour is linear. This is despite the fact that the internal signals are quite nonlinear, such as the capacitor voltage, which is logarithmically related to the output. This is precisely why log domain filters are called 'externally linear, *internally nonlinear*' filters. Notice also that the cutoff frequency of this filter is directly proportional to the bias current, I_{DC}. This is an interesting feature of log domain filters not shared by standard linear filters.

The circuit just described is analogous to the original 'log domain filter' introduced by Adams [1]. In that paper it was suggested that one might be able to generalise the result in a trivial way to higher-order filters. However, Adams himself showed in that paper at least one case where the logarithmic pre-processing and exponential post-processing did not, in fact, work exactly. Later it was shown that this pre- and post-processing would indeed work to provide general, exactly externally linear, log domain filters if the core filters were designed in the right way. As a result, the block diagram of Figure 4.4 can be used to generically describe log domain filters. Before proceeding, it should be understood that the basic goal of log domain filter design is to maximise the dynamic range in transistor circuitry by allowing the usable signal range to greatly exceed the small signal range.

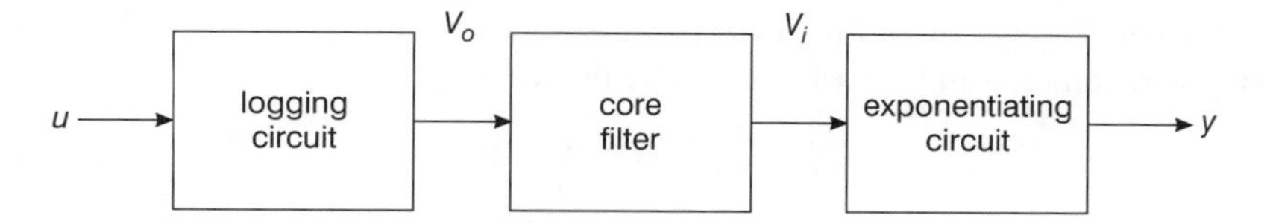

Figure 4.4 General log domain filter block diagram

4.3 Synthesis of log domain filters

4.3.1 Basic theory

Several synthesis approaches have now been proposed for the creation of log domain filters and, as is typically the case, different researchers have adopted their preferences. An attempt will be made to give the reader familiarity with several of these approaches so that he or she may choose that which is deemed most useful. However, to begin it is of value to present a fundamental mathematical framework for log domain filters. The original method [2, 3] proposed for log domain filter synthesis is arguably the most general and unifying approach, since it unifies the explanation of not only log domain filters, but all externally linear filters. This approach is based on the state space description of linear filters. Readers who wish to defer the mathematics may wish to skip this section to get to the synthesis methods directly.

To begin, we assume that the transfer function, $H(s)$, has been specified for some linear filter that we wish to design. Associated with any given linear system is also a linear differential equation relating the input and output of the system (we shall assume that there is only a single input and a single output unless explicitly stated). This differential equation, in general, contains time derivatives of the output up to the Nth order, where N equals the polynomial order of the denominator of the transfer function. It is well known that this single differential equation may always be replaced by a set of N coupled first-order linear differential equations and one linear algebraic equation specifying the output. The N differential equations are called *state equations*, and the algebraic equation is called the *input–output equation*. Collectively these equations are called the *dynamical equations* for the linear filter under consideration. Most elementary circuits and systems textbooks give a good introduction to the specification of dynamical equations for linear circuits. In addition, most modern elementary texts in control theory show how to derive dynamical equations directly from a transfer function in what is generally called the companion form. We will see some examples below of dynamical equations for basic transfer functions.

The dynamical equations for a given practical filter may always be specified in the following standard form:

$$\frac{d}{dt}\overline{\mathbf{x}} = \mathbf{A}\overline{\mathbf{x}} + \overline{\mathbf{b}}u; \quad y = \overline{\mathbf{p}}^T\overline{\mathbf{x}} + du \tag{4.7}$$

where the input, u, and the output, y, are assumed scalars, $\overline{\mathbf{x}} = (x_1, x_2, \cdots, x_N)^T$ is the state vector, $\mathbf{A}$ is the N by N state matrix, $\overline{\mathbf{b}}$ and $\overline{\mathbf{p}}^T$ are N dimensional vectors, and

d is a scalar. The system of Eqn. (4.7) possesses a transfer function, $H(s)$, dependent upon the parameters in Eqn. (4.7) given by the formula,

$$H(s) = \overline{\mathbf{p}}^T (s\mathbf{I} - \mathbf{A})^{-1}\overline{\mathbf{b}} + d \tag{4.8}$$

The reader should note that there is a unique transfer function corresponding to a given system of dynamical equations. However, there are an infinite number of state space descriptions (dynamical equations) for a given transfer function. This fact explains in an abstract way why there are so many filter realisations for a given transfer function.

Starting from a state space description for a linear filter one may derive any of the passive and active filters that are known to engineers. References 4 and 5 give some discussion of this idea that the reader may find useful. An important idea stemming from this observation is that there must be a way to translate the system level description of Eqn. (4.7) into circuit level descriptions, such as Kirchhoff's current and voltage law equations. Otherwise, filter synthesis would not be possible. Log domain filters can be derived by using a special transformation on the dynamical equations. The net result of this transformation is to allow the state equations to become Kirchhoff current law equations for the eventual log domain filter.

To get the basic idea, let us reconsider the first order log domain filter already shown above. Let us design this filter starting from its transfer function. A first-order lowpass filter has the following transfer function:

$$H(s) = \frac{\omega_0}{s + \omega_0} \tag{4.9}$$

Associated with this transfer function are the following dynamical equations:

$$\dot{x} = -\omega_0 x + \omega_0 u; \quad y = x \tag{4.10}$$

Now suppose that we replace the state variable, x, with an exponential function of a new variable, V, and similarly replace the input, u, with an exponential function of another new variable, V_0. This leaves us with the following transformed dynamical equations:

$$x \equiv I_s e^{V/V_t}; \quad u \equiv I_s e^{V_0/V_t}$$

$$\dot{x} = \dot{V}\frac{I_s}{V_t}e^{V/V_t} = -\omega_0 I_s e^{V/V_t} + \omega_0 I_s e^{V_0/V_t}; \quad y = x = I_s e^{V/V_t} \tag{4.11}$$

The constants, I_s and V_t, have been introduced into the transformations to allow flexibility and to make the final results translate to meaningful circuitry. Now let us continue by scaling the state equation above in such a way as to leave only the derivative of the new variable, V, on the left hand side scaled by a new constant, C, which will have obvious significance shortly. Specifically, we have

$$C\dot{V} = -CV_t\omega_0 + CV_t\omega_0 e^{(V_0 - V)/V_t}; \quad y = I_s e^{V/V_t} \tag{4.12}$$

The equations above will take on obvious meaning if we declare that the constants, I_s and V_t, are equal to the reverse saturation current and thermal voltage associated with a bipolar transistor, and the new constant C has units of capacitance. Finally, let us imagine that the variables, V and V_0, are voltages in some network. In this case, the left hand side of the above equation can be thought of as the current flowing in a grounded capacitor of size C. The right hand side can be thought of as the sum of two currents: a constant current (DC source) plus the current through a bipolar transistor junction. These ideas are summarised in the revised version of Eqn. (4.12) shown below:

$$\begin{aligned} C\dot{V} &= I_C = -I_0 + I_0 e^{(V_0 - V)/V_t} \\ &= -I_0 + I_s e^{(V_0 + V_{DC} - V)/V_t} \\ V_0 &= V_t \ln\left(\frac{u}{I_s}\right); \quad y = I_s e^{V/V_t} \\ I_0 &= C V_t \omega_0 \,; \quad V_{DC} = V_t \ln\left(\frac{I_0}{I_s}\right) \end{aligned} \tag{4.13}$$

This revision of the equations captures the full scope of the mathematics introduced in this example. Observe that the voltage, V_0, corresponds to the logarithm of the input, u, and the output, y, is equal to an exponentiated version of the voltage, V. Finally, the voltage, V, is determined through a differential equation that looks like a KCL equation at a node defined by the connection of a grounded capacitor, a current source and an NPN transistor junction with base voltage, $V_0 + V_{DC}$, and emitter voltage, V. Comparing these results with the analysis pertaining to Figure 4.3 reveals that the equations just derived correspond exactly to that circuit, where V_{DC} and I_0 correspond to V_{LS} and I_{DC}, respectively. Thus, we have derived a log domain filter from a transfer function. It now only remains to be seen how this can be generalised to higher-order filters.

The mapping (transformation) idea presented above can be extended in a fairly straightforward way to higher-order filters; however, there are a few pitfalls that one must appreciate before the development of log domain filters can be accomplished in the general sense. First, observe that the exponential mappings (transformations) of Eqn. (4.11) only make sense if the state variable and the input are strictly positive at all times. Otherwise there will not be real values of the voltages, V and V_0, that satisfy these mappings at all times. Since these variables correspond to voltages in a real circuit, this is imperative. Therefore, steps must be taken to ensure this condition by adding a biasing current at the input and possibly at other places in the circuitry.

A corollary to this is that the circuit must possess a suitable DC equilibrium. This is an interesting point. Looking at the circuit of Figure 4.3, it is clear that this circuit possesses a stable DC equilibrium. In particular, in DC equilibrium the capacitor voltage is constant and, therefore, its current is zero. The transistor carries all of the current source current, and thus the circuit operation is quite normal. This observation is borne out by the equation formulation above. In DC equilibrium, all signal derivatives are zero; therefore, the left hand side of Eqn. (4.13) is zero. With the right hand side also equal to zero, we immediately see that the transistor current equals I_0. Note that the DC equilibrium of a circuit is found by solving it with all capacitors open and all inductors shorted.

To see how problems may arise in log domain filter design, let us consider the design of an ideal integrator. Such structures are not generally useful by themselves, but they are quite valuable in multiple feedback loop higher-order filters. Proceeding as before from the transfer function to the dynamical equations and through the mappings, we have the following KCL equation for a log domain integrator:

$$C\dot{V} = I_0 e^{(V_0 - V)/V_t} \tag{4.14}$$

Setting the derivative equal to zero says that the current on the right hand side must be zero. But this time the implication is that the exponential term is zero, which cannot be allowed if the voltages, V and V_0, are to remain finite. This example shows that the issue of DC equilibrium is one which requires serious attention in the design of log domain filters. Of course, DC equilibrium must always be addressed in transistor circuit design.

In order to address this problem in a systematic way, one can manipulate the dynamical equations before using the exponential mappings in such a way that a suitable DC equilibrium is guaranteed. The simplest first step is to take into account the need to offset the input, u. As mentioned above, this must be an always positive quantity. Otherwise, the exponential associated with the input will not yield finite real values for V_0, which represents a voltage in the final circuit realisation. The solution is to define a new input, u_{new}, that is offset from the original input in such a way that the new input is always positive (for all values of u). Specifically, let us define

$$u_{new} = u + u_0 \tag{4.15}$$

where the term, u_0, will be chosen to be a positive constant exceeding the most negative excursion of u for all time. What we are doing, of course, is just adding DC bias to the input. Now, the state equation for the integrator is given by

$$\dot{x} = \omega_0\, u_{new} - \omega_0 u_0 \tag{4.16}$$

Following the steps earlier, applying the exponential mappings to x and the new input, u_{new}, we now have the final result

$$C\dot{V} = I_0 e^{(V_0 - V)/V_t} - I_0 u_0 e^{-V/V_t} \tag{4.17}$$

This time, finite values for both V and V_0 will satisfy the DC equilibrium equation. This equation predicts a fundamental rule in log domain filters – namely, that ideal integrators will always be realised by the interplay of positive and negative log domain transconductors. We will see this in the example given later.

The trick used above to alleviate the DC equilibrium problem works quite well in this case; however, in general cases this is not enough to solve the DC problems created by casually transforming the state equations with the exponential mappings. Fortunately, we may extend the trick to the state variables, thereby solving all of the problems encountered in log domain filter design. The more mathematically sophisticated readers may note that we are about to apply a translation of coordinates to the state space. A more general linear transformation can also be used as discussed in Reference 4. This more basic extension can be written down as follows:

$$\overline{\mathbf{x}}_{new} = \overline{\mathbf{x}} + \overline{\mathbf{x}}_0 \Rightarrow \overline{\mathbf{x}} = \overline{\mathbf{x}}_{new} - \overline{\mathbf{x}}_0 \tag{4.18}$$

Here, each state variable is offset by its own constant (DC) bias. Combining these ideas together results in the transformed dynamical equations given by

$$\begin{aligned} \frac{d}{dt}(\overline{\mathbf{x}}_{new} - \overline{\mathbf{x}}_0) = \frac{d}{dt}\overline{\mathbf{x}}_{new} &= \mathbf{A}\overline{\mathbf{x}}_{new} + \overline{\mathbf{b}}u_{new} - \mathbf{A}\overline{\mathbf{x}}_0 - \overline{\mathbf{b}}u_0 \\ y &= \overline{\mathbf{p}}^T\overline{\mathbf{x}}_{new} + du_{new} - \overline{\mathbf{p}}^T\overline{\mathbf{x}}_0 - du_0 \end{aligned} \tag{4.19}$$

These transformed dynamical equations are the same as before – that is, as shown in Eqn. (4.7) – with the exception that new constant inputs have been added to the equations.

Because of the flexibility involved in specifying the various constants in the translations of the input and the state variables, we now have an approach that will cure all of the DC problems. Clearly, if the original DC equilibrium state vector solution had negative or zero entries, then the new state vector solution can be made to have strictly positive entries by shifting each element of the original state vector appropriately. More important is the fact that this trick will fix all of the potential problems – that is, not just the DC problems – to be encountered in a log domain filter realisation. Specifically, we can look at the excursions of all of the state variables over time for some worst case scenarios and set the individual offsets (translations) in such a way that no state variable or the input can ever be nonpositive. Therefore, the exponential mappings used for the log domain filters will always be well defined.

With these preliminaries we can not only discuss a general synthesis approach for log domain filters, but also we are in a position to unify a number of ideas that should make the understanding of this class of filters much simpler. Even though the use of the transformations above may still seem abstract, let us assume that we know how we want to take a given transfer function into the dynamical equation representation of Eqn. (4.19). For simplicity we will lump all of the new constant terms introduced via translation, as just described, into a constant vector and a scalar added in the state equations and the input-output equation, respectively. In addition we will drop the ‘new’ subscripts, since this only helped us distinguish between some original state space description and the current one. Since we will not need to go back to that, there is no loss of generality at this point in adopting this convention. Thus, we have,

$$\begin{aligned} \frac{d}{dt}\overline{\mathbf{x}} &= \mathbf{A}\overline{\mathbf{x}} + \overline{\mathbf{b}}u - \overline{\mathbf{b}}_0u_0 \\ y &= \overline{\mathbf{p}}^T\overline{\mathbf{x}} + du - d_0u_0 \end{aligned} \tag{4.20}$$

The scaling constants, $\overline{\mathbf{b}}_0$ (a vector) and d_0, have been introduced under the assumption that the constant, u_0, will have the same units as u. In order to specify a log domain filter realisation of these dynamical equations we adopt the following exponential mappings on the inputs and state variables (components of the state vector):

$$(x_1, x_2, \ldots, x_N)^T = (I_s e^{V_1/V_t}, I_s e^{V_2/V_t}, \ldots, I_s e^{V_N/V_t})^T$$
$$u = I_s e^{V_0/V_t}; \quad u_0 = I_s e^{V_{00}/V_t} \tag{4.21}$$

Upon substitution of these relations into Eqn. (4.20) we have a new set of N state equations and the associated input-output equation. The ith state equation has the following form after rearrangement:

$$C_i \dot{V}_i = \sum_{j=1}^{N} C_i V_t A_{ij} e^{(V_j - V_i)/V_t} + C_i V_t b_i e^{(V_0 - V_i)/V_t} + C_i V_t b_{0i} e^{(V_{00} - V_i)/V_t}$$
$$= \sum_{j=1}^{N} I_{ij} e^{(V_j - V_i)/V_t} + I_i e^{(V_0 - V_i)/V_t} + I_{0i} e^{(V_{00} - V_i)/V_t} \tag{4.22}$$

where

$$I_{ij} = C_i V_t A_{ij}, \quad I_i = C_i V_t b_i, \quad I_{0i} = C_i V_t b_{0i}$$

A_{ij} is the ijth element of the state matrix, $\mathbf{A}$, and b_i and b_{0i} are the ith entries of the respective vectors in Eqn. (4.20). This equation now looks like a KCL equation written at a node with a grounded capacitor of size C_i and a number of 'transconductors' connected to it. The exponential terms represent the constitutive laws of these transconductors, and they are clearly nonlinear.

Before discussing the realisation of these transconductors, let us digress for a moment to see how this relates to classic active filter design. Suppose that all of the voltages in the network were accurately described by a DC voltage plus a small perturbation. That is, suppose we assume that

$$V_i = V_{DCi} + v_i \tag{4.23}$$

for each of the $N+1$ voltages ($i = 0, \ldots, N$) in the system. Observe that the voltage, V_{00}, has no AC component, since this is a DC quantity. This 'small signal' assumption allows us to approximate each of the exponential terms above as follows:

$$I_{ij} e^{(V_j - V_i)/V_t} \approx I_{ij} e^{(V_{DCj} - V_{DCi})/V_t} \left[1 + \frac{v_j - v_i}{V_t}\right]$$
$$I_i e^{(V_0 - V_i)/V_t} \approx I_i e^{(V_{DC0} - V_{DCi})/V_t} \left[1 + \frac{v_0 - v_i}{V_t}\right] \tag{4.24}$$
$$I_{0i} e^{(V_{00} - V_i)/V_t} \approx I_i e^{(V_{00} - V_{DCi})/V_t} \left[1 - \frac{v_i}{V_t}\right]$$

Using this approximation to linearise Eqn. (4.22), we can write KCL equations. Specifically,

$$\begin{aligned} C_i\dot{V}_i &= \sum_{j=1}^{N} I_{ij}e^{(V_{DCj}-V_{DCi})/V_t}\left[1+\frac{v_j-v_i}{V_t}\right] + I_i e^{(V_{DCO}-V_{DCi})/V_t}\left[1+\frac{v_O-v_i}{V_t}\right] \\ &\quad + I_{0i}e^{(V_{00}-V_{DCi})/V_t}\left[1-\frac{v_i}{V_t}\right] \\ &= I_{tot-i}\left[1-\frac{v_i}{V_t}\right] + \sum_{j=1}^{N} I_{ij}e^{(V_{DCj}-V_{DCi})/V_t}\frac{v_j}{V_t} + I_i e^{(V_{DCO}-V_{DCi})/V_t}\frac{v_O}{V_t} \end{aligned} \tag{4.25}$$

The term out in front in the last line of Eqn. (4.25) is interesting in that it must always be zero. It can be seen by simply following the algebra in reaching the last result, that

$$I_{tot-i} = \sum_{j=1}^{N} I_{ij}e^{(V_{DCj}-V_{DCi})/V_t} + I_i e^{(V_{DCO}-V_{DCi})/V_t} + I_{0i}e^{(V_{00}-V_{DCi})/V_t} = 0 \tag{4.26}$$

That this term equals zero follows immediately from the fact that the KCL equation must have a stable DC equilibrium – that is, the right hand side of the differential equation must equal zero whenever the AC signals are set to zero. Incorporating this idea into Eqn. (4.25) and recognising that each of the coefficients premultiplying the AC voltages must have units of conductance, we may write

$$\begin{aligned} C_i\dot{V}_i &= \sum_{j=1}^{N} G_{m_{ij}}v_j + G_{m_i}v_O \\ G_{m_{ij}} &= \frac{I_{ij}}{V_t}e^{(V_{DCj}-V_{DCi})/V_t}; \quad G_{m_i} = \frac{I_i}{V_t}e^{(V_{DCO}-V_{DCi})/V_t} \end{aligned} \tag{4.27}$$

This result suggests an equivalent g_m-C filter realisation for the log domain filter under consideration. However, it should be noted that this g_m-C realisation is not truly analogous, topologically speaking, to the log domain filter realisation. In order to find the truly analogous filter we need to rewrite Eqn. (4.25), using Eqn. (4.26), by removing only the constant part of the first term (involving I_{tot-i}) on the right hand side, and by regrouping the AC terms back in with each of the other respective terms. Specifically, we obtain

$$\begin{aligned} C_i\dot{V}_i &= -I_{tot-i}\frac{v_i}{V_t} + \sum_{j=1}^{N} I_{ij}e^{(V_{DCj}-V_{DCi})/V_t}\frac{v_i}{V_t} + I_i e^{(V_{DCO}-V_{DCi})/V_t}\frac{v_O}{V_t} \\ &= \sum_{j=1}^{N} G_{m_{ij}}(v_j - v_i) + G_{m_i}(v_O - v_i) + G_{m_{0i}}(0 - v_i) \\ G_{m_{0i}} &= \frac{I_{0i}}{V_t}e^{(V_{00}-V_{DCi})/V_t} \end{aligned} \tag{4.28}$$

Notice that the sum of the transconductance coefficients on the right hand side of the final result in Eqn. (4.28) equals zero. This follows immediately from Eqn. (4.26), specifically by dividing each term in Eqn. (4.26) by V_t to obtain the transconductance coefficients in Eqn. (4.28). Also notice that the value of each of the transconductances is uniquely determined by the currents, I_{ij}, I_j and I_{0i}, respectively, and the DC operating point solution, specified by V_{DC0} through V_{DCN} and V_{00}, as shown in Eqns. (4.27) and (4.28). The currents, I_{ij}, I_j and I_{0i}, are in turn specified by the capacitor value, C_i, and the parameters in the filter state equations as defined in Eqn. (4.22). Since these currents are defined by the filter state equations that are specified (in general) without concern for the final (DC) operating point (although, they must be specified in such a way as to enable a suitable operating point, as discussed earlier), the constraint given by Eqn. (4.26) will be satisfied by the establishment of the DC operating point in the underlying log domain filter. Indeed, the algebraic solution of the collection of equilibrium equations – that is, Eqn. (4.26) for $i = 1, \ldots, N$ – given values of V_{DC0} and V_{00}, is the DC equilibrium solution for this underlying log domain filter. This same collection of equations is had by setting the voltage derivatives to zero in Eqn. (4.22). In a manner of speaking, the circuit finds a way to make things work out, as long as it is possible. That the voltages, V_{DC0} and V_{00}, are known follows from the fact that they are equilibrium *input* voltages. These voltages are set up by the bias currents used to establish the DC bias on the main input, u, and the extra input, u_0. The designer will specify these currents by choice in consideration of input signal swings, power consumption and dynamic range.

The final result in Eqn. (4.28) is very useful in pointing out the fundamental relationship between log domain filters and g_m-C filters. Eqn. (4.28) is in fact the defining equation for one integrator node of the equivalent g_m-C filter. The collection of all the N nodal equations defines the entire equivalent g_m-C filter realisation for the original transfer function. Because the sum of the transconductances driving each integrator node always equals zero, as mentioned above, only a subset of all g_m-C filters will qualify as possible equivalent filters, which provides a subtle way of comparing log domain filters to g_m-C filters in their most basic terms. Designers will almost certainly observe that the realisation suggested by Eqn. (4.27) is by far the more common g_m-C filter realisation approach, despite the fact that Eqn. (4.28) is equivalent. The key topological difference between the filters is that the realisation based on Eqn. (4.27) requires only single input transconductors, while the realisation based on Eqn. (4.28) requires differential input transconductors. While it is common to use differential transconductors in g_m-C filter design, the differential nature of the transconductors is used to produce a totally differential filter structure. Here, the differential transconductors are needed to produce a single-ended filter structure.

This small signal approach to looking at log domain filters is particularly valuable in comparing these filters to conventional filters. Not only does it highlight the topological similarities of the filter types, but it also puts the fundamental premise of log domain filtering on solid ground. Specifically, every log domain filter can be viewed as a linear filter for small signals, and the small signal transfer function derived from this linear analysis must match that of the log domain filter, even when the signals are not small. Thus we return to the basic reason for considering log domain filters, that is, they provide a systematic means to extend the linear operating regime of a class of transistor based active filters. This analysis also suggests a way to synthesise log domain filters without the mathematical rigour involved in the transformations above.

There is one more feature of log domain filters that distinguishes them from many other filter classes. Specifically, log domain filters are electronically tunable. We saw this property earlier in the derivation of the first order log domain filter where, recalling Eqn. (4.6), the cut-off frequency, ω_0, was directly proportional to the bias current, I_{DC}. This was not a coincidence as suggested by the results in Eqns. (4.22), (4.26) and (4.27). These equations show that the small signal transconductances in a log domain filter are directly proportional to the currents, I_{ij}, I_j and I_{0i}. It is a fundamental property of g_m-C filters that all critical frequencies will be tuned in unison if all the transconductances are scaled together. Hence, if all of the currents, I_{ij}, I_j and I_{0i}, are scaled together, then so will all of the critical frequencies of the log domain filter. Since these currents will be seen to be bias currents in the final circuit realisation, it becomes clear that log domain filters may be tuned electronically by varying the internal biasing currents appropriately. Several researchers have demonstrated tunability over more than two decades in log domain filters. Such a feature makes log domain filters very attractive in applications requiring wide tunability.

4.3.2 A synthesis strategy

We are now in a position to put the above ideas together to form a general methodology for the synthesis of log domain filters. In order to make this synthesis strategy more clear, let us consider an interesting and illuminating example, namely that of a bandpass filter. Suppose that we wished to obtain a log domain filter realisation for the following transfer function:

$$H(s) = \frac{Y(s)}{U(s)} = \frac{s\omega_0}{s^2 + s\omega_0/Q + \omega_0^2} \tag{4.29}$$

A set of dynamical equations corresponding to this transfer function is shown below:

$$\frac{d}{dt}\mathbf{x} = \begin{bmatrix} \dot{x}_1 \\ \dot{x}_2 \end{bmatrix} = \begin{bmatrix} -\omega_0/Q & -\omega_0 \\ \omega_0 & 0 \end{bmatrix} \begin{bmatrix} x_1 \\ x_2 \end{bmatrix} + \begin{bmatrix} \omega_0 \\ 0 \end{bmatrix} u \tag{4.30a}$$

$$y = \begin{bmatrix} 1 & 0 \end{bmatrix} \begin{bmatrix} x_1 \\ x_2 \end{bmatrix} \tag{4.30b}$$

These dynamical equations provide a more interesting example than an ideal integrator of how trouble can result regarding DC equilibrium. By setting derivatives to zero and applying a positive constant at the input, you can quickly verify that the equilibrium value for x_1 must be zero. This will preclude the exponential substitutions, and will therefore preclude a direct log domain realisation of these equations. The solution is to apply the translations to the variables as discussed above. One nice choice of translation yields the following new set of dynamical equations for the system:

$$\frac{d}{dt}\bar{\mathbf{x}} = \begin{Bmatrix} \dot{x}_1 \\ \dot{x}_2 \end{Bmatrix} = \begin{bmatrix} -\omega_0/Q & -\omega_0 \\ \omega_0 & 0 \end{bmatrix} \begin{Bmatrix} x_1 \\ x_2 \end{Bmatrix} + \begin{Bmatrix} \omega_0 \\ 0 \end{Bmatrix} u + \begin{Bmatrix} 0 \\ -\omega_0 \end{Bmatrix} u_0 \tag{4.31a}$$

$$y = \begin{Bmatrix} 1 & 0 \end{Bmatrix} \begin{Bmatrix} x_1 \\ x_2 \end{Bmatrix} - u_0 \tag{4.31b}$$

The translations that resulted in these equations are given by

$$x_{1-new} = x_1 + u_0; \quad x_{2-new} = x_2; \quad u_{new} = u + u_0/Q \tag{4.32}$$

Of course, the 'new' subscripts have been dropped in writing Eqn. (4.31). Using the exponential mappings from above and rearranging as before, we get the KCL equations for the log domain filter realisation shown below:

$$C\dot{V}_1 = -\frac{I_f}{Q} - I_f e^{(V_2 - V_1)/V_t} + I_f e^{(V_0 - V_1)/V_t}$$
$$C\dot{V}_2 = I_f e^{(V_1 - V_2)/V_t} - I_f e^{(V_{00} - V_2)/V_t}$$

$$y = I_s e^{V_1/V_t} - I_f \tag{4.33}$$
$$I_f = C V_t \omega_0$$

Notice that in obtaining the above result we have chosen the constants, C_1 and C_2, to be equal to C. This amounts to requiring equal sized capacitors at the respective nodes in the final implementation. Such a restriction is certainly not necessary; however, equal sized capacitors would certainly be desirable in practice, and this choice makes the mathematics a bit cleaner.

Before attempting the direct log domain realisation of these equations, let us follow the path above in writing the small signal version of Eqn. (4.33). The KCL equations become:

$$C\dot{V}_1 = -G_{m_{11}}(v_1 - v_1) - G_{m_{12}}(v_2 - v_1) + G_{m_1}(v_0 - v_1)$$
$$C\dot{V}_2 = G_{m_{21}}(v_1 - v_2) - G_{m_{02}}(0 - v_2)$$

$$G_{m_{11}} = \frac{I_f}{QV_t}, \quad G_{m_{12}} = \frac{I_f}{V_t} e^{(V_{DC2} - V_{DC1})/V_t}, \quad G_{m_1} = \frac{I_f}{V_t} e^{(V_{DC0} - V_{DC1})/V_t} \tag{4.34}$$
$$G_{m_{21}} = \frac{I_f}{V_t} e^{(V_{DC1} - V_{DC2})/V_t}, \quad G_{m_{02}} = \frac{I_f}{V_t} e^{(V_{00} - V_{DC2})/V_t}$$

As discussed earlier, the transconductance gains are all determined given knowledge of the DC operating point for the circuit, since we already know I_f. This DC operating point is found by solving the algebraic equations obtained from Eqn. (4.33) by setting the derivatives to zero, and using reasonable values for the input voltages, V_0

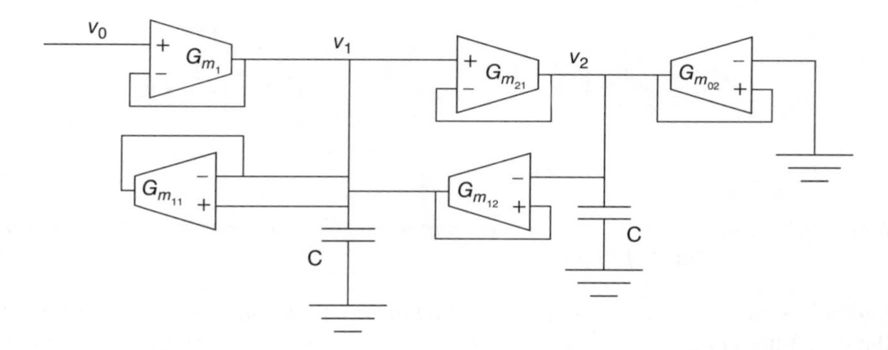

Figure 4.5 g_m*-C core filter realisation for a bandpass filter, where* v_0 *is the input and* v_1 *is the output*

(actually V_{DC0}) and V_{00}. Therefore, we can realise this filter with the circuit shown in Figure 4.5, where v_0 and v_1 are the input and output, respectively. Notice the superfluous transconductor tied to node 1. Although it contributes no current to the node, it will take on significance below. Based on the discussion above, this g_m-C filter must be the small signal equivalent of the log domain filter we are about to design.

The only aspect of this g_m-C filter realisation not covered thus far is the conversion from the overall input and output, u and y, to the core filter input and output, v_0 and v_1, respectively. It is convenient that this conversion may be accomplished using the same transconductors as those used in the core filter. To see this, let us reconsider the input and output relations. We have

$$u = I_s e^{V_0/V_t}, \quad u_0 = I_s e^{V_{00}/V_t}, \quad y = x_1 = I_s e^{V_1/V_t} \tag{4.35}$$

Using the small signal approach used above on these definitions yields

$$u \approx I_s e^{V_{DC0}/V_t}\left(1 + \frac{v_0}{V_t}\right) = I_{DC0} + G_{m_0} v_0 \Rightarrow v_0 = \frac{u - I_{DC0}}{G_{m_0}}$$

where

$$G_{m_0} = \frac{I_{DC0}}{V_t}; \quad I_{DC0} = I_s e^{V_{DC0}/V_t}$$

$$y \approx I_s e^{V_{DC1}/V_t}\left(1 + \frac{v_1}{V_t}\right) = I_{DC1} + G_{m_{out}} v_1 \tag{4.36}$$

where

$$G_{m_{out}} = \frac{I_{DC1}}{V_t}; \quad I_{DC1} = I_s e^{V_{DC1}/V_t}$$

These equations suggest circuit realisations using transconductors as shown in Figure 4.6. Notice that the DC currents only account for the DC offsets resulting from translations applied to ensure always positive signal currents. It is of interest to note that the input circuit is effectively a transresistance amplifier, while the output

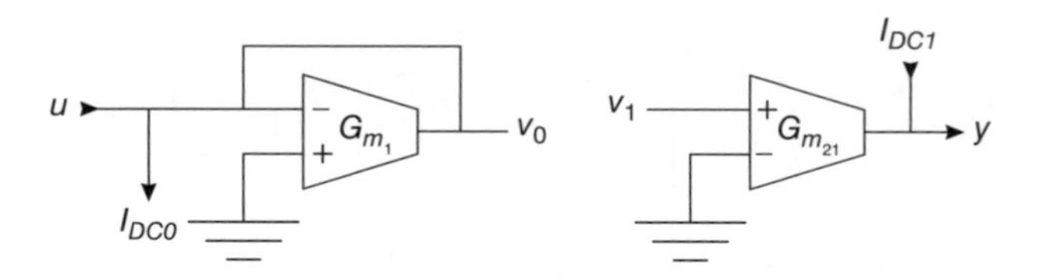

Figure 4.6 *Input transresistance and output transconductance amplifiers for the core filter of Figure 4.5*

circuit is a transconductance amplifier. Using these circuits at the input and output of the core filter completes the g_m-C filter realisation. It is also of interest to note that one may tune the core filter by varying all of the transconductances in that part of the circuit without affecting the input and output blocks. This is equivalent to varying I_f in Eqns. (4.33) and (4.34). Conversely, one may vary the transresistance and transconductance of the input and output blocks, respectively, without affecting the core filter. This is useful in matching the core filter to external circuitry and impedance levels.

So let us move on to the task of designing the log domain filter. To do this, notice that each of the terms in the g_m-C filter equations of Eqn. (4.34) is in one to one correspondence with a respective term in the log domain filter equations of Eqn. (4.33). The difference is that each of the terms in the g_m-C filter equations is realised with a circuit providing a linear transconductance gain on the difference of two input (node) voltages. In the case of the log domain filter, each of the analogous terms is an exponential function of the same two input (node) voltages. Hence, the job of designing the log domain filter is to replace each of the linear transconductance amplifiers in Figure 4.5 by a nonlinear transconductance amplifier obeying the mathematical relations given in Eqn. (4.33). Specifically, we need to make the following association:

$$G_{m_{ij}}(v_j - v_i) \rightarrow G_{m_{ij}} V_t e^{(V_j - V_i)/V_t} \tag{4.37}$$

So the design problem now amounts to the specialised issue of designing a circuit to implement the nonlinear transconductance amplifier. Notice that when $V_j = V_i$ – that is, $i = j$ – the nonlinear transconductor becomes a constant current source. Since this current source is necessary to implement the large signal KCL equations correctly, it must be included. Thus we see the value of leaving the corresponding 'null' transconductor in Figure 4.5. Figure 4.7 shows two simple realisations for this circuit function – one using all NPN transistors and current sources, and one using complementary bipolar transistors and one current source. These circuits are quite adequate to realise the necessary functionality (neglecting base currents); however, they are only applicable if the transconductance – that is, $G_{m_{ij}}$ – is positive. Unlike their linear counterparts, these transconductors do not possess the symmetry to allow trivial implementation of the negative transconductance terms. The inputs can be switched, to be sure; however, the output current will remain positive and drive the wrong node. In fact, the transconductance does become negative, but simple examination of the terms in Eqn. (4.33) – that is, the nonlinear transconductance terms – reveals that a negative transconductance must be accompanied by a negative absolute current.

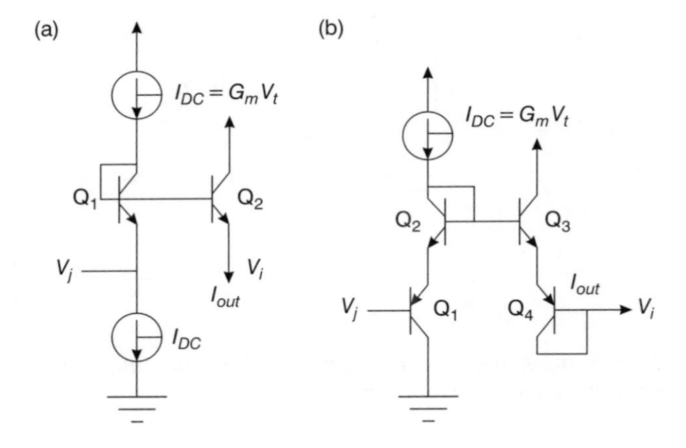

Figure 4.7 Basic positive transconductors for log domain filtering: (a) all-NPN; (b) complementary bipolar

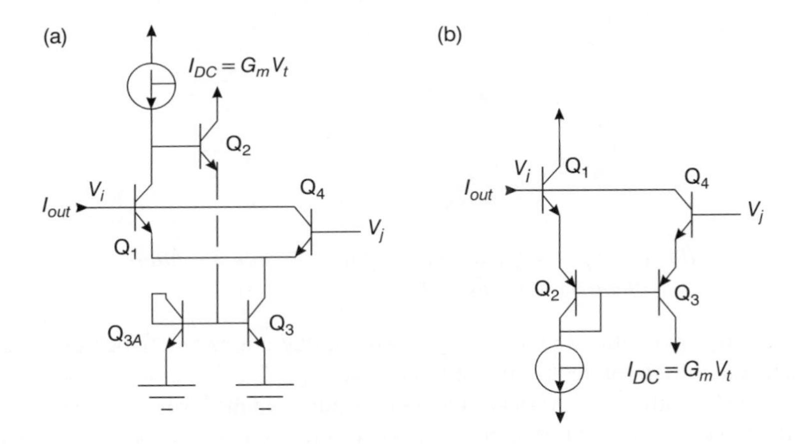

Figure 4.8 Basic negative transconductors for log domain filtering: (a) all-NPN; (b) complementary bipolar. Note that V_j is the input node

Figure 4.8 shows two examples of negative exponential transconductors of the type required. Notice that in the complementary bipolar transistor realisation the change is simple, but in the all-NPN case the change requires significant extra complexity. Nevertheless, the circuits of Figures 4.7 and 4.8 provide the basis for a realisation of the log domain filter suggested by Eqn. (4.33). Only one minor point remains. Recall that when both inputs of the exponential transconductor are the same, the exponential term is constant. Therefore, it may be implemented with only a current source.

As before, we must also implement the interface circuitry between the overall input and output and the core filter input and output. This will involve the design of

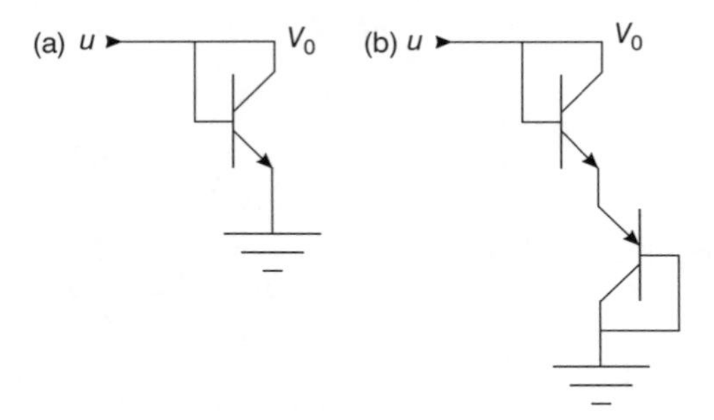

Figure 4.9 *Simple input logging amplifiers for log domain filters: (a) all-NPN; (b) complementary bipolar*

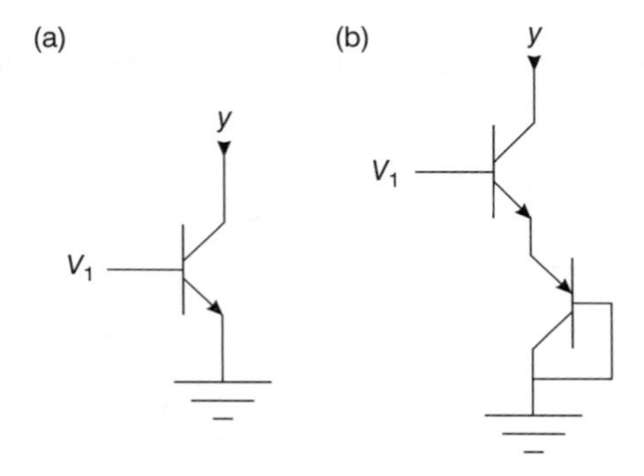

Figure 4.10 *Basic output exponentiators for log domain filters: (a) all-NPN; (b) complementary bipolar*

logging and exponentiating circuitry performing operations analogous to the trans-resistance and transconductance amplifiers in the g_m-C realisation. These interface circuits can be realised in a variety of ways depending on the level shifting used in the core filter. The basic circuitry can be designed directly from the defining equations given in Eqn. (4.35). Figures 4.9 and 4.10 show example circuits for the all-NPN and complementary bipolar cases. These circuits can be combined with level shifting to fit most possible core filter designs. Some examples appear in the circuits shown later.

Figure 4.11 shows the log domain filter realisation of Eqn. (4.33) using complementary bipolar devices. Observe the use of the basic building blocks and how the 'degenerate' block is implemented by the current source, I_f/Q. Also notice the level shifting (down) added to the logging amplifier. An all-NPN version of the filter derived using the alternative circuits from Figures 4.9 and 4.10 is shown in Figure 4.12. The reader should note that the current source, $I_f(1 + 1/Q)$, is the combination of the 'degenerate' block current source and the negative referenced current source associated with the positive transconductor bridging the nodes labelled, V_1 and V_2. The use of level shifting creates subtle advantages in this circuit. Specifically, the log of the

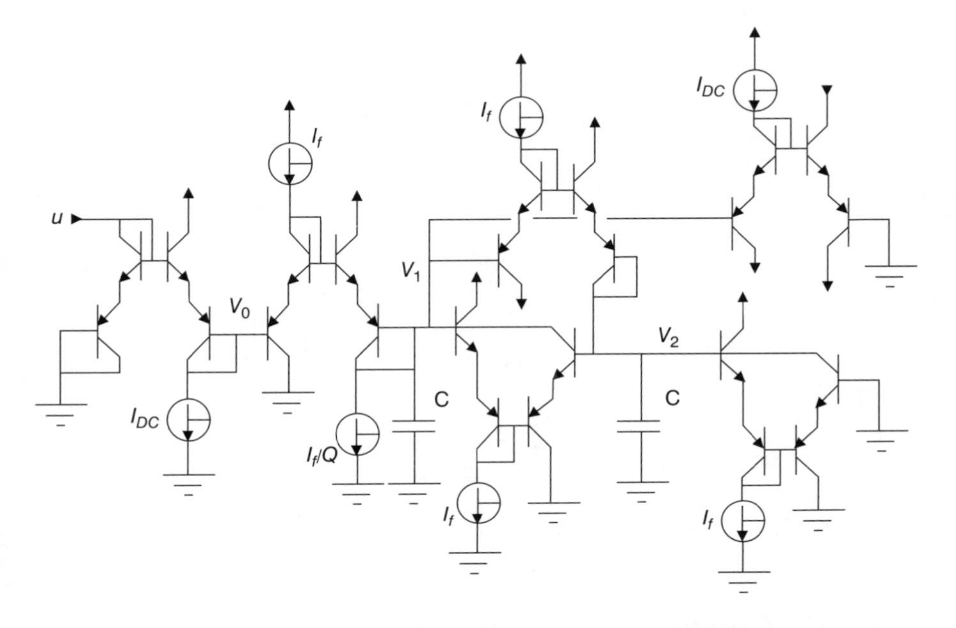

Figure 4.11 *Complete complementary bipolar log domain filter realisation of a bandpass filter*

input is level shifted up by two diode drops, using Q_{LS1} and Q_{LS2}, before application to the core filter. This allows a very effective way of implementing the exponentiator, using Q_{out}, without directly loading the node where V_1 is measured.

At this point we are in a position to design arbitrary log domain filters. The steps to design are clear. One may begin with a transfer function and ultimately derive the KCL equations for the log domain filter via a state space representation and exponential mappings. The last step requires that the exponential terms be implemented using the exponential transconductors shown above. Input logging and output exponentiating circuitry are added as a final step. Alternatively, one may find a g_m-C filter realisation for the filter of the correct form and then simply replace each of the linear differential transconductors with one of the respective exponential transconductors as was done in the above discussion. This second method will probably be the most comfortable for most designers since it does not carry the abstraction of the first approach. Furthermore, a designer may use any of a variety of well known techniques to get to a desirable g_m-C filter realisation. Such techniques can be used to optimise the filter for minimum sensitivity or maximum dynamic range, for example. In this way, one may expect the log domain realisation to inherit the good properties of the g_m-C filter realisation, since the two filters – that is, the log domain and the g_m-C filter – realisations are identical for small signals. Several researchers have used this approach to efficiently design log domain filters [6, 7]. The only caution that must be added to these comments is that a designer may not choose an arbitrary g_m-C filter realisation: specifically, only those g_m-C realisations where the sum of the transconductances

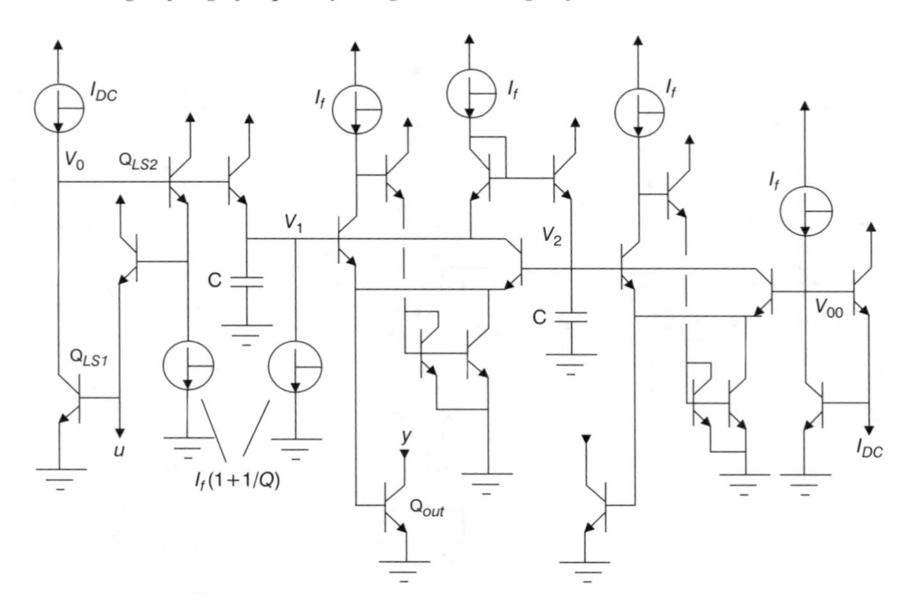

Figure 4.12 Complete all-NPN log domain realisation of a bandpass filter

(coefficients in the respective state equation) driving each node must exactly equal zero. This idea was discussed earlier.

Another generic approach to the design of log domain filters combines the ideas above at a slightly higher level. One may imagine the collection of transconductors driving a given node in the filter plus the associated grounded capacitor as a multiple input integrator. Since it is well known that active filters may be realised with integrators as the fundamental building blocks, this perspective provides a compelling way to design log domain filters. Now, any analytical method which yields effective integrator based topologies for active filters may be used to design log domain filters. Readers familiar with the signal flow graph (SFG) realisation of active filters will immediately appreciate that there is a well established body of literature regarding integrator based design of active filters. Those interested in this approach may consult virtually any modern text in active filter design. For an example of log domain filter design using these ideas the reader may consult Reference 6.

4.3.3 The translinear perspective

The final major synthesis technique applied to the design of log domain filters uses an idea called the 'translinear principle' and its recent extension, the 'dynamic translinear principle'. These ideas allow one to write design equations for log domain filters, and more general 'translinear filters', using a quite different mathematical abstraction. Nevertheless, the ultimate designs are often remarkably similar to those obtained with the already mentioned methods. As with the mapping approach described above,

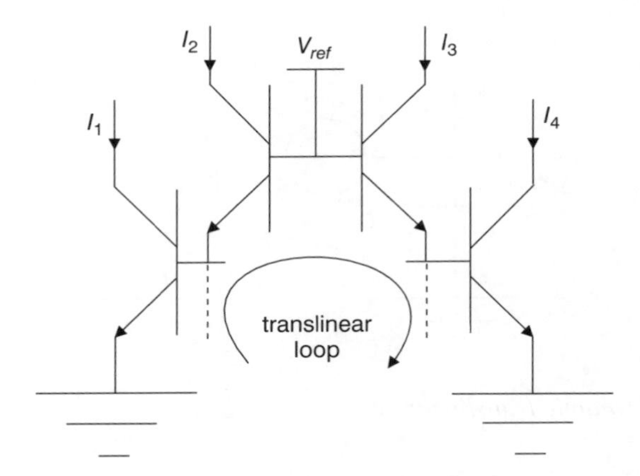

Figure 4.13 A basic 4 transistor translinear loop

this approach begins with differential equations and produces new equations that may be used directly to create log domain filters. The final step in the circuit realisation is conceptually different, however, since it consists of the creation of 'translinear loops' which implement the different terms of the final equations instead of nonlinear transconductors. This approach provides an entirely different approach to the conceptualisation of the final design topology which might allow some new and unobvious log domain filter topologies. At the very least, the practitioners of this approach feel more comfortable with this way of looking at things, which is always important as one attempts to create something new.

The starting point for gaining an understanding of this approach is the classic translinear principle [8, 9] relating currents in circuits comprised solely of bipolar transistors and current and voltage sources. Basically, the translinear principle is a statement that loops of bipolar transistor junctions yield product relationships between the currents flowing in the junctions. Figure 4.13 shows the basic idea using a loop of four junctions, carrying currents of I_1–I_4. The dashed lines indicate the presence of the other circuitry, not shown, that would be needed to establish bias for these devices. The translinear principle states that the product of the currents flowing in junctions of one orientation, e.g. clockwise (I_3 and I_4), in the loop equals the product of currents flowing in the other orientation, e.g. counterclockwise (I_1 and I_2). This rule is easily proven using the logarithmic relationship between voltage and current in the junctions and the assumption of ideal matching between devices. The calculation is identical to that used in Eqn. (4.3). Of course, we are continuing to neglect base currents in this discussion.

The so-called dynamic translinear principle [10] yields another relationship between the product of currents when a capacitor is placed across a transistor in a certain way. Figure 4.14 shows the typical scenario, where the voltage across a capacitor, plus a possible fixed offset, is applied to the base-emitter junction of a

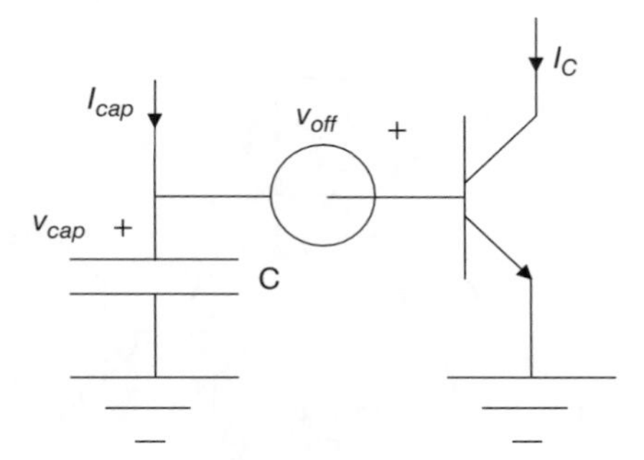

Figure 4.14 Dynamic translinear cell

transistor. The important equations are given by

$$I_C = I_s e^{v_{BE}/V_t}; \quad v_{BE} = v_{cap} + v_{off} \Rightarrow v_{cap} = -v_{off} + V_t \ln\left(\frac{I_C}{I_s}\right)$$
$$\therefore \frac{I_{cap}}{C} = \dot{v}_{cap} = V_t \frac{\dot{I}_C}{I_C} \quad \Rightarrow \quad I_C I_{cap} = C V_t \dot{I}_C \tag{4.38}$$

The final result is the *dynamic translinear principle* that states that the derivative of the collector current, I_C, is proportional to the product of that current and the capacitor current.

The dynamic translinear principle can be used to transform the state equations for a filter into translinear design equations, with the assumption that a translinear circuit will implement the final system. One approach to doing this is as follows. We suppose that each state variable is equivalent to the collector current in a grounded transistor configured as shown in Figure 4.14, where its base-emitter voltage is determined by a capacitor voltage plus a possible DC offset. Then, using the result in Eqn. (4.38), the derivative of each state variable can be replaced by a scaled version of the product of the respective capacitor current and the state variable current. Denoting the input, u, by a current, I_{in}, each of the state variables, x_i, by currents, I_{xi}, and the output, y, by a current, I_{out}, we can rewrite the dynamical equations of Eqn. (4.7) as

$$\frac{1}{CV_t}\begin{bmatrix} I_{x1}I_{cap1} \\ I_{x2}I_{cap2} \\ \vdots \\ I_{xN}I_{cap\,N} \end{bmatrix} = \mathbf{A}\begin{bmatrix} I_{x1} \\ I_{x2} \\ \vdots \\ I_{xN} \end{bmatrix} + \overline{\mathbf{b}}I_{in}\,; \quad I_{out} = \overline{\boldsymbol{p}}^T \begin{bmatrix} I_{x1} \\ I_{x2} \\ \vdots \\ I_{xN} \end{bmatrix} + dI_{in} \tag{4.39}$$

Eqn. (4.39) is interesting in that it characterises a dynamical circuit with purely algebraic equations. Of course, the dynamics is built in by the presence of the N capacitor currents. With some manipulation, as shown below, these equations suggest translinear circuit configurations which implement the underlying filter.

Let us now consider the issue of circuit realisation by using Eqn. (4.39). This can be done by making the reasonable assumption that the circuit topology at each

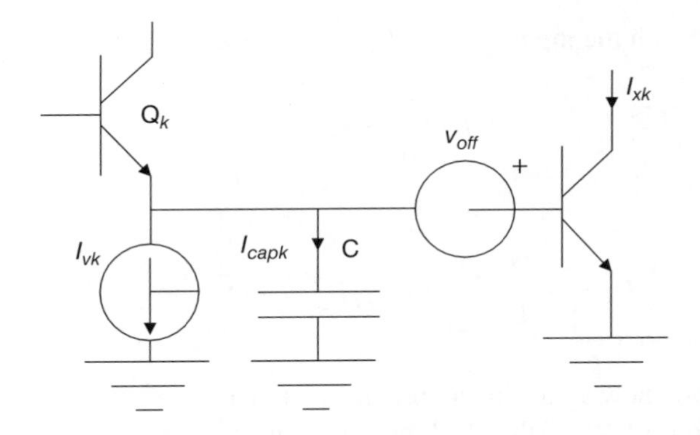

Figure 4.15 Generic integrating node for translinear analysis

capacitor node has the general structure shown in Figure 4.15, which is clearly related to Figure 4.14. Any additional circuitry which may actually appear connected to the capacitor in a final implementation might be elaborate, but its net current contribution may always be treated as if it were the current source, labelled I_{vk}. The validity of this statement is proven via the substitution theorem from circuit theory, and for our purposes the current, I_{vk}, will be used as a slack variable to obtain a desired mathematical form for each equation. Now observe that the net current entering the capacitor, I_{capk}, is given by the difference of I_{Ek}, the emitter current in transistor Q_k driving the node, and I_{vk}, the current acting as the slack variable.

Incorporating the information provided by the assumed topology in Figure 4.15, the kth transformed state equation from Eqn. (4.39) may be rewritten as

$$I_{xk}(I_{Ek} - I_{vk}) = \sum_{j=1}^{N} CV_t A_{kj} I_{xk} + CV_t b_k I_{in} = \sum_{j=1}^{N} I_{DC-kj} I_{xk} + I_{DC-k0} I_{in}$$

where

$$I_{DC-kj} = CV_t A_{kj}; \quad I_{DC-k0} = CV_t b_k \tag{4.40}$$

$$\Rightarrow \quad I_{xk} I_{Ek} = I_{xk} I_{vk} + \sum_{j=1}^{N} I_{DC-kj} I_{xk} + I_{DC-k0} I_{in}$$

Eqn. (4.40) states that the product of the current in transistor Q_k driving the kth capacitor node and the 'output' current – that is, state variable current, I_{xk} – relative to this node is equal to the sum of products of currents.

Recalling the basic translinear principle stated earlier, a loop of transistor junctions leads to a relation where one product of currents equals another product of currents, and not a sum of products of currents. With this as motivation, suppose that we use the slack variable, I_{vk}, to eliminate all but one of the current products on the right hand side of Eqn. (4.40). As a simple example, suppose the remaining current product was that associated with the input current, I_{in}. Then we would have the following:

$$I_{xk} I_{Ek} = I_{DC-k0} I_{in}$$

where

$$I_{vk} = -\frac{1}{I_{xk}} \sum_{j=1}^{N} I_{DC-kj} I_{kj} = -\sum_{j=1}^{N} \frac{I_{DC-kj} I_{kj}}{I_{xk}} \tag{4.41}$$

This result shows how a translinear circuit equation can be obtained from the general equation of Eqn. (4.40) if the slack current is made equal to just the right sum of terms. At first, these terms may appear to be difficult to implement, but each of them in its own right can be implemented using the translinear principle on simple circuit topologies.

An example of this synthesis procedure is had by considering again the bandpass filter of Eqn. (4.31) which has been augmented with the extra DC input, u_0, which we will call I_{DC} in this example, to allow for a realisable final circuit as before. It should be noted that with the addition of this extra input all state equations have terms with both positive and negative coefficients, which is also a necessary constraint for the translinear technique to be applicable. The reader should appreciate that the translinear approach to log domain filter realisation must start from a suitable state space formulation for the desired transfer function. For example, there must be a stable equilibrium. No approach to log domain filter realisation can circumvent the basic issues discussed thus far. On the other hand, some approaches are better than others at managing the problems from a conceptual point of view. Finally, it should be noted that some researchers have come up with other ways – that is, not starting from a state space filter description – to implement the translinear design approach that is shown here. For more information the reader should see References 10 and 11.

Now let us continue with the example. Applying the dynamic translinear principle as above, and following the development that led to Eqn. (4.40), the system equations become:

$$\begin{aligned} I_{E1} I_{x1} &= I_{v1} I_{x1} - I_Q I_{x1} - I_f I_{x2} + I_f I_{in} \\ I_{E2} I_{x2} &= I_{v2} I_{x2} + I_f I_{x1} - I_f I_{DC} \\ I_{out} &= I_{x1} \\ I_f = C V_t \omega_0 \,; &\quad I_Q = \frac{1}{Q} I_f \end{aligned} \tag{4.42}$$

These equations have only one positive coefficient in each of the state equations which must remain after the selection of slack variables in order to allow for a realisable circuit. Recall that the translinear principle only relates positive products of currents to one another. Hence, only the following choices for the slack variables and the accompanying translinear design equations can be had:

$$I_{E1}I_{x1} = I_f I_{in}$$
$$I_{E2}I_{x2} = I_f I_{x1} \tag{4.43}$$
$$I_{out} = I_{x1}$$

where

$$I_{v1} = I_Q + \frac{I_f I_{x2}}{I_{x1}}$$
$$I_{v2} = \frac{I_f I_{DC}}{I_{x2}}$$
$$I_f = CV_t\omega_0; \quad I_Q = \frac{1}{Q}I_f$$

The transformed state equations suggest a pair of simple 4-transistor loops, where the constraints on the slack variables suggest an additional current source and a pair of divider structures. Specifically, I_{v2} is defined through a translinear equation where I_{v2} is one of the currents in a translinear loop. I_{v1} is the sum of a current in a translinear loop, involving I_f, I_{x1} and I_{x2}, and the constant current, I_Q. A translinear design corresponding to the system of Eqn. (4.43) is shown in Figure 4.16, where it should be noted that the transistor used to create the current, I_{x2}, is not needed in practice, since this current is not actually required. Transistors used solely for level shifting purposes and their biasing circuitry have been replaced by batteries in Figure 4.16 to aid in the understanding of this circuit. Each battery corresponds to a

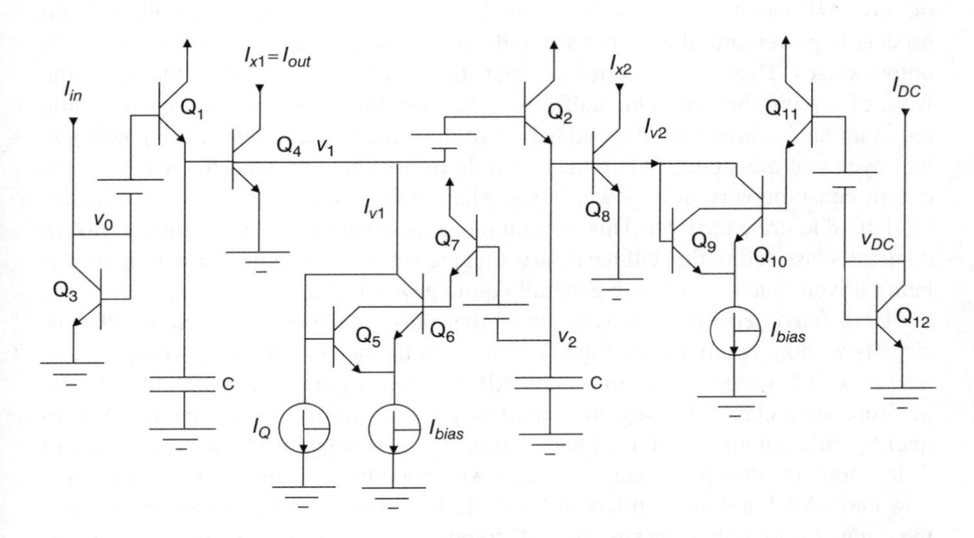

Figure 4.16 Log domain bandpass filter based on the translinear design approach

level shifting device with a nominal current of I_f flowing in it. Thus, the batteries may be treated as transistors carrying I_f amps in writing translinear loops. The translinear loop comprising Q_1, Q_3, Q_4 and a battery establishes the first equation of Eqn. (4.43) involving I_{E1} (the emitter current of Q_1). The translinear loop comprising Q_2, Q_4, Q_8 and a battery establishes the second equation of Eqn. (4.43) involving I_{E2} (the emitter current of Q_2). The translinear loop comprising Q_4, Q_5, Q_6, Q_7, Q_8 and a battery sets up the translinear part of the equation defining I_{v1} in Eqn. (4.43). Even though this translinear loop involves six transistors, it still results in a relation similar to four transistor loops – that is, the product of the currents with clockwise orientation in the loop equals the product of currents with counterclockwise orientation in the loop. Besides, the currents in Q_5 and Q_7 are equal and in opposite orientation in the loop, which cancels their contribution. Finally, the translinear loop comprising Q_8, Q_9, Q_{10}, Q_{11}, Q_{12} and a battery sets up the equation defining I_{v2} in Eqn. (4.43). Here the currents in Q_9 and Q_{11} offset one another in the translinear loop. The design of Figure 4.16 is very similar to the log domain filter design shown in Figure 4.12, developed using the mapping strategy. In fact, one may find translinear loops in the circuit of Figure 4.12 that exactly correspond to the translinear equations of Eqn. (4.43). This demonstrates that the approaches are substantially equivalent for use in designing this class of log domain filters. Nevertheless, their different perspectives provide interesting alternative views.

4.4 Class AB log domain filters

Because of their internally nonlinear behaviour, log domain filters can be designed with features that no purely linear filter could ever possess. One such feature is that of class AB operation. Circuit designers familiar with the design of amplifiers, and especially power amplifiers, are used to the idea of so-called 'push-pull' or class AB output stages. These are essentially output stages that have two halves that share the work of driving the load. One half drives the load during positive half cycles while the other half is idle, then the two halves of the circuit switch roles during negative half cycles of the signal. When there is little or no signal present, the halves of the circuit consume very little power. Only when the signal is large does the circuitry work hard to drive the load. This operation results in very power efficient designs. In contrast, class A circuits will have parts that are never idle whether or not the load is being driven. Such circuits are generally quite power inefficient.

Until fairly recently, it was assumed that only 'memoryless' circuits – that is, circuits without reactive elements – could enjoy the benefits of class AB operation. Seevinck [12] proposed the first class AB dynamical circuit in 1990. At that time he proposed a class AB integrator based on the translinear design principle. Subsequently, this circuit was refined and, ultimately, was shown to be a special case of A log domain filter possessing the class AB property [13]. In order to understand how class AB log domain filters differ from those shown so far, let us observe that the circuits shown thus far are 'class A' log domain filters. That is, they are circuits which must have sufficient DC bias so that under no circumstances may the composite signal current drop to zero. To ensure this, the quiescent currents are typically chosen considerably larger than is necessary for average signal levels. Therefore, under typical conditions the DC bias levels are significantly larger than the AC signals present. Since the intrinsic noise in bipolar transistors is proportional to the square root of the total emitter current (assuming negligible contributions from the resistive parasitics), it follows that larger than necessary DC bias levels translate to larger than necessary noise. Thus, we might wish that the standing bias currents would always be as small as possible to minimise the overall noise in the filter circuitry.

The ultimate approach to realising this goal is to borrow the 'push-pull' concept from classic amplifier designs. Unfortunately, this is not a straightforward task. Nevertheless, it can be done. To begin, it is necessary to create two circuit halves in an analogous way to class AB output stages. Then it is necessary to create an input 'splitter' that will apply just the right kind of signal to these two halves to achieve the desired class AB operation. Figures 4.17 and 4.18 show a basic current splitter and the current waveforms it produces in response to a sinusoidal input current, u. V_{ref} establishes the quiescent currents for u_1 and u_2 when $u = 0$. This reference voltage is usually derived by forcing a reference current in a pair of diode connected transistors in series tied to ground. Note how the two outputs are waveforms that are large during opposite half cycles and never go to zero during their respective 'idle' half cycles. The only constraint that these waveforms must satisfy is that their difference is exactly equal to the input signal.

This last point is very important. It is worth digressing for a moment to appreciate it more fully. Suppose that an arbitrary input signal, $u(t)$, is applied to a linear system whose transfer function is given by $H(s)$. Letting $U(s)$ denote the Laplace transform of the input, the output transform response will be given by $H(s)\,U(s)$. Now suppose that the input, $u(t)$, is preprocessed into two signals, $u_1(t)$ and $u_2(t)$, such that $u(t) = u_1(t) - u_2(t)$. Then let each of these signals be applied as input to a pair of linear systems whose respective transfer functions are each given by $H(s)$. Finally, let us derive an output, $y(t)$, by subtracting the two outputs, $y_1(t)$ and $y_2(t)$, from

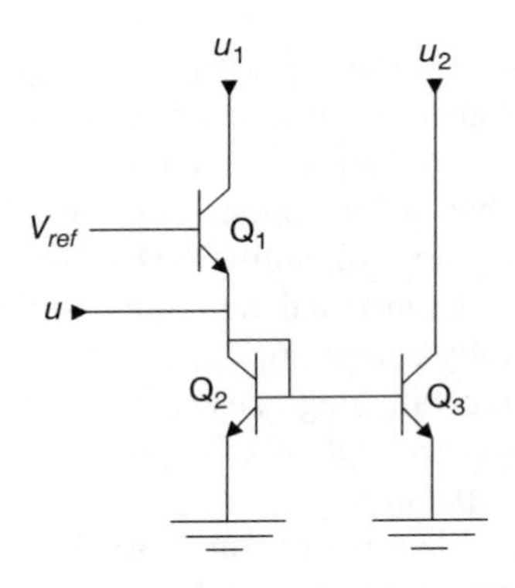

Figure 4.17 Basic class AB current splitter

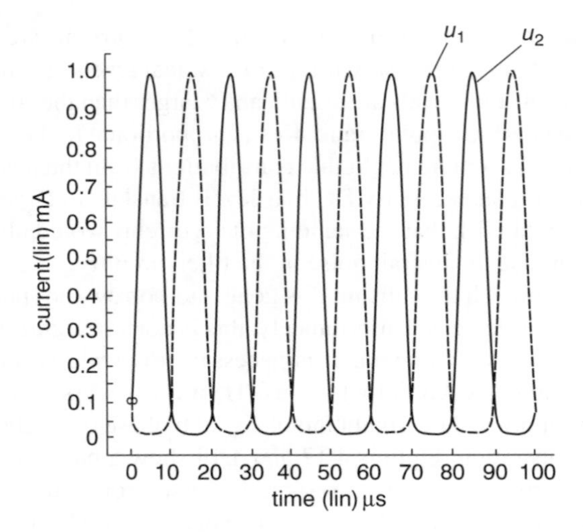

Figure 4.18 Current splitter output waveforms

these respective linear systems. The total operation is summarised below:

$$
\begin{aligned}
U(s) \longleftrightarrow u(t) = u_1(t) - u_2(t); \quad U_1(s) \longleftrightarrow u_1(t),\ U_2(s) \longleftrightarrow u_2(t) \\
y_1(t) \longleftrightarrow Y_1(s) = H(s)U_1(s),\ y_2(t) \longleftrightarrow Y_2(s) = H(s)U_2(s) \\
y(t) = y_1(t) - y_2(t) = H(s)(U_1(s) - U_2(s)) = H(s)U(s)
\end{aligned} \tag{4.44}
$$

The final line makes the key point – namely, that the final output is the same as that which would have resulted from applying the original input to the single filter. The interesting aspect of using the splitter and the pair of filters is that the two inputs can be created using an arbitrary nonlinear operation as long as their difference is linearly related to the input. An extreme example of this is to use half wave rectifiers to preprocess the input.

The splitter of Figure 4.17 is more graceful in its operation than is a pair of half wave rectifiers. This highlights the difference between class B amplifiers, whose operation would resemble that of half wave rectifiers, and class AB amplifiers whose operation is analogous to that of the splitter in Figure 4.17. The outputs from the splitter in Figure 4.17 never go to zero, unlike the half wave rectifiers; therefore, these signals tend to work much better in actual transistor circuitry where very small signals at any instant in time generally compromise the performance of the transistors – for example, their bandwidth typically degrades at very low currents. Furthermore, the class AB output signals are compatible with log domain filters, since the inputs may never go to zero for proper operation.

There is still a subtle problem in creating class AB log domain filters using the idea above. This problem arises from the fact that filters of higher order than one often exhibit overshoot in the time domain at various nodes within the filter, including the

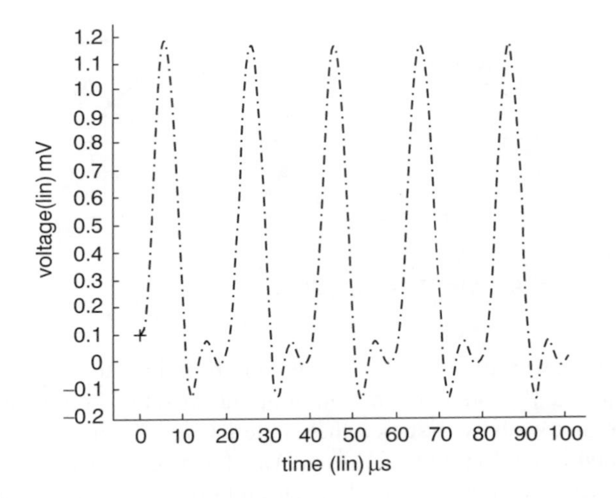

Figure 4.19 Filtered current splitter output

output. As a result, the highly nonlinear waveforms from a current splitter, which resemble step functions at the transition times between idle and active half cycles, can cause internal waveforms in the linear filters to which these signals are applied to have negative excursions. Figure 4.19 shows the waveform that results from applying one of the output currents from the current of Figure 4.17 to a second-order filter. Notice how the ringing produced by the second-order filter results in a negative going signal at the end of each 'active' half cycle. Despite the fact that the overall output of the differential system will be unaffected by this phenomenon, it becomes clear that the internal signals no longer possess the true class AB property whereby signals are idle for a full half cycle. More important is the fact that these internal signals now take on negative values, making them incompatible with log domain filters, since many of the internal signals – for example, state variables – must remain strictly positive in log domain filters (recall the exponential mappings). Hence, something more than just a class AB current splitter is required if class AB log domain filters are to be developed.

Seevinck offered a solution to this problem by proposing a class AB integrator in his contribution to translinear filters. By using a circuit like his as the basic building block it is possible to maintain the class AB quality of the signals all through a filter despite the potential for overshoot created by higher-order filters. The key to this circuit and those which have been subsequently proposed is to cross-couple internal points in the two linear filters to be used to process the class AB input. This cross-coupling is best done at each integration node in the circuitry, which amounts to the creation of cross-coupled integrators of the type proposed by Seevinck [12]. Figure 4.20 shows his core integrator idea. Notice the pair of single ended log domain integrators implemented by Q_1 and Q_2 along with capacitors C_1 and C_2. The cross-coupling transistors Q_3 and Q_4 perform two functions. First, they enable the class

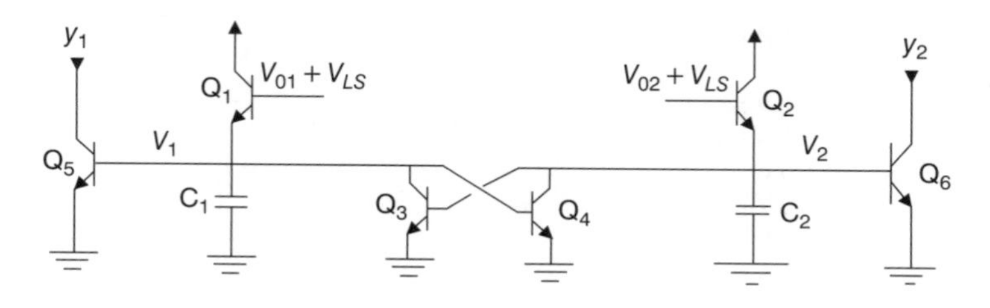

Figure 4.20 Seevinck's core class AB filter

AB operation by never allowing the currents in Q_1 and Q_2 to go to zero, assuming the inputs, V_{01} and V_{02}, are derived from logging operations on a pair of class AB currents. Note that we have assumed that some DC level shifting, denoted by V_{LS}, has been applied to the log inputs, V_{01} and V_{02}, for proper operation. Secondly, Q_3 and Q_4 provide a way to quiescently bias Q_1 and Q_2 so that current sources are not required. This makes the transfer function from the input to the output that of an ideal integrator, which is a very nice feature. Finally, the output currents that can be measured in Q_5 and Q_6 turn out to be class AB currents which are appropriate for the application to further stages of class AB processing.

While a full treatment of the mathematics behind class AB filters is beyond the scope of this text, a basic appreciation for the ideas can be had by the following derivation regarding the operation of the integrator of Figure 4.20. We begin by writing the KCL equations at each of the integrating nodes. This is done below under the simplifying assumption that both capacitors are equal to C. We assume that level shifted versions – that is, shifted up by V_{LS} as defined below – of the log inputs, V_{01} and V_{02}, are applied to the core filters:

$$C\dot{V}_1 = I_s e^{(V_{01}+V_{LS}-V_1)/V_t} - I_s e^{V_2/V_t}$$
$$C\dot{V}_2 = I_s e^{(V_{02}+V_{LS}-V_2)/V_t} - I_s e^{V_1/V_t}$$

where

$$V_{LS} = V_t \ln\left(\frac{I_f}{I_s}\right)$$

$$\Rightarrow \begin{array}{l} C\dot{V}_1 = I_f e^{(V_{01}-V_1)/V_t} - I_s e^{V_2/V_t} \\ C\dot{V}_2 = I_f e^{(V_{02}-V_2)/V_t} - I_s e^{V_1/V_t} \end{array} \tag{4.45}$$

In Eqn. (4.45), I_f is assumed to be some DC biasing current. Rearranging these results, we have:

$$\begin{aligned} \frac{\dot{V}_1}{V_t} e^{V_1/V_t} &= \frac{d}{dt}[e^{V_1/V_t}] = \frac{I_f}{CV_t} e^{V_{01}/V_t} - \frac{I_s}{CV_t} e^{(V_1+V_2)/V_t} \\ \frac{\dot{V}_2}{V_t} e^{V_2/V_t} &= \frac{d}{dt}[e^{V_2/V_t}] = \frac{I_f}{CV_t} e^{V_{02}/V_t} - \frac{I_s}{CV_t} e^{(V_1+V_2)/V_t} \end{aligned} \tag{4.46}$$

Next we may subtract these equations and scale by I_s, with the result,

$$\frac{d}{dt}[I_s e^{V_1/V_t} - I_s e^{V_2/V_t}] = \omega_0[I_s e^{V_{01}/V_t} - I_s e^{V_{02}/V_t}] \tag{4.47}$$

where

$$\omega_0 = \frac{I_f}{CV_t}$$

Now suppose that we redefine the mappings on the input and state variable introduced earlier by using the difference of two exponentials in each case. Specifically, suppose we define

$$u \equiv I_s e^{V_{01}/V_t} - I_s e^{V_{02}/V_t}; \quad x \equiv I_s e^{V_1/V_t} - I_s e^{V_2/V_t} \tag{4.48}$$

With these substitutions Eqn. (4.47) becomes the equation for an ideal integrator, where we assume that the output is equal to the state variable, x. Although this derivation does not dwell on the mathematical subtleties, it does show how one may extend all of the ideas above to the context of class AB filtering.

All we need to employ this concept now is a way to make the basic integrators capable of accepting multiple inputs of either polarity and of being made lossy – that is, equivalent to a first-order lowpass filter. Loss can be added in the same way that it is in single-ended log domain filters – that is, using DC current sources. Multiple inputs are added in exactly the same way as for single-ended log domain filters. Specifically, we can connect pairs of positive and negative transconductors to the pairs of grounded capacitors, creating pairs of log domain integrators as before. The key difference will be in the cross-coupling transistors connected between the capacitor nodes, which will enable the class AB performance. As a result of these observations the synthesis of class AB log domain filters is no more difficult than single-ended class A log domain filters. Therefore, all of the above discussion related to integrator based synthesis is applicable here. For more options in creating class AB filters, the reader may consult References 7 and 13.

4.5 Log domain building blocks and nonideal behaviour

Having discussed the synthesis of log domain filters at the top level, it is now appropriate to discuss some of the more interesting designs for log domain building blocks. We have already done some of this by showing some simple transconductor and integrator structures. The circuits shown thus far are useful in a variety of cases where there are not serious constraints on precision or power supply requirements. Suppose that we consider modifications or alternative circuits from this perspective. To begin, suppose we wish to improve the accuracy of the basic positive transconductor of Figure 4.7(a). Up to now, we have assumed that the transistor current gain, β, is very large. In practice this will not be the case. Therefore, the base current of Q_2 will appear as an error both at the input to this transconductor and the output, since only the collector current of a bipolar transistor obeys the ideal exponential law in theory. We can cure at least part of this error by adding a third 'β helper' transistor as shown

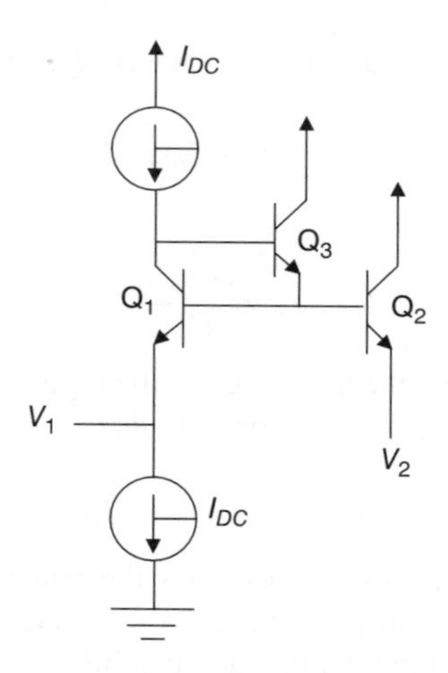

Figure 4.21 *Positive transconductor with reduced sensitivity to base current*

in Figure 4.21. This reduces the base current error seen at the input; however, the output error is still there. Fortunately, the input error is usually more of a problem in log domain filter performance than the output error.

Looking at the negative transconductor of Figure 4.8(a) one can see that the base current of Q_4 is still a problem at the input, but not at the output. This time the base current could be reduced by making Q_4 a Darlington pair and compensating in the translinear loop with another Darlington pair, but this solution does not produce good quality log domain transconductors because of uncertainties in the transistor βs and the performance degradation of the transistors running at decreased current levels. This performance degradation regarding frequency response is exacerbated by the 'hidden node' introduced between the transistors making up the Darlington pair. Another source of error is the base current of Q_1 which is connected to the output node. This base current is constant since it is in the level shifter path which runs at constant current. Nevertheless, this constant current acts like a fixed current attached to the output node, which in turn will change the tuning of the filter at the output node, since current sources at log domain integrator nodes program the pole location of the lossy integrators. Note that if the integrator were designed to be ideal, this current would make it lossy, almost certainly changing the performance of the overall filter. Fortunately, this problem can be cured, or at least greatly reduced, by adding a compensating current at the output node.

We can see by the above discussion that log domain filter functional blocks possess nonidealities just as do any active building blocks. Mitigating these effects is always at the heart of good circuit design. It is worth mentioning the nonideal effects introduced by the transistors, and how they contribute to performance degradation in log domain filters. As mentioned above, transistor base current can cause two types of performance degradation. To see how this works, let us write the nodal equations for the second-order filter of Figure 4.12, this time taking into account a few of the nonzero base current errors. The new nodal equations become:

$$
\begin{aligned}
C\dot{V}_1 &= -\frac{I_f}{Q} - I_f e^{(V_2-V_1)/V_t} + I_f e^{(V_0-V_1)/V_t} + \frac{I_f}{\beta} e^{(V_0-V_1)/V_t} - \frac{I_f}{\beta} \\
C\dot{V}_2 &= I_f e^{(V_1-V_2)/V_t} + I_f e^{(V_{00}-V_2)/V_t} - \frac{I_f}{\beta} I_f e^{(V_2-V_1)/V_t}
\end{aligned}
\tag{4.49}
$$

The new terms added are, of course, those divided by β. To see the effect they have on the filter performance, let us apply the exponential mappings in reverse to these equations. We already know that if the new terms were not added then we would recover the original state equations for the filter. The result this time is the original state equations with new terms added. Specifically, we have:

$$
\begin{aligned}
\dot{x}_1 &= -\frac{\omega_0}{Q} x_1 - \omega_0 x_2 + \omega_0 u + \frac{\omega_0}{\beta} u - \frac{\omega_0}{\beta} x_1 \\
\dot{x}_2 &= \omega_0 x_1 - \omega_0 u_0 - \frac{\omega_0}{\beta} \frac{x_2^2}{x_1}
\end{aligned}
\tag{4.50}
$$

From here we can easily see how each base current error affects the system. The new term in x_1 in the first equation amounts to a Q shift, while the other new term in the first equation produces only an input gain error. Neither of these terms causes nonlinear distortion. On the other hand, the new term in the second equation of Eqn. 4.50 introduces the term, x_2^2/x_1, which certainly introduces distortion. Thus we see that log domain filters suffer in a variety of ways from the finite current gain of the transistors.

We may consider other nonideal effects as well. A lack of ideal log conformity in the transistors – that is, a deviation from the ideal exponential current/voltage relation – will contribute distortion and an apparent shift in the gain or pole frequencies of a filter. The mathematics needed to prove this is more involved, but follows the same general approach as that above. Log conformity in bipolar transistors is usually excellent until the emitter current gets too large. This is because the intrinsic ohmic resistance in the emitter, and to some extent in the base, causes significant additional voltage drops between the base and the emitter of the transistor that are only linearly, and not exponentially, related to the emitter current. Another common error in bipolar circuitry is caused by the offset between the different transistors resulting from the fact that the reverse saturation current of the different devices varies randomly. While there is typically quite good matching of devices on a single integrated circuit, the mismatch is still in the order of 1 mV across a circuit. Fortunately, in log domain filters these errors only cause errors in the constant coefficients of each of the exponential terms in the nodal equations, which ultimately results in errors in the coefficients of the state equations. These are linear errors which affect the frequency response and gain of the filters but do not contribute distortion.

To gain an appreciation for this, recall the g_m-C filter comparison given above. There it was shown that the DC equilibrium voltages control the g_m of each transconductor. Since these g_m values directly affect the filter response, it follows that the log domain filter response is affected by shifts in DC equilibrium caused by such factors as transistor mismatch. That shifts in DC equilibrium do not produce distortion is a fact that can be proven using the exponential mapping ideas already introduced. The Early effect in the transistors causes a similar effect to that of offset in the base-emitter voltages of the transistors, and as a result is generally not a concern. The Early effect might be more significant in causing errors and possibly distortion if the signal swings at the various nodes in the filter were large, but in log domain filters these swings are typically quite small, that is, 100 mV or less.

Another nonideal property of transistors that degrades performance in log domain filters, but is more difficult to characterise, is limited bandwidth. The capacitance of the base-emitter junction of the transistors causes distortion, as does any more subtle delay effect at high frequencies. Stray capacitance to ground typically appears in parallel with capacitors intentionally placed in the circuit, resulting in only an error in the frequency response. However, parasitic capacitance bridging between nodes in a log domain filter, such as base-emitter junction capacitance, is a problem. To see this, let us return to the second-order filter of Figure 4.12. This time let us assume that all things are ideal except that a small capacitance, C_B, is placed between the two integration nodes. It is a simple matter to show that the nodal equations are now:

$$\begin{aligned} C\dot{V}_1 &= -\frac{I_f}{Q} - I_f e^{(V_2-V_1)/V_t} + I_f e^{(V_0-V_1)/V_t} + C_B(\dot{V}_2 - \dot{V}_1) \\ C\dot{V}_2 &= I_f e^{(V_1-V_2)/V_t} + I_f e^{(V_{00}-V_2)/V_t} - C_B(\dot{V}_2 - \dot{V}_1) \end{aligned} \tag{4.51}$$

To see how things have really changed, it is necessary to resolve these equations for the individual capacitor current terms. This has been done below:

$$\begin{aligned} C\dot{V}_1 &= -\frac{I_{f1}}{Q} - I_{f1} e^{(V_2-V_1)/V_t} + I_{f1} e^{(V_0-V_1)/V_t} + I_{f2} e^{(V_1-V_2)/V_t} + I_{f2} e^{(V_{00}-V_2)/V_t} \\ C\dot{V}_2 &= I_{f1} e^{(V_1-V_2)/V_t} + I_{f1} e^{(V_{00}-V_2)/V_t} - \frac{I_{f2}}{Q} - I_{f2} e^{(V_2-V_1)/V_t} + I_{f2} e^{(V_0-V_1)/V_t} \end{aligned} \tag{4.52}$$

where

$$I_{f1} = \frac{C + C_B}{C + 2C_B} I_f\ ; \quad I_{f2} = \frac{C_B}{C + 2C_B} I_f$$

Now let us apply the mappings in reverse, as done earlier. This yields the new state equations given below:

$$\begin{aligned} \dot{x}_1 &= -\frac{\omega_{01}}{Q} x_1 - \omega_{01} x_2 + \omega_{01} u + \omega_{02} \frac{x_1{}^2}{x_2} - \omega_{02} \frac{x_1 u_0}{x_2} \\ \dot{x}_2 &= \omega_{01} x_1 - \omega_{01} u_0 - \frac{\omega_{02}}{Q} x_2 - \omega_{02} \frac{x_2{}^2}{x_1} + \omega_{02} \frac{x_2 u}{x_1} \end{aligned} \tag{4.53}$$

where

$$\omega_{01} = \frac{C + C_B}{C + 2C_B} \omega_0\ ; \quad \omega_{02} = \frac{C_B}{C + 2C_B} \omega_0$$

Clearly, a variety of changes have occurred which not only perturb the linear behaviour of the filter, but also introduce distortion terms. Thus, anything that can be done to reduce the bridging capacitance between nodes in log domain filters is a necessity if the best high frequency performance is desired. The design tradeoff resulting from this observation regards the sizing of transistors. Using the smallest possible devices is always the best way to reduce parasitic junction capacitance. On the other hand, the smaller the devices the worse the log conformity of those devices at larger currents. Larger currents are generally desirable to improve speed and signal to noise ratio, so this trade-off is an important one in the design of high frequency log domain filters. Note that, since one of the main advantages of log domain filtering is in the design of high frequency filters, this trade-off will often become important in practical designs. One factor which brightens the picture in this regard is the fact that log domain filters usually require a relatively large integrating capacitor due to the low impedance of the active signal path. Therefore, capacitive parasitics, both due to transistor parasitics and layout parasitics, are usually small compared to the integrating capacitors.

4.6 Very low voltage log domain filter design

Having taken a look at the major nonideal effects in log domain filters which result from transistor nonidealities, let us revisit the issue of designing basic building blocks. Clearly, the best building blocks are those which exhibit the least amount of nonideal behaviour due to the transistors. The circuits shown thus far, and those shown below, are amongst the best discovered in this regard; however, log domain filtering is a new area of electronic design, so better circuit topologies may yet be discovered. Let us continue the discussion now with topologies aimed at very low voltage operation.

Log domain filters are excellent candidates for systems requiring low power. This is because the transconductors involved require very little headroom due to their simplicity. Thus, the design of log domain filters to operate on 3V supply rails is quite straightforward. Power can be further conserved if class AB designs are used. If even lower voltage power supplies must be used, log domain filters can still be used by adopting the more modern folded cascode designs which are prevalent elsewhere. To achieve the lowest supply operation, let us make some basic observations. Designs based on the complementary bipolar transconductors of Figures 4.7(b) and 4.8(b) require too much headroom, since every 'junction' really comprises two transistor junctions, which almost doubles the power supply requirements for a given topology. Using such stacked transistor topologies limits supply rails to a minimum of nearly 5 V, which of course may be fine in some applications, but not here. Since level shifting up and down in log domain filters must be accomplished using similar devices, the use of complementary devices in the signal path is quite difficult without doubling up devices. To date, the author knows of no good solutions for this problem. Hence, the lowest possible supply rail operation seems to be achievable with all NPN transistors in the signal path. Of course, all PNP designs are also possible; however, NPN transistors are usually the highest performance devices available on a given process, so there is typically no good reason for choosing PNP transistors over NPN transistors.

However, PNP transistors are valuable in designing the current sources off the positive rail, as must be done in any log domain design.

Now that we have restricted attention to all NPN designs (neglecting current sources), let us consider the supply that must be used with any of the designs discussed thus far. If one considers the essential elements of log domain filters we can estimate the minimum supply possible. Clearly, at least one diode drop is necessary for the logging operation at the input. Considering the transconductors shown thus far, one additional diode drop is required for the positive transconductor, and at least a 'saturation voltage' extra is required for the negative transconductor shown. This is because this transconductor requires a diode drop down from the nominal integrating node voltage plus the drop ('saturation voltage') across the current source below this. In the design for the all NPN second-order filter shown in Figure 4.12, this 'saturation voltage' has been made equal to another diode drop, allowing a simpler design. Finally, there must be one more 'saturation voltage' drop added for current sources off the top rail. This yields at least two diode drops and two 'saturation voltage' drops, which results in an estimate of about 2 V for the minimum supply rail. This is certainly very good, but is not good enough for 1.8 V operation, and certainly rules out 1.2 V operation.

To further reduce the supply requirements, researchers have introduced folded topologies. Furthermore, by reversing the orientation of the level shifters in the positive transconductors it is possible to eliminate a diode drop in the design, allowing operation down to about 1.2 V. This is truly impressive for a bipolar transistor active circuit. To see how this is done, consider the input logging circuit of Figure 4.22(a). Here the input, u, is forced to flow in a device whose base voltage is fixed. Thus, the emitter is at a voltage that drops with increasing input, as opposed to the logging circuits shown earlier where the input log voltage becomes more positive with increasing input signals. Conceptually, this is no different from the cases discussed above; however, it does require a rethinking of all the log domain building blocks. Mathematically, this scenario arises if the exponential mappings used to transform the state equations have a minus sign in the exponent. To demonstrate this, the derivation of the first-order log domain filter given earlier in Eqn. (4.11) is redone. Specifically,

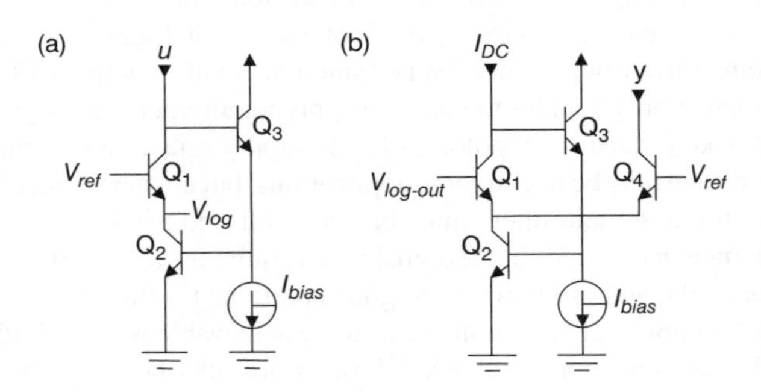

Figure 4.22 Very low voltage input logging and output exponentiating circuits

we have the following transformation of the dynamical equations:

$$x \equiv I_s e^{-V/V_t}; \quad u \equiv I_s e^{-V_0/V_t}$$

$$\dot{x} = -\dot{V}\frac{I_s}{V_t}e^{-V/V_t} = -\omega_0 I_s e^{-V/V_t} + \omega_0 I_s e^{-V_0/V_t}; \quad y = x = I_s e^{-V/V_t} \tag{4.54}$$

Rearranging as before, we have

$$C\dot{V} = I_0 - I_0 e^{(V-V_0)/V_t} \tag{4.55}$$

This almost trivial change in the formulation suggests a design based on a reversal of the sense of all the signals in the earlier designs. The reversal of the current source, I_0, is of no consequence; however, the transconductors must now be redesigned to exploit this point of view. The level shifting idea associated with the negative transconductors shown earlier can be used to advantage in designing both transconductors this time. Figure 4.23 shows circuits that can be used for positive and negative transconductors in this new context. Note that now a current mirror (Q_5 and Q_6) is necessary to implement the negative transconductor. This is somewhat of a liability, since both accuracy and bandwidth are compromised by current mirrors; however, this may be a small price to pay in applications where absolute minimum power supply operation is a necessity. Using these new transconductors allows the creation of the full range of log domain filters starting from a building block approach just as before. Finally, the output exponentiator of Figure 4.22(b) may be used to obtain the final output, y. Thus, we are able to design very low voltage log domain filters with the same facility as before. Recall that one of the key advantages that log domain filters have is in maximising dynamic range in applications where voltage swings do not permit large dynamic range.

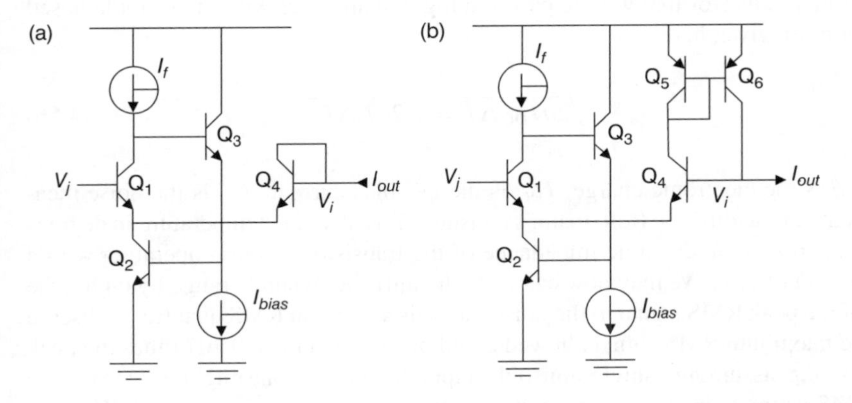

Figure 4.23 Very low voltage positive and negative transconductors

4.7 Noise in log domain filters

We now turn to one of the most complex issues in understanding log domain filters – that is, noise. As we have seen from the start, log domain filters permit the use of as much as 20 dB of additional headroom in transistor-based active filters. However, this increase in headroom is offset by the fact that signal swings are limited relative to the noise floor. To understand the real trade-offs involved regarding noise in log domain filters, let us consider a single transistor that might be part of a log domain filter. We restrict attention to class A filters to begin. The DC bias current in that transistor must be chosen such that under peak signal swings the total current can never drop to zero. Thus, we may assume that the DC bias current is approximately equal to the peak signal swing, since there is no value in having significantly larger bias currents, because this would only waste power and, as we shall see, increase the noise floor. However, it should be noted that in practice one would want to make the DC bias a little larger than the expected maximum signal swing so that the total current in the transistor would never fall below some minimum current chosen to ensure reasonable bandwidth and matching requirements. This observation stems from the fact that transistor bandwidth and current gain are severely degraded at very low emitter currents.

Having made a preliminary estimate for maximum signal swing relative to the DC bias current, we are in a position to derive the fundamental upper bound on dynamic range in log domain filters. This calculation can be made by making the reasonable assumption that the noise in the transistor under consideration is due primarily to shot noise. One might be tempted to assume that white noise due to parasitic base and emitter resistance should also be included, but the reader should recall that log domain filters must be biased in such a way that log conformity is not compromised in any of the transistors. A natural consequence of this design choice is that the net resistive parasitic reflected to the emitter is much less than the dynamic impedance of the emitter. It then follows that the noise due to the resistors will be small compared to that due to shot noise. Therefore, we can accurately approximate the quiescent noise current in a transistor that will be part of a log domain filter with the formula based on shot noise given here:

$$i_n = \sqrt{2qI_{DC}\Delta f} = \sqrt{2kT\Delta f/r} \tag{4.56}$$

where q is the electronic charge, I_{DC} is the DC bias current, Δf is the noise measurement bandwidth, k is Boltzmann's constant, T is absolute temperature in degrees Kelvin, and r is the dynamic impedance of the transistor junction operating with a bias current of I_{DC}. We may now calculate the intrinsic dynamic range by taking the ratio of the peak RMS signal to the noise (noise is always an RMS quantity). Observe that the maximum RMS signal allowed could be assumed to be 0.707 times the peak signal swing, assuming a single sinusoidal input. In practice, one might want to reduce this RMS estimate using the assumed crest factor of the signals expected. We shall assume the sinusoidal case for simplicity. Thus we obtain the following estimate for maximum signal to noise ratio:

$$\text{DR(dynamic range)} = \frac{I_{DC}/\sqrt{2}}{\sqrt{2qI_{DC}\Delta f}} = \frac{\sqrt{I_{DC}}}{2\sqrt{q\Delta f}} = \frac{\sqrt{kT}}{2q\sqrt{\Delta f}}\frac{1}{\sqrt{r}} \tag{4.57}$$

This result is interesting in that it proves that dynamic range (DR) is proportional to the square root of DC bias current, or inversely proportional to the square root of the dynamic impedance, which is, of course, determined by the DC bias current. We conclude that improvements in dynamic range will come with a significant penalty in power dissipation – that is, a 3 dB increase in DR results from a doubling of power consumption. Alternatively, DR is increased by reducing the impedance level in the circuit, which is intrinsically related to the dynamic impedance, r. Specifically, a reduction by a factor of 2 in the impedance levels will increase the DR by 3 dB. Therefore, one may conclude that ultimately all capacitor sizes *and* power dissipation in a given filter will have to be doubled in order to improve the DR by 3 dB without affecting the filter critical frequencies. This is a more serious trade-off than that necessary in standard voltage mode filters, for example.

To get a feel for the kind of dynamic range available in a log domain filter, let us evaluate the formula above for a special case. Suppose we consider a case where the DC bias current is 100 μA and the measurement bandwidth is 1 MHz. Assuming room temperature, this yields a DR of 82 dB. Clearly, this figure for DR is better than one would expect in an actual filter, since this only applies for a single device. A reasonable first cut estimate would scale the noise power by the number of transistors in the signal path. For example, a circuit with ten transistors in the signal path might hope to achieve a 72 dB DR. Of course, such a crude estimate would have to be tempered by the knowledge that filter characteristics would play an important role in the actual filter dynamic range.

To get a better feel for the overall noise in a log domain filter, let us go back to the simplified block diagram for these filters shown earlier in Figure 4.4. This picture highlights the three main contributors of noise in a log domain filter, namely the front-end logging circuitry, the core filter circuitry, and the back-end exponentiation circuitry. It is valuable to separate the overall circuit this way because the different sections of the circuit will often work at different current levels. Specifically, the input logging and output exponentiating circuitry typically operate at currents that are commensurate with the interface needs of the filter. For example the filter may need to drive and/or be driven from a specific source and load; for example, each might have 75 Ω of characteristic impedance. On the other hand, the core filter may be designed to operate at current levels consistent with other requirements, such as total capacitance or power consumption. Moreover, the core filter current levels may be varied to tune the filter. As a result, the computation of noise for each part of the overall filter must be done assuming different quiescent currents. Designing the system with these degrees of freedom, that is, the choice of DC bias in the different parts of the circuit, can be quite challenging.

A good understanding of the trade-offs involved here may be had by returning to the small-signal perspective of the circuit operation. This perspective can be misleading, as will be discussed later, but for class A circuits it is quite efficient in obtaining noise estimates. From a small-signal perspective, the front-end logging stage is a transresistance stage, the core filter is a voltage mode filter, and the exponentiating circuitry is a transconductance amplifier. Thus we can estimate the output of the logging amplifier to be a voltage signal, Ri_{in}, where R is the small-signal transresistance of the input stage plus a noise component approximately equal to the noise voltage of a resistor, R. Actually, the noise is equivalent to that of a resistor whose size is $R/2$, because of the fact that it is shot noise, and not thermal noise, that gives rise to this. Next, this composite signal plus noise is applied to the core filter whose noise can be computed assuming small-signal noise models for each of the transistors. This analysis is not trivial, but at least it is something that any circuit simulator can provide. Thus, at the output of the core filter we can assume there to be a total noise resulting from the filtering of the input noise from the logging amplifier and the noise from the core filter. Finally, this composite noise is amplified by the final transconductance amplifier. Of course, the transconductance amplifier adds noise of its own. Its equivalent input noise voltage is roughly equal to that of a resistor whose size is half that of the reciprocal of the transconductance. Once again the half comes from the fact that shot noise is causing this added noise as opposed to thermal noise.

Having laid out the picture, let us consider a few simple scenarios. Suppose the input logging and output exponentiating circuits have a DC bias current of 1 mA, while the internal core filter operates with typical bias currents of 100 μA. Then the transresistance of the front-end will be about 25 Ω (at room temperature), and the transconductance of the output stage will be about 40 mΩ^{-1}. The voltage noise contributed by these two stages will be quite small, in the order of 0.5 nVHz$^{-1/2}$. However, the core filter will, due to its higher impedance levels, contribute significantly more noise voltage. Thus the core filter noise will be much more significant than the other noise sources. Suppose the core filter output noise were about 20 nVHz$^{-1/2}$. Then the final output noise would be about 800 pAHz$^{-1/2}$. Considering the output bias current this would yield a dynamic range of about 119 dBHz$^{-1/2}$ of measurement bandwidth. If the measurement bandwidth were 1 MHz, the filter dynamic range would be 59 dB.

We may wish to improve this rather low DR by adjusting the current levels in the various stages. Assuming no constraints on input and output current levels, we may simply increase all the currents in the system by a factor of ten, gaining an additional 10 dB of dynamic range based on the discussion above. This would be quite inefficient, though. Instead, let us recognise that the noise due to the core filter was by far the most significant. Therefore, let us increase the core filter current by a factor of ten, leaving the input and output stages unchanged. Assuming the core noise drops by 10 dB and the other noise remains the same, we get a new overall dynamic range estimate of almost exactly 69 dB. Thus, by recognising where the noise is coming from we get a much more efficient solution. In fact, a little thought will demonstrate that the basic trade-offs here are not very different from those in any multistage design.

Were the present discussion about classic linear filters, we would have said everything of significance regarding noise. However, since log domain filters are internally nonlinear, there is more to the story. The main consequence of this fact is that log domain filters exhibit modulation noise. Specifically, since the noise in the individual transistors is due primarily to shot noise, the overall noise in a log doman filter depends on the signal current. This is because the shot noise in each transistor at any instant in time is due to the total current flowing in the device at that instant. Therefore, the instantaneous noise of each transistor is constantly changing in magnitude. While the mathematics is rather cumbersome, it can be shown that the effective RMS noise in each transistor can be computed by using the average value of the total current flowing in the device to compute the shot noise. In a completely linear system the average value would simply equal the DC bias. For signals that are 6 dB or more below the maximum in class A log domain filters this remains an excellent approximation; however, as the signal levels approach the maximum, the average current can be shown to be a little higher than the DC bias, due to the internal nonlinearity. The difference is large enough that, at peak level signals, the shot noise can be about 1 dB higher than the DC bias current would predict. This deviation is not particularly significant in many applications, but it should be taken into account, since it diminishes the maximum available signal to noise ratio by 1 dB.

While this modulation noise phenomenon is of little significance in class A log domain filters, it is of paramount importance in class AB filters. In these filters the maximum signal can be many times the DC bias current level, and as a result the noise observed in these filters is heavily dependent on the signal strength. In fact, it can be shown that due to the combined action of all three stages in the filter, the signal to noise ratio will saturate to some maximum value as signal levels get large enough. In contrast, linear systems will display an ever increasing signal to noise ratio as the signals get larger. Clipping is neglected in both these cases. We assume there is some peak input level beyond which the system will become overloaded in any filter. The net result of these observations about class AB filters is that one should design these filters in such a way that the maximum signal to noise ratio is adequate to meet the needs of the overall design. The literature shows several examples of very interesting class AB log domain filter designs (see Reference 7, for example).

Before concluding the discussion of noise, it is worth mentioning another unique limitation that log domain filters have relative to their linear counterparts. The same phenomenon that leads to modulation noise in log domain filters also leads to a 'blocking' phenomenon. This phenomenon is somewhat familiar to RF designers and is related to the fact that when large 'blocking' signals are present in addition to relatively weak desired signals, the blocking signals can cause interference of the weaker signals, making them more difficult to recover than might be expected. A simple extreme example of this occurs when the blocking signal causes the system to clip. This severe nonlinearity can easily contaminate the weak signals that may also be present, making it impossible to accurately recover them. Typically, a blocking signal will cause only mild distortion; however, this may still be enough to irrevocably contaminate the weaker signals, depending on the disparity in their strength.

In log domain filters this phenomenon also exists and is more complicated. Imagine a scenario where a large out of band signal is present in addition to a desired in-band signal. This combined signal may be applied to a log domain filter in order to reject the out of band signal. But notice that before the succeeding stages of the filter can significantly eliminate the 'blocking' signal, this signal will have caused the overall currents in some of the transistors to significantly increase over their DC bias level. As a result, there will be an increase in the broadband, and therefore in-band, noise. This increased noise will degrade the weaker in-band signal component in the process. The extent to which this will occur, of course, is contingent upon the magnitude of the 'blocking' signal relative to the DC bias. In class A filters, one might not expect too much effect; however, in class AB filters the effect could be catastrophic. A complete treatment of these modulation noise issues is beyond the scope of this chapter, but the reader should be aware of this phenomenon in considering the benefits of using log domain filters. References 14 and 15 give a great deal more discussion on these matters.

4.8 Other applications of the log domain filtering idea

Now we have seen how log domain filters work and how they can be designed and understood. The design context has been solely that of bipolar transistor designs. One may rightly ask if these concepts can be applied to MOS transistor circuits. The answer is yes. The simplest application to MOS design is had by replacing the bipolar transistors of the designs discussed by FET transistors operating in their weak inversion region. In that region the drain current follows an exponential relation relative to the gate to source voltage. This idea is very useful. First, circuits based on FET transistors operating in weak inversion use very little power. In applications such as hearing aids this can be a wonderful alternative to other analogue or digital signal processing. Even better is the fact that these transistors draw no base current, so all of the performance degradation due to base current effects is gone. As a result, researchers have given this class of log domain filters a fair amount of attention [16]. One drawback of using sub-threshold operation of MOS transistors is that the body effect can be a significant problem, making it difficult to achieve high accuracy in the log conformity of the transistors.

Other applications of the ideas presented in this chapter include the use of CMOS transistors in saturation to directly replace the bipolar transistors. It has been shown [17] that, while this is analytically incorrect, the idea produces filters in CMOS with good performance. Research in this direction is only in its infancy. Another promising extension of the log domain filters is in the creation of dynamically biased syllabic companding filters. This idea uses the well known concept of companding to enhance the overall system performance of the filters in much the same way as have classic noise reduction systems, such as the Dolby system. An in-depth discussion of this intriguing design direction is given in References 18–23.

4.9 Conclusions

Log domain filtering is a very new area of active filter design. As a result, there are no generally accepted design procedures. The discussion in this chapter has therefore centred on familiarising the reader with the basic ideas and concepts related to this new idea. An attempt has been made to summarise several perspectives on the design of these filters, with the intent to give the reader an ability to not only appreciate how to go about designing log domain filters, but also access the literature for more information and perspective.

Hopefully, the discussion has also highlighted the main advantages and disadvantages of log domain filters. The main advantages of log domain filters involve their potential for new solutions to the hard problems associated with low voltage wide bandwidth filter design and in the new perspective the concept itself offers. Clearly, to the extent that log domain filters add 10–20 dB of dynamic range to active filters, they are of great interest. On the other hand, this increase in dynamic range is applicable only to a special class of transistor filters with already limited dynamic range compared to more classic voltage mode designs. Nevertheless, in cases where supply voltage, tunability or bandwidth necessitate a deviation from the old solutions, log domain filters are a very interesting alternative. From the perspective of pure science, the log domain concept is one of the most revolutionary ideas to be proposed in recent history. It is quite intriguing to suppose that nonlinearity itself can be used to advantage in creating linear circuits. That the mathematics can be posed in such a fundamental way is truly a gift of understanding into what may someday be a broad range of nonlinear circuits implementing useful electronic functions.

4.10 References

1 ADAMS, R. W.: 'Filtering in the Log Domain', Preprint #1470, presented at the 63rd AES Conference, New York, May 1979.

2 FREY, D. R.: 'Log Domain Filtering: an Approach to Current Mode Filtering', *IEE Proceedings G*, Vol. 140, No. 6, pp. 406–416, December 1993.

3 FREY, D. R.: 'Exponential State Space Filters: A Generic Current Mode Design Strategy', *IEEE Trans. Circuits Syst., I: Fund. Theory and Appl.*, Vol. 43, No. 1, pp. 34–42, January 1996.

4 FREY, D. R.: 'State Space Synthesis and Analysis of Log Domain Filters', *IEEE Trans. Circuits Syst., II*, Vol. 45, No. 9, pp. 1205–1211, September 1998.

5 FREY, D. R.: 'Future Implications of the Log Domain Paradigm', *IEE Proceedings Circuits, Devices and Systems*, Vol. 147, No. 1, pp. 65–72, February 2000.

6 PERRY, D. and ROBERTS, G. W.: 'The Design of Log-domain Filters Based on the Operational Simulation of LC Ladders', *IEEE Trans. Circuits and Syst., II*, Vol. 43, No. 11, pp. 763–774, November 1996.

7 PUNZENBERGER, M. and ENZ, C. C.: 'A 1.2V Low-power BiCMOS Class AB Log-domain Filter', *IEEE Journal of Solid-State Circuits*, Vol. 32, No. 12, pp. 1968–1978, December 1997.

8 GILBERT, B.: 'Translinear Circuits: a Proposed Classification', *Electronics Letters*, Vol. 11, No. 1, pp. 14–16, 1975.

9 GILBERT, B.: 'Current-mode Circuits from a Translinear Viewpoint: a Tutorial', in TOUMAZOU, C., LIDGEY, F. J. and HAIGH, D. G.: 'Analogue IC Design: The Current Mode Approach', Peter Peregrinus, London, 1990.
10 MULDER, J., VAN DER WOERD, A. C., SERDIJN, W. A. and VAN ROERMUND, A. H. M.: 'General Current Mode Analysis Method for Translinear Filters', *IEEE Trans. Circuits Syst., I*, Vol. 44, pp. 193–197, March 1997.
11 DRAKAKIS, E. M., PAYNE, A. J. and TOUMAZOU, C.: 'Log-domain Filters, Translinear Circuits, and the Bernoulli Cell', *Proc. IEEE Int. Symp. Circuits and Systems*, pp. 501–504, Hong Kong, June 1997.
12 SEEVINCK, E.: 'Companding Current-mode Integrator: a New Circuit Principle for Continuous-time Monolithic Filters', *Electronics Letters*, Vol. 26, No. 24, pp. 2046–2047, November 1990 .
13 FREY, D. and TOLA, A.: 'A State Space Formulation for Externally Linear Class AB Dynamical Circuits', *IEEE Trans. Circuits Syst., I*, Vol. 46, No. 3, pp. 306–314, 1999.
14 MULDER, J., KOUWENHOVEN, M. H. L. and VAN ROERMUND, A. H. M.: "Signal x Noise Intermodulation in Translinear Filters', *Electronics Letters*, Vol. 33, pp. 1205–1207, July 1997.
15 TÓTH, L., TSIVIDIS, Y. P. and KRISHNAPURA, N.: 'On the Analysis of Noise and Interference in Instantaneously Companding Signal Processors', *IEEE Trans. Circuits Syst., II*, Vol. 45, No. 9, pp. 1242–1249, September 1998.
16 TOUMAZOU, C., NGARNNIL, J. and LANDE, T. S.: 'Micropower Log-domain Filter for Electronic Cochlea', *Electronics Letters*, Vol. 30, pp. 1839–1841, October 27 1994.
17 FREY, D.: 'C-Log Domain Filters', Proc. of IEEE Intl. Symp. Circuits and Systems, Geneva, June 2000.
18 TSIVIDIS, Y., GOPINATHAN, V. and TÓTH, L.: 'Companding in Signal Processing', *Electronics Letters*, Vol. 26, pp. 1331–1332, August 1990.
19 TSIVIDIS, Y. and LI, D.: 'Current-mode Filters Using Syllabic Companding', Proc. IEEE ISCAS, Vol. 1, pp. 121–124, Atlanta, 1996.
20 MULDER, J., SERDIJN, W. A., VAN DER WOERD, A. C. and VAN ROERMUND, A. H. M.: 'A Syllabic Companding Translinear Filter', Proc. 1997 IEEE Int. Symp. Circuits and Systems, Vol. 1, pp. 101–104, Hong Kong, 1997.
21 FREY, D. and TSIVIDIS, Y.: 'A Syllabically Companding Log Domain Filter', *Electronics Letters*, Vol. 33, No. 18, pp. 1506–1507, 1997.
22 KRISHNAPURA, N., TSIVIDIS, Y. and FREY, D. R.: 'Simplified Technique for Syllabic Companding in Log-Domain Filters', *Electronics Letters*, Vol. 36, No. 15, pp. 1257–1259, 2000.
23 FREY, D., TSIVIDIS, Y. P., EFTHIVOULIDIS, G. and KRISHNAPURA, N.: 'Syllabic-Companding Log Domain Fitters', *IEEE Trans. Circuits Syst. II*, Vol. 48, No. 4, pp. 329–339, April 2001.

Chapter 5
Low voltage techniques for switched-current filters

John Hughes, Apisak Worapishet and Rungsimant Sitdhikorn

5.1 Introduction

The switched-current technique (SI) was introduced in the late 1980s [1–5] at a time when switched capacitors (SC) could not be easily implemented with the *digital* CMOS IC processes available at that time. This came about for two main reasons. First, SC operated by charge *transfer* and this could only be performed efficiently with the linear floating capacitors that came with special process options (e.g. double-polysilicon layers). SI, on the other hand, used charge *storage* and this required only grounded, nonlinear capacitances of the sort that occur naturally at the gate of any MOS transistor. Second, SI used current rather than voltage to represent its signals. This made it much better placed to contend with the diminishing power supply voltages needed for emerging digital circuits.

Early SI circuits were very simple. The basic current memory required only a pair of complementary MOS transistors, one as the storage device and the other to provide its bias current, and a few MOS switches to effect sampling. The cell merged the properties of storage and buffering into the same physical device (the memory transistor), whereas SC required separate storage (usually linear floating capacitors) and buffering (closed-loop OTAs). So, as well as the economies resulting from the use of the most basic process, further cost savings could be expected through reduction of chip area.

The simplicity of the circuits also suggests that SI could be capable of higher sampling frequencies than SC, which may be limited by compensation and slewing in its closed-loop OTAs. This promised SI a good future for video and higher frequency applications [6]. More recently, class AB cells have been developed [7–9] which retain the traditional advantages accruing from the class A cell's 'MOSFET-only' circuit style while giving considerably better circuit performance.

Current memories are the most fundamental of all switched-current cells. They are used in feedback pairs to make the integrators for leapfrog or state-variable filters [10] or in the modulators of sigma-delta data converters [11]. They are also used in pairs as the unit-delay cells of FIR filters [6] and in cascades for pipeline data converters [12]. So, the current memory has been the natural choice for circuit development over the past decade.

Despite their early promise, it is only recently that the benefits of SI over SC have been rigorously demonstrated. In Section 5.2, we show that, in its primitive form, the current memories can achieve levels of performance (signal-to-noise ratio, power consumption and clock frequency), to rival those of comparable SC cells. What's more, that performance, unlike that of their SC counterparts, is maintained at low supply voltage.

Less fundamental, but equally important, is the cell's precision. Lack of precision results from the MOSFET's nonideal circuit behaviour and limits the filter's precision and linearity. These aspects are reviewed in Section 5.3 for single-ended class A and class AB cells.

Primitive cells do not have adequate precision, in either SI or SC cells, and circuit enhancements are usually required to produce performance of practical value. For SC, this takes the form of using an operational (transconductance) amplifier. For SI, extra voltage gain has similar benefits and has been applied in a variety of ways. The challenge to the circuit innovator has been to provide enhanced precision without unduly penalising the other performance vectors or the low voltage working of the cells. Section 5.4 reviews the use of balanced circuit structures and Section 5.5 describes how neutralisation techniques can be used to give extra precision. The combined use of balanced structures and neutralisation is shown to produce good precision in both class A and class AB cells.

In Section 5.6, we demonstrate that the low voltage techniques developed in the earlier sections give good performance in practice. Switched-current cells, including a memory, an integrator and a second-order bandpass filter are designed and simulated performance is given.

5.2 Performance comparison of SI and SC

A performance comparison of SI and SC has recently been made in [13, 14]. In this, primitive SC and SI cells (both class A and class AB) were analysed for maximum clock sampling frequency (F_c), power consumption (P) and signal-to-noise ratio (SNR). Overall performance was then compared by combining these vectors into a single figure-of-merit (FoM), defined as

$$FoM = F_c \frac{SNR}{P} \tag{5.1}$$

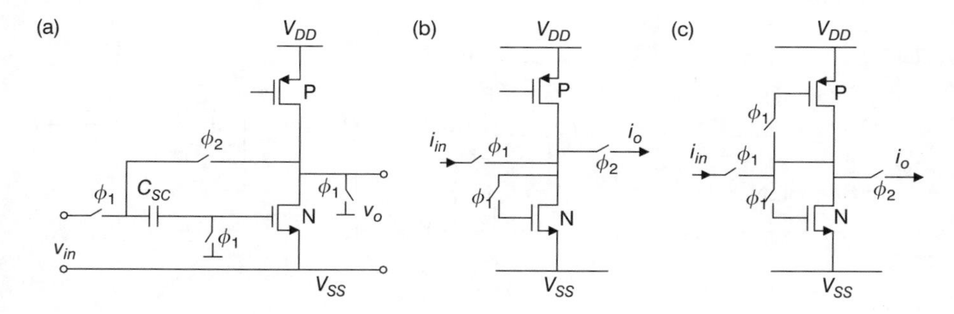

Figure 5.1 Primitive memory cells used for performance comparison: (a) SC; (b) class A SI; (c) class AB SI

The circuit vehicles used for this comparison are shown in Figure 5.1. For SC, the basic class A stage shown in Figure 5.1(a) was chosen as this has a simple MOS complementary pair as its buffer amplifier. For SI, the basic class A and class AB stages shown in Figures 5.1(b) and (c) were chosen because they use the same simple MOS complementary pair for buffering and memory. Such cells are to be found in both filters and data converters and have the virtue of simplicity which enabled the comparisons to focus on fundamental issues, without the complications involved in more sophisticated cells.

First, the operation of the test cells is described with reference to Figure 5.2. The SC circuit (Figure 5.2(a)) has a sampling or input phase ϕ_1 during which the buffer amplifier (N, P) is open loop while the sampling capacitor C_{SC} samples the input voltage v_{in}. On the hold or output phase ϕ_2, the buffer amplifier's feedback loop is closed by the connection of C_{SC} and the loop settles to produce the output voltage v_o, which is close to v_{in}.

The SI circuits (Figures 5.2(b) and (c)) have a sampling phase ϕ_1 during which the buffer amplifier is in a closed loop due to the closure of the memory switch S. The voltage on its sampling capacitor(s), C_g, settles to a value determined by the input current i_{in} and the transconductance of the memory transistor(s), G_m. On the hold phase ϕ_2, the buffer amplifier is open loop and the voltage held on C_g, together with the same transconductance, produces an output current i_o which is close to i_{in}.

Clearly, the cells' operation demonstrates the duality that exists between voltage and current mode circuits. The SC circuit presents a high input impedance with its buffer amplifier's loop open on its sampling phase and a low output impedance with the loop closed and settling on its output phase. On the other hand, the SI circuits present a low input impedance with their buffer amplifier's loop closed and settling on their sampling phase and a high output impedance with their loops open on their output phase. The allowed clock frequency will be determined by the settling behaviour when the buffer amplifier's loop is closed, i.e. during the SC's ϕ_2 phase or during the SI's ϕ_1 phase.

The individual performance vectors, clock sampling frequency (F_c), power consumption (P) and signal-to-noise ratio (*SNR*), were analysed and then combined to form analytic expressions for the *FoM*'s of each cell. Then these expressions

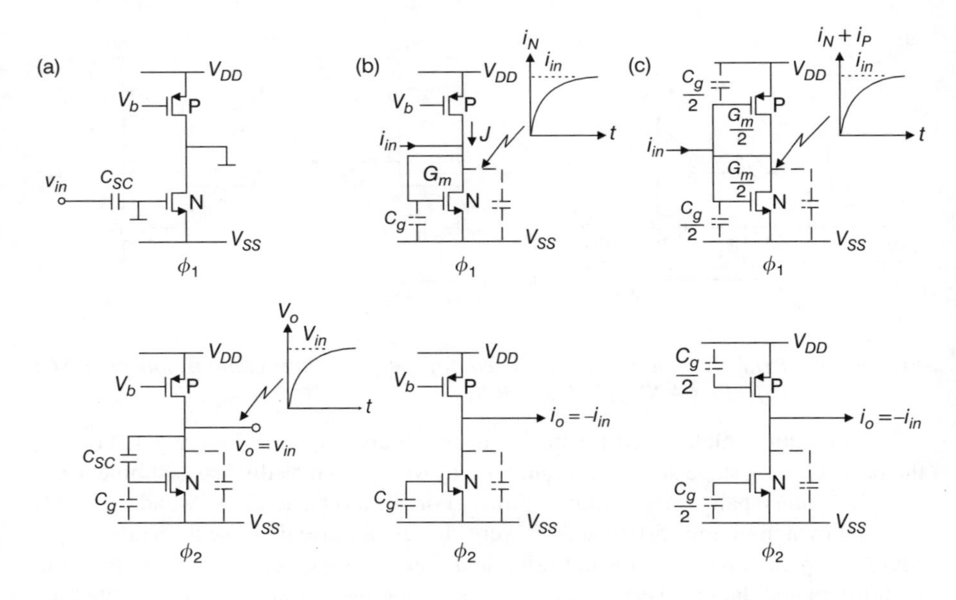

Figure 5.2 Operation of the memory cells: (a) SC; (b) class A SI; (c) class AB SI circuits

Table 5.1 Forecast of CMOS technology generations

Year	Technology (μm)	Process supply voltage (V)	Threshold voltage (V)
1991	0.80	5.0	0.80
1993	0.50	3.3	0.60
1995	0.35	3.3	0.50
1997	0.25	2.5	0.45
1999	0.18	1.8	0.35
2002	0.13	1.5	0.30
2005	0.10	1.2	0.25
2008	0.07	0.9	0.25
2011	0.05	0.6	0.20

were numerically evaluated to give a performance comparison with CMOS processes over the period 1991–2011. To do this, the process development as forecast by the 1999 Semiconductor Industries Association (SIA) roadmap was used and this is summarised in Table 5.1.

Figure 5.3 shows the resulting calculated performances of the three cells. We see that, although SC has enjoyed a performance superiority over SI through the 1990s, it is steadily falling and we can expect it to be surpassed by SI during the next decade.

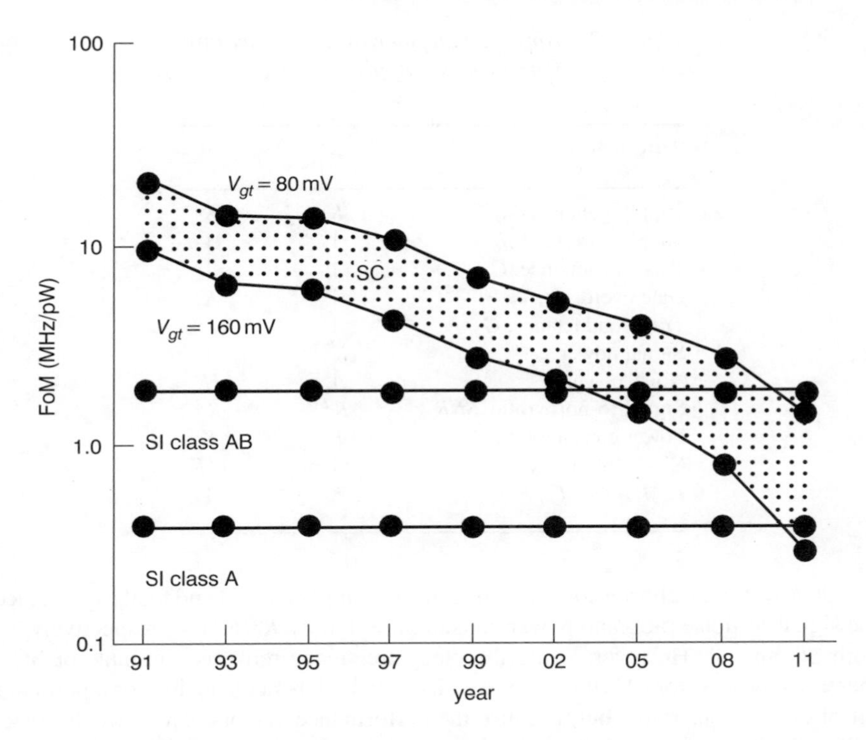

Figure 5.3 Figure-of-merits in SC and SI memory cells at different CMOS technology generations in Table 5.1

This performance behaviour can be explained intuitively with the help of the approximate comparison given in Table 5.2. We consider the CMOS process V_{DD} falling from 5 V to 0.6 V (K falling from, say, unity to 0.12) with the constraint of constant supply current, I_{DD}, and total capacitance, C. For SC, the gate overdrive V_{gt} stays constant at its minimum value and so the transconductance, G_m, also stays constant. The signal power falls by K^2 because the signal voltage falls by K but the noise power stays constant because G_m stays constant. The signal-to-noise ratio falls by K^2 and the power consumption by K. The clock frequency stays constant because both G_m and C remain constant. Consequently, the figure-of-merit falls by K.

For SI, if we assume simplistically that the threshold voltage falls linearly with the process supply voltage (reasonable until V_{DD} reaches 0.9 V in 2008), then the gate overdrive, V_{gt}, falls by K to ensure saturated operation and so the transconductance, G_m, increases by 1/K. The signal power remains constant because the supply current remains constant but the noise power increases by $1/K^2$ because the increase in G_m increases both the noise power spectral density and the noise bandwidth by $1/K$. The signal-to-noise ratio falls by K^2 and the power consumption falls by K. The clock frequency increases by $1/K$ because G_m increases by $1/K$ and C remains constant. Consequently, the figure-of-merit stays constant.

Table 5.2 Approximate performance scaling with process supply voltage, V_{DD}

Performance	SC	SI
Supply voltage, V_{DD}	K	K
Supply current, I_{DD}	1	1
Total capacitance, C	1	1
Gate overdrive, V_{gt}	1	K
Transconductance, G_m	1	$1/K$
Signal power	K^2	1
Noise power	1	$1/K^2$
Signal-to-noise ratio, *SNR*	K^2	K^2
Power consumption, P	K	K
Clock frequency, F_c	1	$1/K$
FoM, $SNR \cdot F_c/P$	K	1

Clearly, for the chosen constraint of constant supply current and total capacitance, the signal-to-noise ratio and power consumption fall by K^2 and K, respectively, for both SC and SI. However, while the clock frequency remains constant for SC it increases by $1/K$ for SI. Of course, we have reached this conclusion for a particular set of circuit constraints but, because the performance vectors can be freely traded within the bounds of constant *FoM*, the result is perfectly general. The performance superiority enjoyed by SC at higher supply voltage is steadily eroded and, as the supply voltage approaches 1V, class AB SI performance can be expected to match and then surpass that of SC.

Ultimately, it will be enhanced versions of these primitive cells, designed with properly engineered trade-offs and safety margins, that will settle the argument. These may suffer extra problems (e.g. SC may suffer slew limited settling) not encountered in our primitive cells, and this will colour our comparison. Nevertheless, the result is a strong one and indicates several things: on the one hand, it explains why SI has been outperformed by SC in older, higher voltage CMOS processes, while on the other it indicates that SI, in due course, should offer both cost and performance advantages.

5.3 Single-ended SI circuits

Having established in the last section the viability of SI for low supply voltage operation, we now turn our attention to aspects of precision and how these are influenced by transistor nonideal behaviour [15]. The signal transmission between cells is only perfect if all the signal current sampled and held in one memory is received exactly by a second memory. If this were possible, filters with infinitely high Q-factors and with perfect linearity would be possible. In practice, transistor nonideal behaviour produces systematic errors and these fall into three catagories: conductance errors, settling errors and charge injection errors. Conductance errors occur because the memory cell has neither infinite input conductance nor zero output conductance, with the result that some of the signal current transmitted from one cell to another leaks to ground. Settling errors are caused because the loop formed by the closure of the cell's switches settles exponentially during the sampling phase. Charge injection errors occur because, when the memory switch is opened, a proportion of its channel and drain-gate overlap charge is forced onto the memory capacitance, causing a gate voltage disturbance which translates to a current error in the output signal. The combined effect of these errors is to degrade precision through the generation of offsets, gain errors and nonlinearity.

5.3.1 Single-ended prototype cells

To compare the precisions of the class A and class AB cells, we first define a pair of prototype cells and these are shown in Figure 5.4. The class A cell shown in Figure 5.4(a) has bias current J, controlled by the bias voltage V_b, and an nMOS memory with width W and length L to give a quiescent transconductance G_m and gate overdrive voltage V_{gt}. The signal current i defines the modulation index $m = i/J \leq 1$. The pMOS bias transistor has width and length scaled to give the same quiescent transconductance and gate overdrive voltage. The memory switch has a quiescent on-conductance, G_s, defined by its width W_s and length which is set to L_{min}. We assume for simplicity that the gate capacitance, C_g, is determined by the gate area (WL) and the drain and drain-gate capacitances, C_d and C_{dg}, by W. The quiescent drain conductances are assumed equal with a value G_{ds} controlled by the value of the ratio J/L.

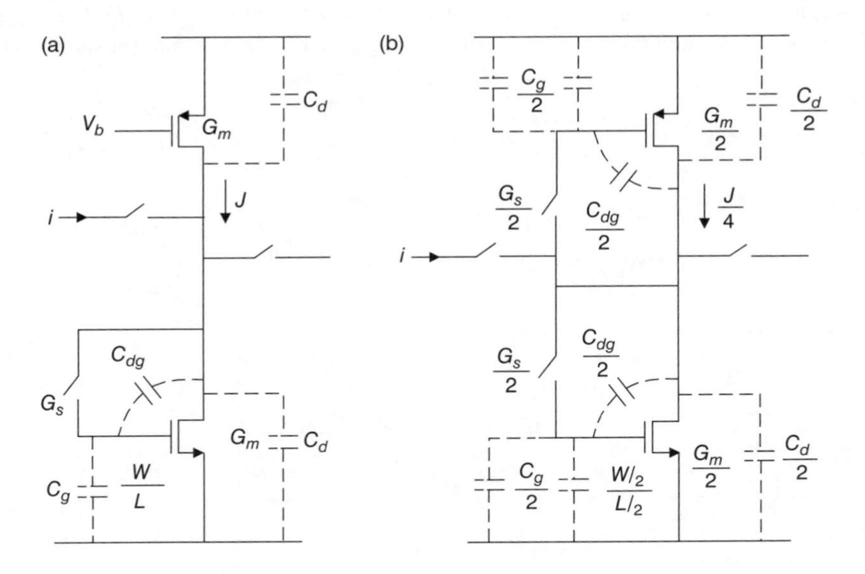

Figure 5.4 Prototype single-ended memory cells with scaled parameters: (a) class A; (b) class AB

The class AB cell shown in Figure 5.4(b) has bias current $J/4$ and an nMOS memory with width $W/2$ and length $L/2$ to give a quiescent transconductance $G_m/2$ and a gate overdrive voltage $V_{gt}/2$. The pMOS memory has width and length scaled to produce the same quiescent transconductance and gate overdrive voltage. As explained later, this cell can handle the same signal current, i, as the class A cell even though it has only a quarter of its bias current. The memory switches have a quiescent on-conductance of $G_s/2$ defined by switch widths $W_s/2$ and lengths L_{min}. The gate capacitances are made equal to $C_g/2$ partly from the gate-oxide capacitances of the scaled memory transistors and partly from added gate-oxide capacitances. The drain and drain-gate capacitances are $C_d/2$ and $C_{dg}/2$ and quiescent drain conductances are $G_{ds}/2$.

So, we see that the adopted regime has resulted in a class AB prototype which has parameters for each memory with half those of the class A prototype, and consequently the two cells have nearly identical bandwidth and settling characteristics. While other scaling regimes may give more optimal precision, the adopted regime has the merit of simplifying the analyses which follow.

5.3.2 *Single-ended class A*

Operating range Best performance is achieved when the quiescent gate overdrives are set to their maximum allowable values. This maximum is determined by the need to keep the MOSFETs operating in their saturation region.

Figure 5.5(a) shows two class A cells where N_1 is storing bias and signal current, $J - i$, and N_2 is sinking $J + i$. It is easily shown that, for equal threshold voltages ($V_{tn} = V_{tp} = V_t$) and gate overdrive voltages ($V_{gtn} = V_{gtp} = V_{gt}$), all transistors stay

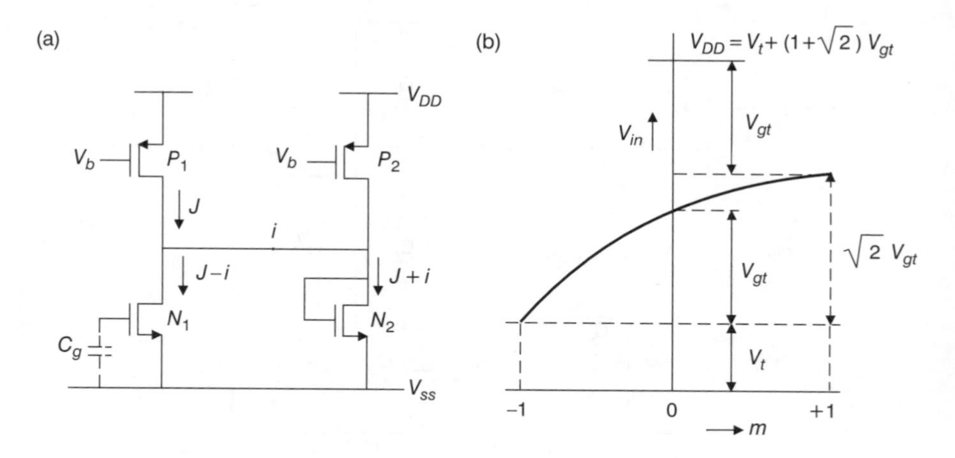

Figure 5.5 *Transmission between primitive single-ended class A SI memory cells: (a) circuit arrangement; (b) input voltage variation with signal with* $\hat{m} = 1$

saturated so long as the quiescent gate overdrive voltage, V_{gt}, satisfies the condition

$$V_{gt} \leq \frac{V_t}{\sqrt{1+\hat{m}} - \sqrt{1-\hat{m}}} \tag{5.2}$$

and the supply voltage is set to

$$V_{DD} \geq V_t + V_{gt}\left(1 + \sqrt{1+\hat{m}}\right) \tag{5.3}$$

where $\hat{m}$ is the peak modulation index $(\hat{i}/J)$. In principle, $\hat{m} \leq 1$ but a more realistic design might be for $V_{gt} = V_t$, for which $\hat{m} \leq 0.87$ and $V_{DD} \geq 3.37V_t$. However, as we will see later, this can produce poor precision and so, consequently, $\hat{m}$ may have to be limited to a lower value.

Figure 5.5(b) shows that the input voltage of the class A cell varies with the square root of the input current. This property will be shown to seriously impair the linearity of the single-ended class A cell.

Conductance errors Referring to Figure 5.5, the signal transmitted from one memory cell to another is, in practice, not perfect because of nonideal behaviour of the transistors. First, the driving cell, N_1P_1, and the receiving cell, N_2P_2, have transconductances, $G_{mn1}(m)$ and $G_{mn2}(m)$, which are signal-dependent, i.e. they are dependent on the instantaneous modulation level, $m = i/J$:

$$G_{mn1}(m) = G_m\sqrt{1-m} \approx G_m\left(1 - \frac{m}{2} - \frac{m^2}{8}\ldots\right) \tag{5.4}$$

$$G_{mn2}(m) = G_m\sqrt{1+m} \approx G_m\left(1 + \frac{m}{2} - \frac{m^2}{8}\ldots\right) \tag{5.5}$$

where G_m is the quiescent value of the transconductance.

Second, transistor N_1 has an output conductance resulting from both channel length modulation and feedback from its drain to its open gate capacitor C_g via its drain-gate capacitor C_{dg}, and transistors P_1, N_2, P_2 have signal-dependent output conductances resulting from channel length modulation. So, the output conductances are:

$$G_{dsn1}(m) = G_{ds}(1-m) + \frac{C_{dg}}{C_{dg}+C_g}G_m\sqrt{1-m} \tag{5.6}$$

$$G_{dsp1}(m) = G_{ds} \tag{5.7}$$

$$G_{dsn2}(m) = G_{ds}(1+m) \tag{5.8}$$

$$G_{dsp2}(m) = G_{ds} \tag{5.9}$$

where G_{ds} is the quiescent drain conductance of each transistor. The effective output conductance is $G_{oA}(m)$, where

$$\begin{aligned} G_{oA}(m) &= G_{dsn1}(m) + G_{dsp1}(m) + G_{dsn2}(m) + G_{dsp2}(m) \\ &= 4G_{ds} + \frac{C_{dg}}{C_{dg}+C_g}G_m\sqrt{1-m} \end{aligned} \tag{5.10}$$

and this leaks some of the signal (δi) away from the receiving cell. The signal-dependent conductance error is

$$\begin{aligned}\epsilon_{gA}(m) &= \frac{\delta i}{i} \approx \frac{G_{oA}(m)}{G_{mn2}(m)} \\ &= \frac{4G_{ds} + C_{dg}G_m\sqrt{1-m}/(C_{dg}+C_g)}{G_m\sqrt{1+m}} \\ &\approx \frac{4G_{ds}}{G_m}\left(1 - \frac{m}{2} + \frac{3m^2}{8}\right) + \frac{C_{dg}}{C_{dg}+C_g}\left(1 - m + \frac{m^2}{2}\right), \quad m \ll 1\end{aligned} \tag{5.11}$$

Clearly, the large-signal conductance error of the single-ended class A cell is highly signal-dependent and this gives rise to gain errors and both even and odd harmonic distortion. As m approaches unity this error becomes very large and so, in practice, the modulation index, $\hat{m}(=\hat{i}/J)$, is often limited to about 0.5 for the single-ended class A cell.

Settling errors The closed loop of the class A cell has the properties of a second order system. The class A cell has a small-signal Q-factor of

$$QF_A = \frac{2G_mC_dC_g/G_s}{2C_d + C_g} \tag{5.12}$$

The loop is normally designed for small-signal critical damping ($QF_A = 0.5$) to minimise the settling error.

The large-signal settling behaviour for the single-ended class A cell is depicted in Figure 5.6. Large-signal settling behaviour tends to be underdamped for positive signals and overdamped for negative signals because of the variation of transconductance with drain current, and this leads to signal-dependent settling and consequent distortion.

Charge injection errors Charge injection in the single-ended class A cell is shown in Figure 5.7. The charge injected onto the memory capacitor when the memory switch is turned off is given by [15]

$$q_A \approx C_{OL}(V_H - V_L) + \alpha\left[V_H - \left(1 + \frac{\gamma}{3}\right)v_X - V_t\right]C_{CH} \tag{5.13}$$

where C_{OL} and C_{CH} are the switch's drain-gate overlap and the channel capacitances, V_H and V_L are the high and low levels of the clock voltage, γ is the backgate parameter, α is the fraction of the switch's channel charge entering the gate capacitor, C_g, and v_X is the signal-dependent voltage at the switch. In the class A cell, $v_X = V_{gs}$, and it can be shown (Chapter 4 of [15]) that the charge injection error in the output signal is of the form

$$\delta i_A = \frac{Q_A}{C_g}\sqrt{2\beta I_d} - \frac{C_A}{C_g}I_d = \frac{Q_A}{C_g}G_m\sqrt{1+m} - \frac{C_A}{C_g}J(1+m) \tag{5.14}$$

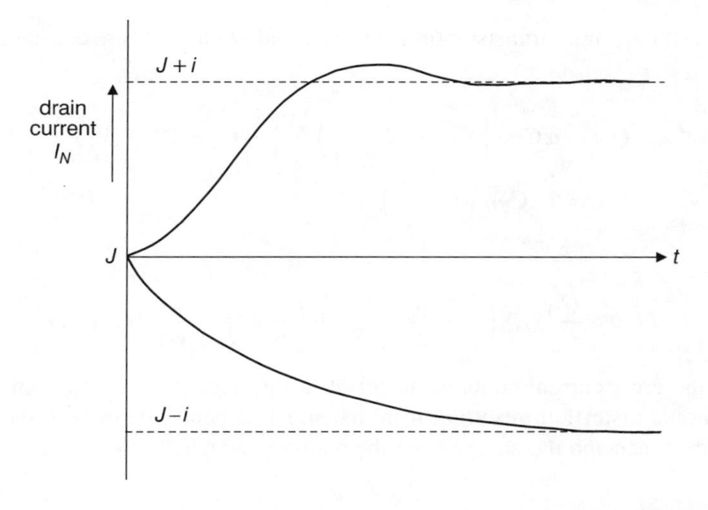

Figure 5.6 Large-signal settling behaviour of the single-ended class A memory cell with positive and negative signals

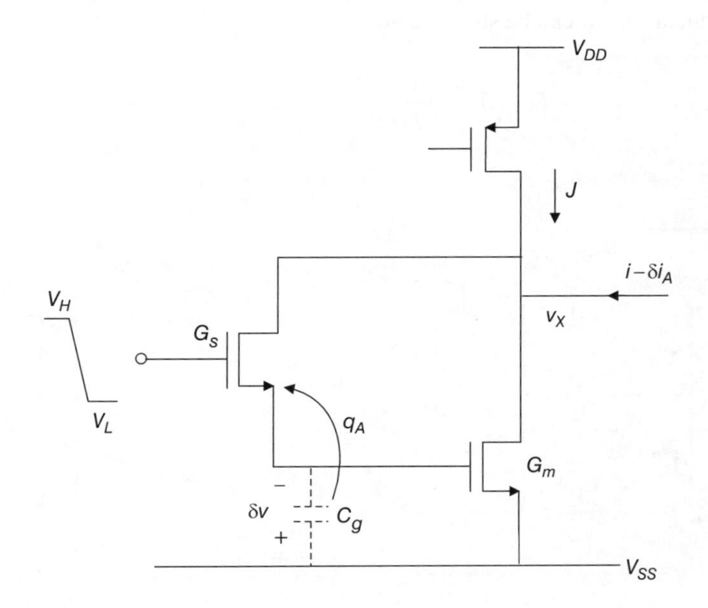

Figure 5.7 Charge injection in single-ended class A memory cell

where I_d is the memory transistor drain current, and Q_A and C_A are constants defined by:

$$Q_A = \alpha C_{CH}\left[V_H - \left(2 + \frac{\gamma}{3}\right)V_t\right] + C_{OL}(V_H - V_L) \tag{5.15}$$

$$C_A = 2\alpha C_{CH}\left(1 + \frac{\gamma}{3}\right) \tag{5.16}$$

So,

$$\delta i_A \approx \frac{Q_A}{C_g} G_m \left(1 + \frac{m}{2} - \frac{m^2}{8} + \frac{m^3}{16} \cdots\right) - \frac{C_A}{C_g} J(1+m) \tag{5.17}$$

Clearly, the error current contains an offset, a linear gain error and both even and odd harmonic distortion, resulting from the signal-dependence of both the memory transconductance and the charge from the memory switch.

5.3.3 Single-ended class AB

Operating range Figure 5.8(a) shows transmission of signal current i between two class AB cells. The two memories, N_1 and P_1, are storing I_n and I_p, respectively, to produce an output current $i = I_p - I_n$. This signal then splits between the two memories N_2 and P_2 which sink I_p and I_n, respectively. For transistors with equal transconductances, it can be shown that

$$I_{n,p} = J_{AB}\left(1 \pm \frac{i}{4J_{AB}}\right)^2 \quad -4J_{AB} \le i \le +4J_{AB} \tag{5.18}$$

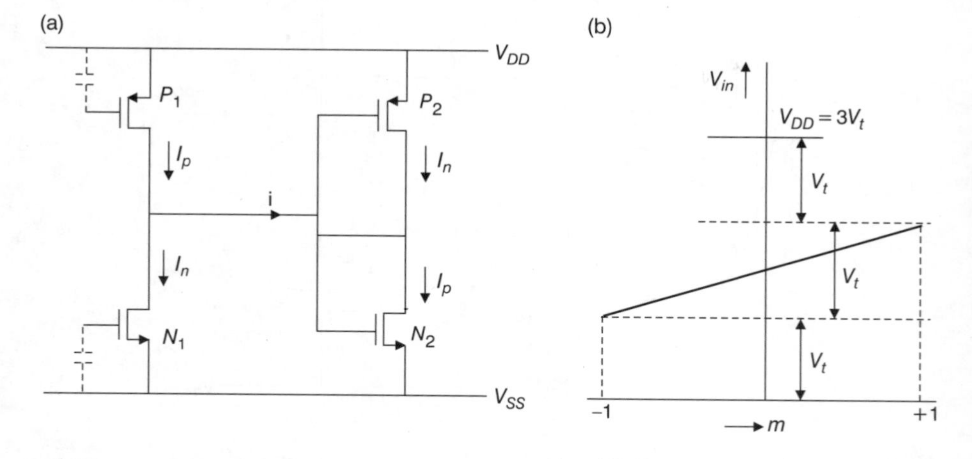

Figure 5.8 Transmission between primitive single-ended class AB SI memory cells: (a) circuit arrangement; (b) input voltage variation with signal for $\hat{m} = 1$

which for $J_{AB} = J/4$ reduces to

$$I_{n,p} = \frac{J}{4}(1 \pm m)^2 \quad -1 \leq m \leq +1 \tag{5.19}$$

where $m = i/J$ is the class AB instantaneous modulation level. For equal threshold voltages, all transistors stay in saturation over the whole signal range, $-1 \leq m \leq +1$, so long as

$$V_{gtAB} \leq \frac{V_t}{2\hat{m}} \tag{5.20}$$

and the supply voltage is set to

$$V_{DD} = 2(V_{gtAB} + V_t) = V_t \left(2 + \frac{1}{\hat{m}}\right) \tag{5.21}$$

So, for $\hat{m} = 1$, we may design with:

$$V_{gtAB} = \frac{V_t}{2} \tag{5.22}$$

$$V_{DD} = 3V_t \tag{5.23}$$

Figure 5.8(b) shows that the input voltage varies linearly with signal in the class AB cell. We will see later that this property gives an added benefit to class AB operation.

Whereas the bias current for the class A cell is determined by the pMOS transistor, this is not the case for the class AB cell. As V_{DD} varies or the transistor parameters vary with process spreads and temperature, special control of the bias current is needed. This is usually achieved by controlling the current in a reference circuit similar to the memory cell by either regulating the supply voltage or adjusting the n-well bias to regulate the threshold voltage [7] as described in Section 5.6.

Conductance errors Referring to Figure 5.8, the transistors N_1, P_1 of the driving cell and N_2, P_2 of the receiving cell have signal-dependent transconductances $G_{mn1}(m)$, $G_{mp1}(m)$ and $G_{mn2}(m)$, $G_{mp2}(m)$, where for equal nMOS and pMOS transconductance parameters, it can be shown, using Eqn. (5.19), that:

$$G_{mn1}(m) = \frac{G_m}{2}(1 - m) \tag{5.24}$$

$$G_{mp1}(m) = \frac{G_m}{2}(1 + m) \tag{5.25}$$

$$G_{mn2}(m) = \frac{G_m}{2}(1 + m) \tag{5.26}$$

$$G_{mp2}(m) = \frac{G_m}{2}(1 - m) \tag{5.27}$$

where $G_m/2$ is the quiescent value of the transconductance of each transistor, as defined in Figure 5.4(b). So, the transconductances of the driving and receiving cells are $G_{m1}(m)$ and $G_{m2}(m)$, where:

$$G_{m1}(m) = G_{mn1}(m) + G_{mp1}(m) = G_m \tag{5.28}$$

$$G_{m2}(m) = G_{mn2}(m) + G_{mp2}(m) = G_m \tag{5.29}$$

So, the transconductance of the class AB cell is not signal-dependent, remaining fixed at the quiescent value (which is consistent with the linear variation of input voltage with signal as shown in Figure 5.8).

Transistors N_1, P_1 have output conductances resulting from both channel length modulation and feedback from their drain to their open gate capacitors $C_g/2$ via their drain-gate capacitors $C_{dg}/2$, and transistors N_2, P_2 have signal-dependent output conductances resulting from channel length modulation. So, the output conductances are:

$$G_{dsn1}(m) = \frac{G_{ds}}{2}(1-m)^2 + \frac{C_{dg}}{C_{dg}+C_g}G_{mn1}(m) \tag{5.30}$$

$$G_{dsp1}(m) = \frac{G_{ds}}{2}(1+m)^2 + \frac{C_{dg}}{C_{dg}+C_g}G_{mp1}(m) \tag{5.31}$$

$$G_{dsn2}(m) = \frac{G_{ds}}{2}(1+m)^2 \tag{5.32}$$

$$G_{dsp2}(m) = \frac{G_{ds}}{2}(1-m)^2 \tag{5.33}$$

where $G_{ds}/2$ is the quiescent drain conductance of each transistor. The effective output conductance is $G_{oAB}(m)$, where

$$\begin{aligned} G_{oAB}(m) &= G_{dsn1}(m) + G_{dsp1}(m) + G_{dsn2}(m) + G_{dsp2}(m) \\ &= 2G_{ds}(1+m^2) + \frac{C_{dg}}{C_{dg}+C_g}G_m \end{aligned} \tag{5.34}$$

and this leaks some of the signal (δi) away from the receiving cell. The signal-dependent conductance error is

$$\begin{aligned} \epsilon_{gAB}(m) &= \frac{\delta i}{i} \approx \frac{G_{oAB}(m)}{G_{m2}(m)} \\ &= \frac{2G_{ds}(1+m^2) + C_{dg}G_m/(C_{dg}+C_g)}{G_m} \\ &= \frac{2G_{ds}}{G_m}(1+m^2) + \frac{C_{dg}}{C_{dg}+C_g} \qquad m \le 1 \end{aligned} \tag{5.35}$$

Note that this expression is exact for square-law MOS operation and there is no approximation as for the class A analysis. The first term is signal-dependent, but as there is no first-order dependence (compare with Eqn. (5.11) for class A) there is no even harmonic distortion, even with extreme modulation ($\hat{m} = 1$). The second term

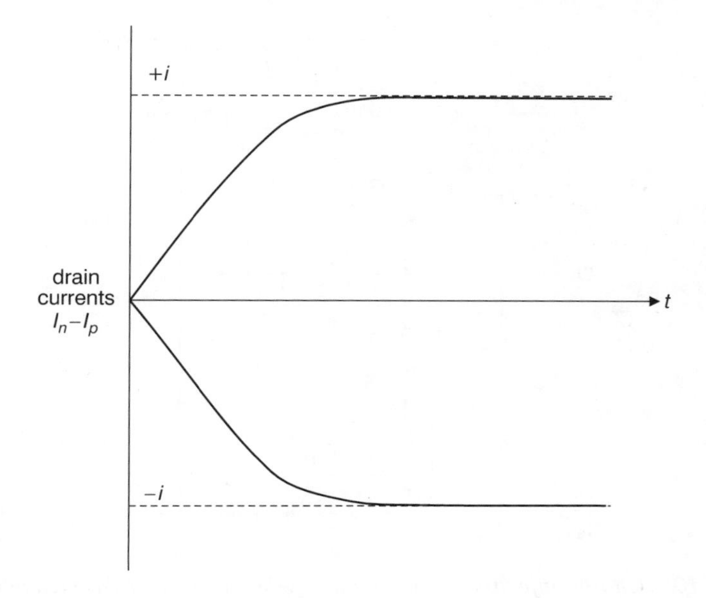

Figure 5.9 *Large-signal settling behaviour of the single-ended class AB memory cell for positive and negative signals*

is constant and produces only a gain error. Consequently, conductance errors in the single-ended class AB cell produce less distortion than in the single-ended class A cell.

Settling errors The single-ended class AB cell has a Q-factor of

$$QF_{AB} = \frac{\sqrt{(G_{mn} + G_{mp})C_g C_d / G_s}}{C_g + C_d} \tag{5.36}$$

because the nMOS and pMOS loops settle in parallel. As $G_{mn} + G_{mp} = G_m$ is signal-independent, the settling error is fairly constant, resulting in only a gain error. The large-signal settling behaviour for single-ended class AB cells is depicted in Figure 5.9. Whereas the class A cell large-signal settling behaviour (Figure 5.6) tends to be underdamped for positive signals and overdamped for negative signals because of the variation of transconductance with total current, the class AB large-signal settling behaviour is substantially signal-independent with consequent benefit to linearity.

Charge-injection errors In the class AB cell (Figure 5.10), charge is injected from each switch onto its respective memory capacitor, and the resulting gate voltage disturbances are translated into error currents in the drains of each memory transistor to give a composite error signal at the cell's output. The switch voltage, v_X, is linearly dependent on the input signal current

$$v_X = V_X + \frac{i}{G_{inAB}(m)} \tag{5.37}$$

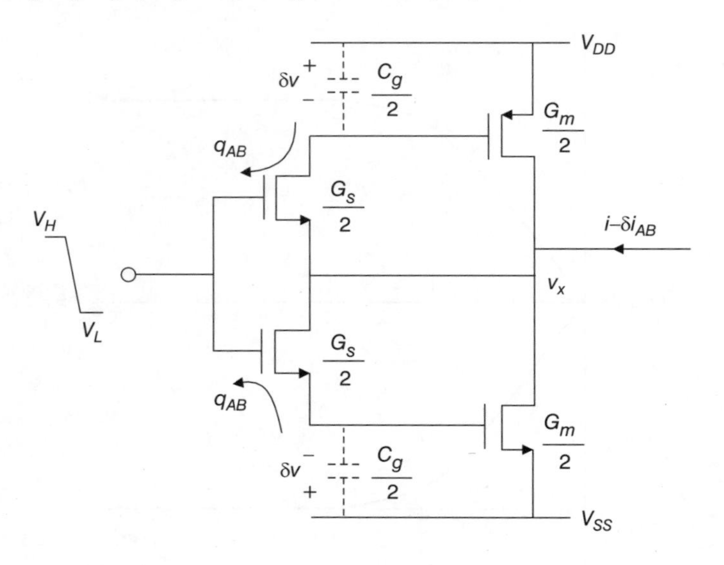

Figure 5.10 Charge-injection errors in the single-ended class AB memory cell

where $G_{inAB}(m) = G_m$ is the signal-independent input conductance and $V_X = V_{DD}/2$ (for $V_{SS} = 0$) is the quiescent voltage at the switches. So, for equal memory capacitances $C_g/2$ and equal memory switches, it can be shown that the charge injected by each switch is given by

$$q_{AB} = Q_{AB} - C_{AB}\frac{i}{G_{inAB}(m)} \tag{5.38}$$

where Q_{AB} and C_{AB} are constants which, because the class AB switches are half the size of the class A switches, are given by:

$$Q_{AB} = \frac{Q_A}{2} \tag{5.39}$$

$$C_{AB} = \frac{C_A}{2} \tag{5.40}$$

The gate voltage disturbances are

$$\delta v_p = \delta v_n = \delta v = \frac{2q_{AB}}{C_g} \tag{5.41}$$

These gate voltage disturbances cause an output signal current error given by

$$\delta i_{AB} = \delta v(G_{mp}(m) + G_{mn}(m)) = \frac{Q_A G_m}{C_g} - \frac{C_A J}{C_g}m \tag{5.42}$$

So, the single-ended class AB cell transmits a charge injection error which is an offset error of $Q_A G_m / C_g$ and a linear gain error given by

$$\epsilon_{qAB} = -\frac{C_A}{C_g} \tag{5.43}$$

Neither of these errors causes distortion.

5.3.4 Summary

In Section 5.2, we showed that the single-ended class AB primitive SI cell had about an eightfold higher figure-of-merit than the single-ended class A primitive SI cell. This gives the class AB cell higher signal-to-noise ratio and clock frequency, and lower power consumption. In this section, we have examined how the common transistor defects (conductance, settling and charge injection) affect precision. It was found that the signal-independent transconductance of the class AB cell has resulted in signal-independent settling and charge-injection errors as well as a reduced conductance error. Consequently, the single-ended class AB cell has considerably better linearity than the class A cell.

5.4 Balanced SI circuits

The single-ended SI cells discussed in Section 5.3 do not have sufficiently high precision for demanding applications. Balanced circuits have the well-known advantage for a mixed signal environment that they reject impulsive noise on the substrate. Further, signal inversion can be achieved simply by interchanging the signal pair. In this section, we examine the additional benefits of employing balanced structures for class A and class AB SI cells.

5.4.1 Prototype balanced cells

As with the single-ended cells we define prototype balanced class A and class AB cells, and these are shown in Figure 5.11. They are simply two half-sized single-ended cells (Figure 5.4) each handling one of the signal pair. They have no input common-mode rejection but this may be unnecessary in filter circuits, for example when using double-sampling integrators [16]. The circuits have the same power consumption, signal-to-noise ratio and bandwidth [15] as their single-ended counterparts. However, the balanced cells have much better precision. The parameters of the balanced circuits are scaled from their single-ended cells in the same manner as before. The (trans)conductances and currents (both signal and bias) of the half circuits are half those of the single-ended cells, and the voltage swings are identical.

5.4.2 Balanced class A cells

Conductance errors As the half circuits of the balanced cell are uniformly scaled versions of the single-ended cell, the conductance error of each half circuit is also the

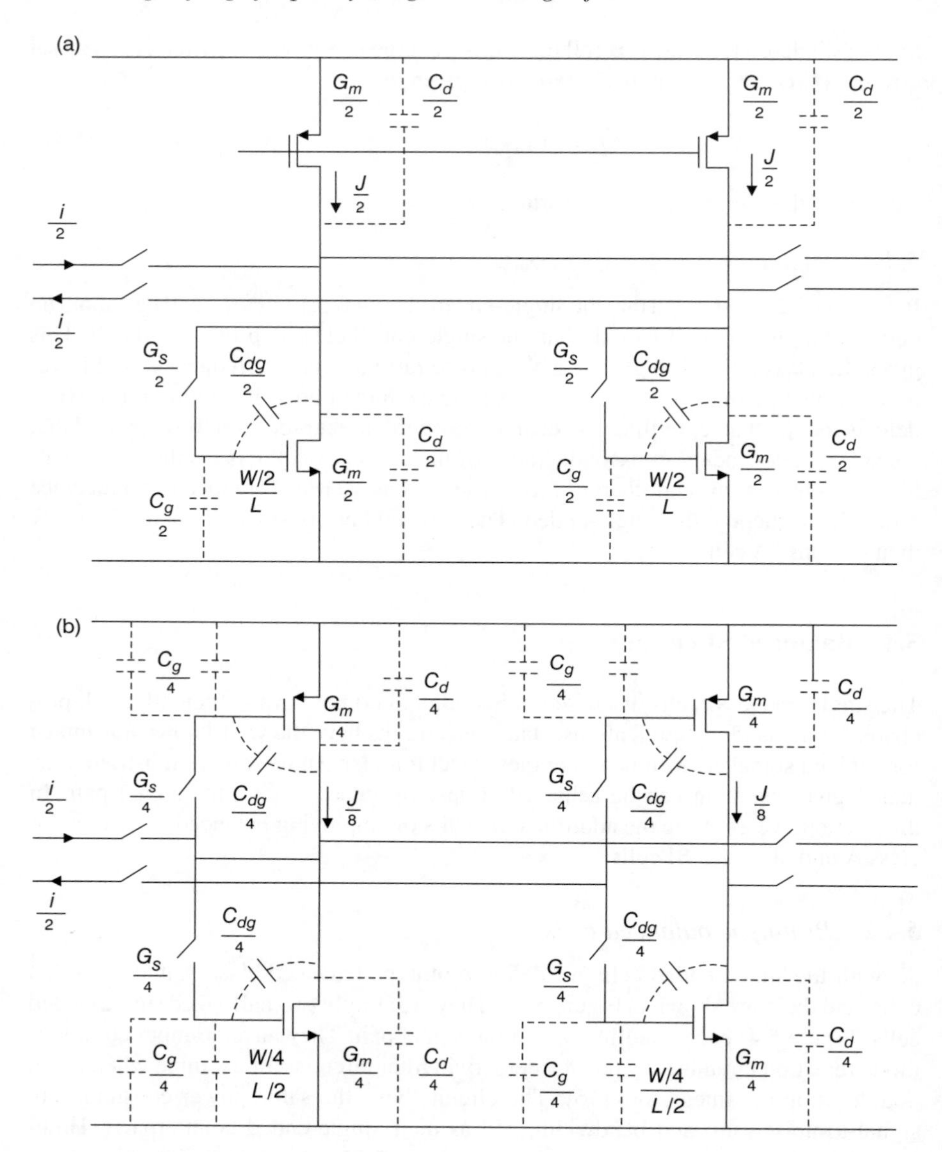

Figure 5.11 Prototype balanced memory cells with scaled parameters: (a) class A; (b) class AB

same. So, these two half cell errors are:

$$\epsilon_{gA}^{+}(m) \approx \frac{4G_{ds}}{G_m}\left(1 + \frac{m}{2} + \frac{3m^2}{8}\right) + \frac{C_{dg}}{C_{dg} + C_g}\left(1 + m + \frac{m^2}{2}\right) \tag{5.44}$$

$$\epsilon_{gA}^{-}(m) \approx \frac{4G_{ds}}{G_m}\left(1 - \frac{m}{2} + \frac{3m^2}{8}\right) + \frac{C_{dg}}{C_{dg} + C_g}\left(1 - m + \frac{m^2}{2}\right) \tag{5.45}$$

Consequently, the differential conductance error, $\epsilon_{gA\Delta}(m)$, is given by

$$\begin{aligned}\epsilon_{gA\Delta}(m) &= \frac{\epsilon_{gA}^{+}(m) + \epsilon_{gA}^{-}(m)}{2} \\ &\approx \frac{4G_{ds}}{G_m} + \frac{C_{dg}}{C_{dg} + C_g} + \left[\frac{6G_{ds}}{G_m} + \frac{1}{2}\frac{C_{dg}}{C_{dg} + C_g}\right] m^2 \quad m \ll 1\end{aligned} \tag{5.46}$$

So, the odd-order terms have cancelled (and will re-appear as common-mode errors), while the even-order terms remain unchanged. For a differential sinusoidal input signal, $i = \hat{i}\sin(\omega t)$, this produces a differential modulation at the input of $m = \hat{m}\sin(\omega t)$ and at the output of $m_{out} = \hat{m}[1 - \epsilon_{gA\Delta}(m)]\sin(\omega t)$. It can be shown that the distortion of the output signal is given by:

$$HD_{gA\Delta(1)} = \frac{4G_{ds}}{G_m} + \frac{C_{dg}}{C_{dg} + C_g} \tag{5.47}$$

$$HD_{gA\Delta(2)} = 0 \tag{5.48}$$

$$HD_{gA\Delta(3)} \approx \left[\frac{6G_{ds}}{G_m} + \frac{1}{2}\frac{C_{dg}}{C_{dg} + C_g}\right]\frac{\hat{m}^2}{4} \tag{5.49}$$

where $HD_{gA\Delta(1)}$ is simply the linear gain error and $HD_{gA\Delta(2)}$ and $HD_{gA\Delta(3)}$ are the second- and third-harmonic distortions resulting from conductance errors. Balanced operation of the class A cell has resulted in linearised performance because both the input and the output differential conductances have extended linear ranges.

Settling While the small-signal bandwidth of the cells is the same as the single-ended cell, the large-signal settling behaviour is quite different. Figure 5.12 shows this large-signal settling. We see the two memory currents of the balanced cell exhibiting the same underdamped response for increasing current and overdamped response for decreasing currents that we described in the single-ended cell (Figure 5.6). However, the net differential current in the cell is $i_{A\Delta} = (i^{+} - i^{-})/2$, and this is shown in Figure 5.12(b), where we see a response close to being critically damped.

This may also be explained by the differential transconductance of the balanced memory cells, which is given by

$$G_{mA\Delta}(m) = \frac{G_{mn}^{+}(m)G_{mn}^{-}(m)}{G_{mn}^{+}(m) + G_{mn}^{-}(m)} \tag{5.50}$$

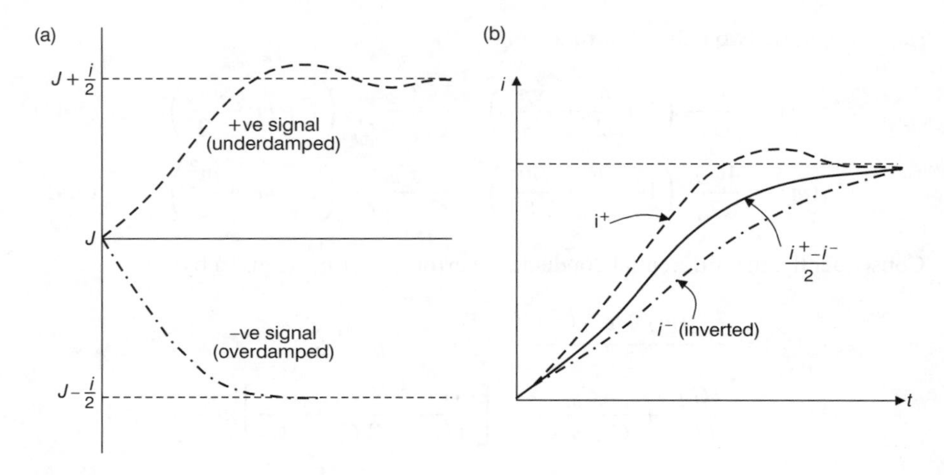

Figure 5.12 Large-signal settling behaviour in balanced class A cells

where $G^{+}_{mn}(m)$ and $G^{-}_{mn}(m)$ are the signal-dependent transconductances of the cell's memory transistors defined by:

$$G^{+}_{mn}(m) = \frac{G_m}{2}\sqrt{1+m} \approx \frac{G_m}{2}\left(1 + \frac{m}{2} - \frac{m^2}{8} + \frac{m^3}{16}\cdots\right) \tag{5.51}$$

$$G^{-}_{mn}(m) = \frac{G_m}{2}\sqrt{1-m} \approx \frac{G_m}{2}\left(1 - \frac{m}{2} - \frac{m^2}{8} - \frac{m^3}{16}\cdots\right) \tag{5.52}$$

So, the differential transconductance, $G_{mA\Delta}(m)$, is given by

$$G_{mA\Delta}(m) \approx \frac{G_m}{4}\left(1 - \frac{3m^2}{8}\right) \tag{5.53}$$

So, the more linear settling response is due largely to the balanced cell's linearised differential transconductance given by Eqn. (5.53). Comparing this with Eqn. (5.5) it can be seen that, as there is no first-order term, the settling error does not produce second-harmonic distortion but the second-order term implies that there is still third-harmonic distortion.

Charge-injection Because the balanced class A cells have half sized memory switches which deliver half the charge and half the gate capacitances of their single-ended cells, the error currents at the outputs of each half cell resulting from charge injection are:

$$\delta i^{+}_{A} = \frac{Q_A}{C_g}\frac{G_m}{2}\sqrt{1+m} - \frac{C_A}{C_g}\frac{J}{2}(1+m) \tag{5.54}$$

$$\delta i^{-}_{A} = \frac{Q_A}{C_g}\frac{G_m}{2}\sqrt{1-m} - \frac{C_A}{C_g}\frac{J}{2}(1-m) \tag{5.55}$$

This gives a differential charge injection error current of

$$\delta i_{A\Delta} = \frac{\delta i_A^+ - \delta i_A^-}{2} \approx \left[\frac{Q_A}{C_g} \frac{G_m}{2J} \left(1 + \frac{m^2}{8}\right) - \frac{C_A}{C_g} \right] \frac{i}{2} \tag{5.56}$$

So, the charge injection error is

$$\epsilon_{qA\Delta} = \frac{\delta i_{A\Delta}}{\frac{i}{2}} \approx \frac{Q_A}{C_g} \frac{G_m}{2J} \left(1 + \frac{m^2}{8}\right) - \frac{C_A}{C_g} \tag{5.57}$$

This produces further distortion of the output signal:

$$HD_{qA\Delta(1)} = \frac{Q_A}{C_g} \frac{G_m}{2J} - \frac{C_A}{C_g} \tag{5.58}$$

$$HD_{qA\Delta(2)} = 0 \tag{5.59}$$

$$HD_{qA\Delta(3)} = \frac{Q_A}{C_g} \frac{G_m}{2J} \frac{\hat{m}^2}{32} \tag{5.60}$$

This shows that charge injection in the balanced class A cell produces a further linear gain error and further third-harmonic distortion but no even-harmonic distortion.

5.4.3 Balanced class AB cells

Conductance errors The balanced class AB conductance error is much the same as the single-ended case. This is because each half circuit has the same input and output conductance relationships as the single-ended cell. Alternatively, it can be seen from Eqn. (5.35) that the single-ended class AB conductance error expression has only a second-order term in m and so, even though the half circuits of the balanced cell pass opposite polarities of signal, they produce the same error as the single-ended cell. So, the balanced class AB differential conductance error, $\epsilon_{gAB\Delta}(m)$, is given by

$$\epsilon_{gAB\Delta}(m) = \epsilon_{gAB}(m) = \frac{2G_{ds}}{G_m}(1 + m^2) + \frac{C_{dg}}{C_{dg} + C_g} \tag{5.61}$$

This produces harmonics of the output signal given by:

$$HD_{gAB\Delta(1)} = \frac{2G_{ds}}{G_m} + \frac{C_{dg}}{C_{dg} + C_g} \tag{5.62}$$

$$HD_{gAB\Delta(2)} = 0 \tag{5.63}$$

$$HD_{gAB\Delta(3)} = \frac{G_{ds}}{G_m} \frac{\hat{m}^2}{2} \tag{5.64}$$

where $HD_{gAB\Delta(1)}$ is simply the class AB cell's linear gain error and $HD_{gAB\Delta(3)}$ is its third-harmonic distortion resulting from conductance errors.

Settling errors The balanced class AB cell has the same signal-independent settling response as the single-ended cell which, being linear, gets no further improvement from balanced operation.

Charge-injection errors For the balanced class AB cell, the charge-injection errors for each half circuit can be described by the single-ended behaviour (Eqn. (5.42)). Each half circuit has an input conductance of $G_m/2$ and a signal current of $\pm i/2$. So, the two half circuit errors are:

$$\delta i_{AB}^{+} = \frac{Q_A}{C_g}\frac{G_m}{2} - \frac{C_A}{C_g}\frac{i}{2} \tag{5.65}$$

$$\delta i_{AB}^{-} = \frac{Q_A}{C_g}\frac{G_m}{2} + \frac{C_A}{C_g}\frac{i}{2} \tag{5.66}$$

giving a differential charge-injection error, $\epsilon_{qAB\Delta}$,

$$\epsilon_{qAB\Delta} = \frac{\delta i_{AB}^{+} - \delta i_{AB}^{-}}{i} = -\frac{C_A}{C_g} \tag{5.67}$$

and a signal-independent common-mode charge injection error current, $i_{qAB\Sigma}$,

$$i_{qAB\Sigma} = \frac{\delta i_{AB}^{+} + \delta i_{AB}^{-}}{2} = \frac{Q_A}{C_g}\frac{G_m}{2} \tag{5.68}$$

So, the balanced operation of the class AB cell produces a common-mode and linear gain error but with neither offset nor distortion.

5.4.4 *Summary*

The use of balanced circuits with half-sized transistors in each half circuit preserves the *FoM* of the single-ended cell as well as providing a degree of immunity from substrate noise and a simple means of inverting signals. The simple balanced cell does not give common-mode rejection but when used in filters employing double-sampling integrators this is unimportant as common-mode signals at their inputs are rejected by the input sampling.

Neither class A nor class AB cells produced offsets or even-harmonic distortion, as expected for balanced cells. Balanced class A cells produce linear gain errors and odd-harmonic distortion through conductance, settling and charge injection errors but with reduced magnitude. Balanced class AB cells produce linear gain errors from conductance, settling and charge injection errors but only conductance errors resulted in odd-harmonic distortion.

5.5 Neutralised SI circuits

As well as the benefits to linearity provided by balanced circuits, their complementary signal pairs offer an opportunity for further error reduction. This arises through the existence of open gate nodes in the memory transistors when they are delivering their output signals. In the normal balanced memory cells, voltage disturbances at the drains are coupled to the memory transistor's gates via their parasitic drain-gate overlap capacitance and this produces a *loss* error in the output signal. Now suppose that extra capacitances are added with one end connected to the memory gates and the other ends cross-connected to their partners' drains. Voltage disturbances at the drains are now coupled with inversion to the memory transistor's gates via the extra capacitances and this produces a *gain* error in the output signal. Clearly, we have a means to neutralise any residual errors [17, 18].

5.5.1 Neutralised class A cells

Figure 5.13 shows the balanced class A memory cell with extra capacitors C_n cross-coupled between drains and gates. On the input phase (Figure 5.13(a)), the currents in the memory transistors develop drain voltages v_p and v_n. On the output phase (Figure 5.13(b)), the driving memory cell is connected to the receiving memory cell and, because of the signal inversion in the driving cell, the drain voltages change to v_n and v_p. This change of drain voltage feeds back to the open gates of the driving memory to produce gate voltages of $v_p + \delta v_p$ and $v_n + \delta v_n$. Charge conservation at the gate nodes gives

$$\delta v_p = -\delta v_n = (v_p - v_n)\frac{C_n - \frac{C_{dg}}{2}}{\frac{C_g}{2} + C_n + \frac{C_{dg}}{2}} \tag{5.69}$$

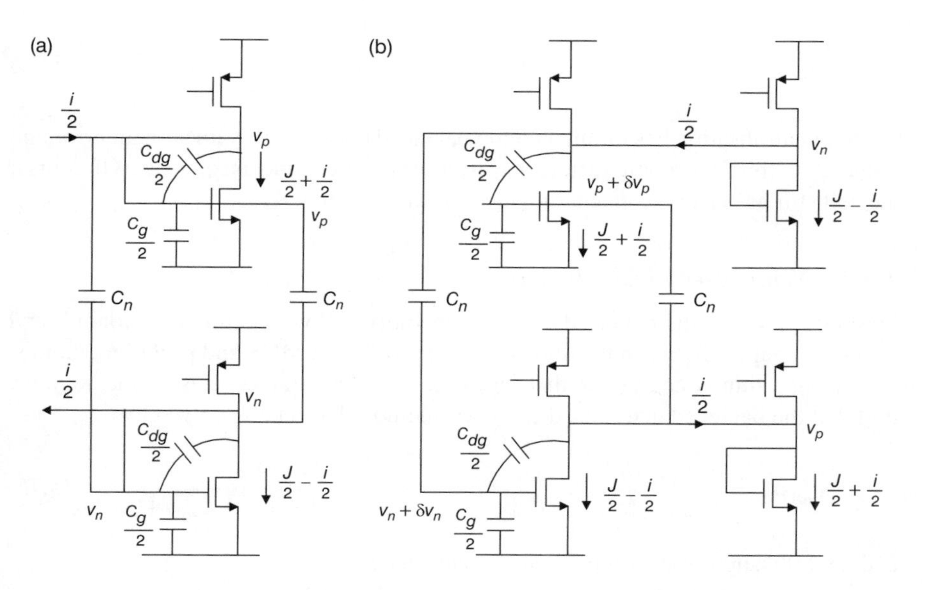

Figure 5.13 Error neutralisation in balanced class A memory cell: (a) input phase; (b) output phase

Following the same procedure for deriving the conductance error given in Section 5.4.2, it can be shown that the conductance error of the neutralised class A cell is given by

$$\epsilon_{gA\Delta}^{N} = \frac{G_{oA\Delta}(m)}{G_{mA\Delta}(m)}$$
$$\approx \frac{4G_{ds}}{G_m} + \frac{C_{dg} - 2C_n}{C_{dg} + 2C_n + C_g} + \left[\frac{6G_{ds}}{G_m} + \frac{1}{2}\frac{C_{dg} - 2C_n}{C_{dg} + 2C_n + C_g}\right] m^2 \quad (5.70)$$

If we make

$$\frac{4G_{ds}}{G_m} = -\frac{C_{dg} - 2C_n}{C_g + 2C_n + C_{dg}} \quad (5.71)$$

or, equivalently,

$$C_n \approx \frac{C_{dg}}{2} + \frac{C_g}{2}\frac{4G_{ds}}{G_m} \quad (5.72)$$

then the signal-independent error is neutralised, giving

$$\epsilon_{gA\Delta}^{N} \approx \left[\frac{4G_{ds}}{G_m}\right] m^2 \quad (5.73)$$

and this produces third-harmonic distortion, $HD_{gA\Delta(3)}^{N}$:

$$HD_{gA\Delta(3)}^{N} \approx \frac{G_{ds}}{G_m}\hat{m}^2 \quad (5.74)$$

So, neutralisation has eliminated the balanced class A cell's linear gain error and reduced its third-harmonic distortion resulting from conductance errors. Of course, the cell still has errors from charge injection and settling.

5.5.2 Neutralised class AB cells

Figure 5.14 shows the balanced class AB memory cell with extra capacitors C_n and C_p cross-coupled between the drains and gates of both nMOS and pMOS memories. Following a similar analysis to that for the balanced neutralised class A memory, we find that the balanced neutralised class AB memory has an error given by

$$\epsilon_{gAB\Delta}^{N} = \frac{G_{oAB\Delta}(m)}{G_{inAB\Delta}(m)} \approx \frac{2G_{ds}}{G_m}(1 + m^2) + \frac{C_{dg}/4 - C_n}{C_g/4 + C_n + C_{dg}/4} \quad (5.75)$$

and neutralisation of the output conductance occurs when

$$C_n = C_p \approx \frac{C_{dg}}{4} + \frac{C_g}{4}\frac{2G_{ds}}{G_m} \quad (5.76)$$

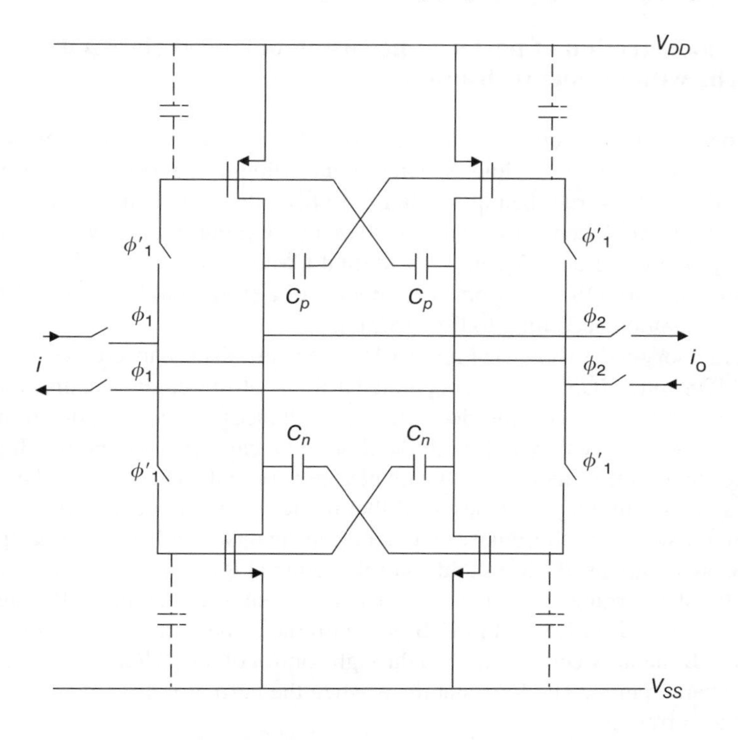

Figure 5.14 Error neutralisation in balanced class AB memory cell

The neutralised conductance error is then

$$\epsilon^N_{gAB\Delta} \approx \frac{2G_{ds}}{G_m} m^2 \tag{5.77}$$

and this gives third-harmonic distortion, $HD^N_{AB\Delta(3)}$, of

$$HD^N_{AB\Delta(3)} \approx \frac{G_{ds}}{G_m} \frac{\hat{m}^2}{2} \tag{5.78}$$

5.5.3 *Summary*

The addition of cross-coupled feedback capacitors to the balanced class A and class AB memories provides an effective means to create zero effective output conductance under quiescent conditions. In the balanced class A cell, neutralisation removed the signal-independent conductance error. In the balanced class AB cell, it removed the signal-independent error while leaving the odd-harmonic distortion unaffected. So, the neutralised class A cell is left with odd-harmonic distortion resulting from conductance, settling and charge injection errors while the neutralised class AB cell has odd-harmonic distortion only from conductance errors.

5.6 Demonstration of performance using balanced class AB cells with full neutralisation

To demonstrate the effectiveness of the class AB low voltage techniques described in the previous sections, various SI circuits including a memory cell [18], an integrator and a second-order bandpass filter have been designed and simulated. All the circuit designs have been optimised for 100 MHz sampling frequency (50 MHz clock frequency, F_c) using a 0.35 μm, 3.3 V, digital CMOS process with $V_t \approx 0.5$V for both pMOS and nMOS transistors. The supply voltage V_{DD} has been set to 1.6 V (for optimum operation according to Eqn. (5.21)).

The memory cell is shown in Figure 5.15. The neutralising capacitors were realised using pMOS and nMOS transistors operated in their cut-off region, a realisation which gives close matching with the drain-gate capacitances in the memory transistors. When sized optimally they null both the drain-gate capacitances and the drain conductances to produce a memory cell with near-zero output conductance. The memory switches include dummy switches to further reduce charge injection errors.

To stabilise the circuit's quiescent bias current against supply voltage, temperature and process variations, the threshold control circuit in Figure 5.16 was employed. This is a master-slave arrangement in which the bias current in a reference cell, comprising a diode-connected pMOS and nMOS pair with the same transistor as those used in the class AB memory cell, is adjusted through control of the pMOS threshold voltage using a charge pump. The loop stabilises when the current in the reference cell, J_r, equals the current J_{AB}.

The simulations performed in this section all used transistor-level models with spectreRF, a commercial switching circuit simulator with a noise analysis capability. In all cases, the signal-to-noise ratios (SNR) and distortions (THD and IMD) were simulated with the maximum peak differential output current of $4J_{AB}$, and noise powers were calculated by integrating the differential noise power spectral densities over the Nyquist band (25 MHz for the memory and 50 MHz for the double-sampling integrator and filter).

For the balanced SI memory cell, the simulated transmission errors, when driven by an ideal differential current source and terminated by a balanced memory of the same type, are summarised in Table 5.3 for different extremes of processing, temperature and input current. The overall transmission error is less than 0.2% of the full-scale signal or, equivalently, the accuracy is better than 8 bits. Also given in the table are the voltage levels of V_{bk} generated by the threshold control for each case. When simulated with a typical process, the memory exhibits an SNR of 69.6 dB, a THD of -49 dB (for 5 MHz input frequency) and a power dissipation of 0.12 mW, yielding a worst-case figure-of-merit ($FoM = SNR \cdot F_c/P$) of 0.87 MHz/pW.

The circuit schematic diagram of the balanced bilinear SI integrator constructed from the neutralised class AB memories is shown in Figure 5.17. Based on the architecture in Reference 16, the integrator samples at twice the clock frequency, F_c, and has a capability to reject common-mode signals. The simulated amplitude and phase responses under typical process conditions are shown in Figure 5.18. It can be seen that the Q-factor has been tuned to a high value (by optimising the memory neutralisation)

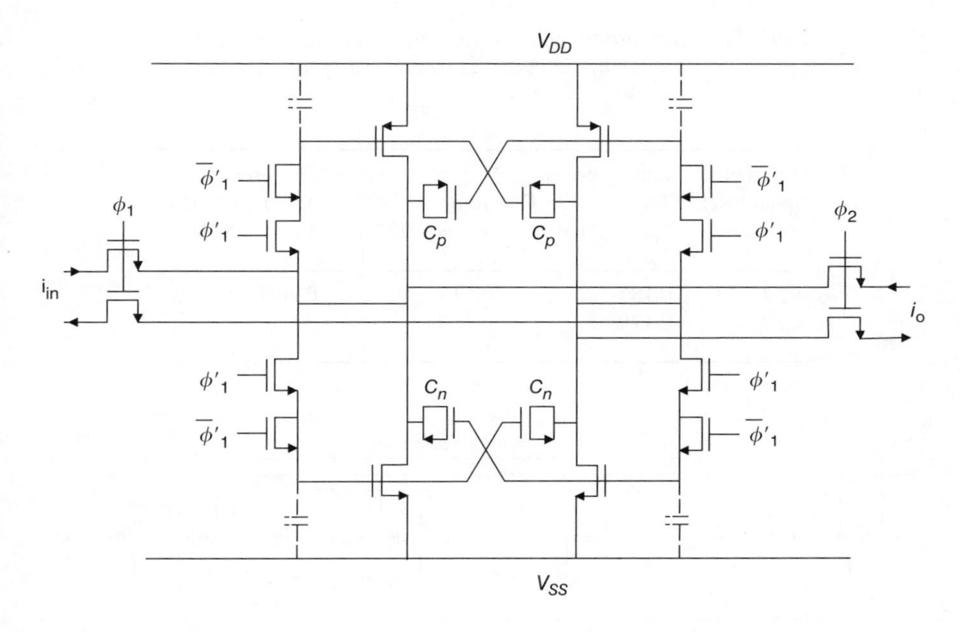

Figure 5.15 Practical class AB balanced memory with full neutralisation

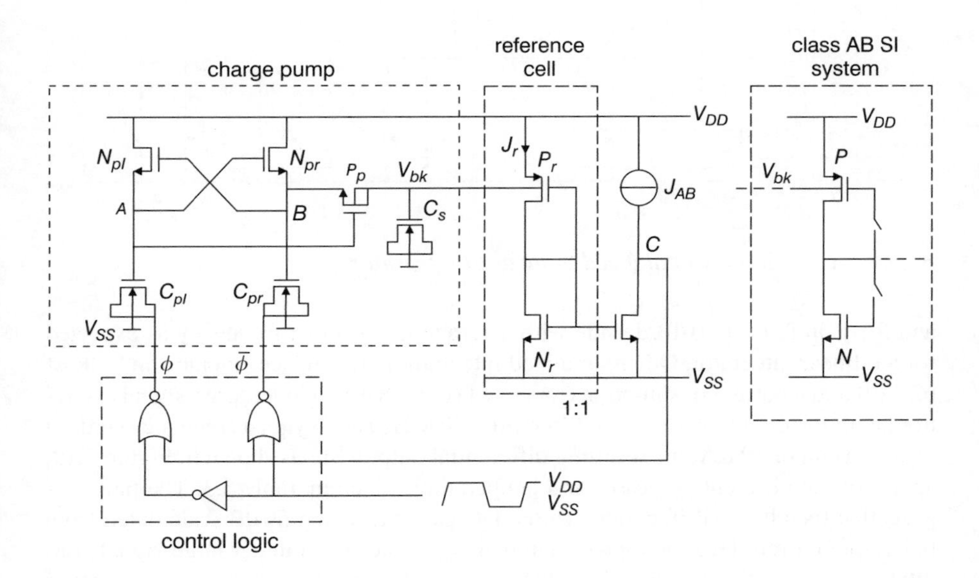

Figure 5.16 Arrangement for stabilising the bias current

Table 5.3 Simulated transmission errors of the class AB balanced memory at 100 MHz sampling frequency and $V_{DD} = 1.6$ V

Differential signal level	Slow process Temp. = 125°C $V_{bk} = 2.1$ V	Typical process Temp. = 27°C $V_{bk} = 2.37$ V	Fast process Temp. = −40°C $V_{bk} = 2.81$ V
0.2 J	0.18%	0.14%	0.28%
4.0 J	0.17%	0.11%	0.1%

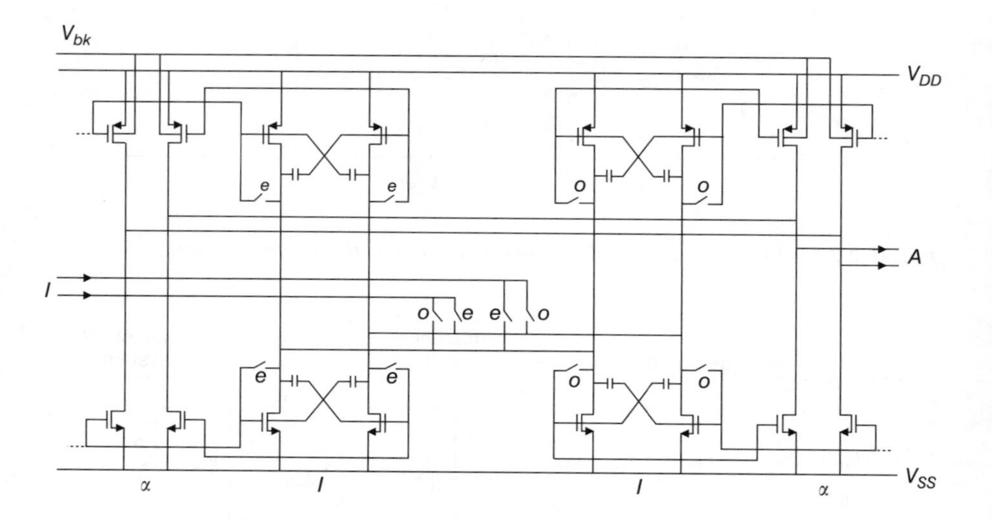

Figure 5.17 Class AB balanced bilinear SI integrator

and the gain falls at 20 dB/decade with a zero at the Nyquist frequency as expected for a bilinear integrator. Other simulated integrator performances include an SNR of 62.1 dB and a power consumption (for $\alpha = 1$) of 0.48 mW. Under large-signal excitation, with the circuit driven by a differential 50 MHz (i.e. Nyquist frequency) pulsed input current of 10 μA, the resulting differential outputs are as shown in Figure 5.19, again for the different processes and temperatures given in Table 5.3. The plot suggests that the class AB integrator with a DC gain exceeding 50 dB (calculated from the output current level of approximately $4J_{AB}$) is achieved under large-signal conditions. Again, this corresponds to the associated memories having an accuracy of better than 8 bits.

Finally, an SI second-order bandpass filter with a centre frequency of 25 MHz and a quality factor of 10 has been designed using the bilinear integrator, following a design process similar to that outlined in Reference 19. The resulting signal flow

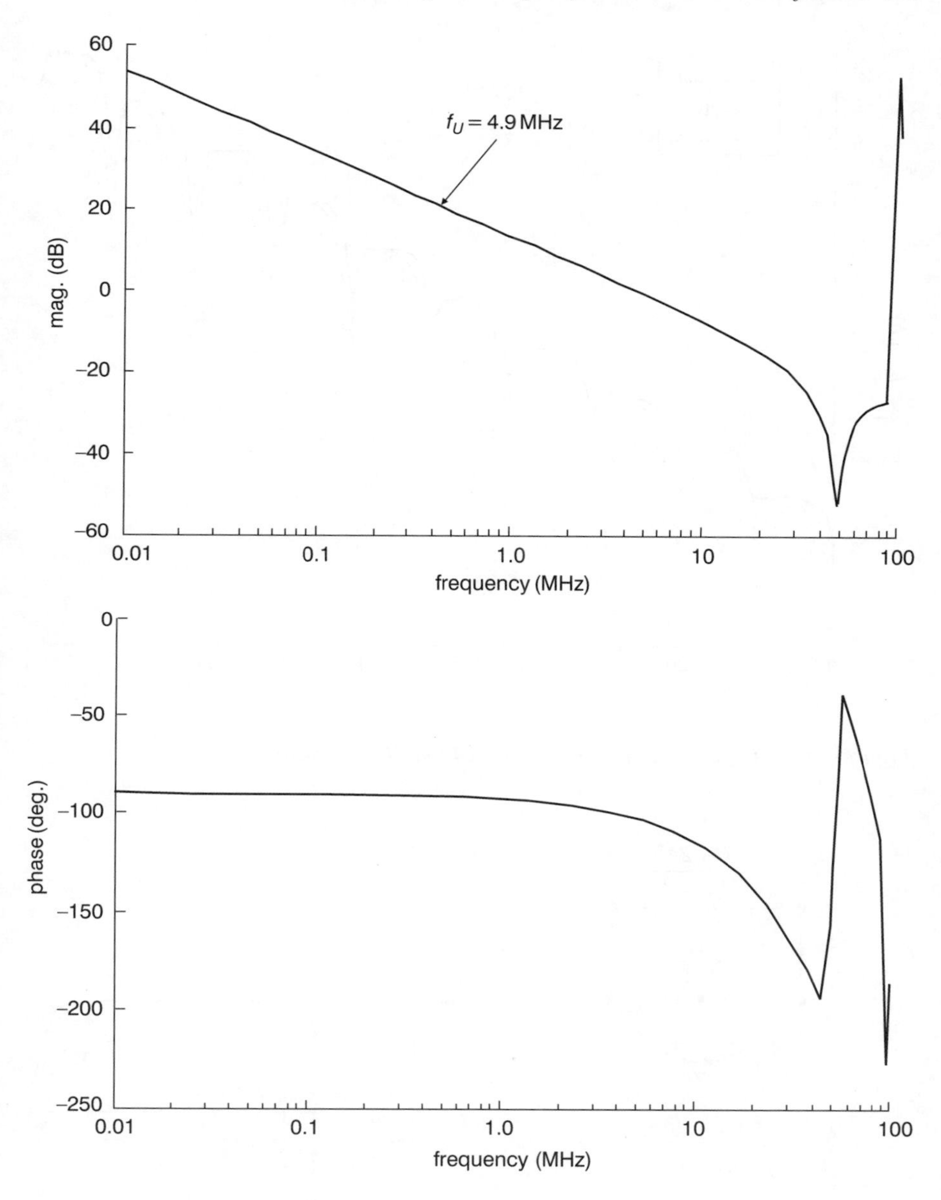

Figure 5.18 Simulated amplitude response of the integrator

graph and architecture with the calculated coefficients is shown in Figure 5.20(a), (b). Figure 5.21 shows the simulated amplitude response of the bandpass filter where the expanded passband response indicates a centre frequency of 24.8 MHz and a Q-factor of 10.1. The filter's SNR is approximately 48.5 dB, close to the predicted value.

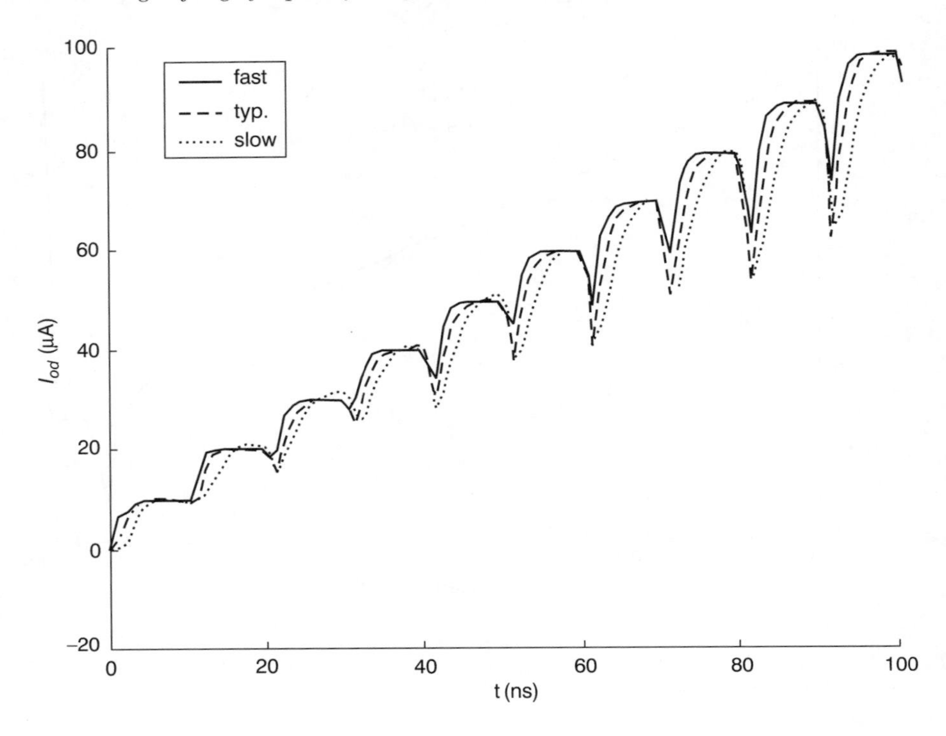

Figure 5.19 Differential output current of the SI integrator

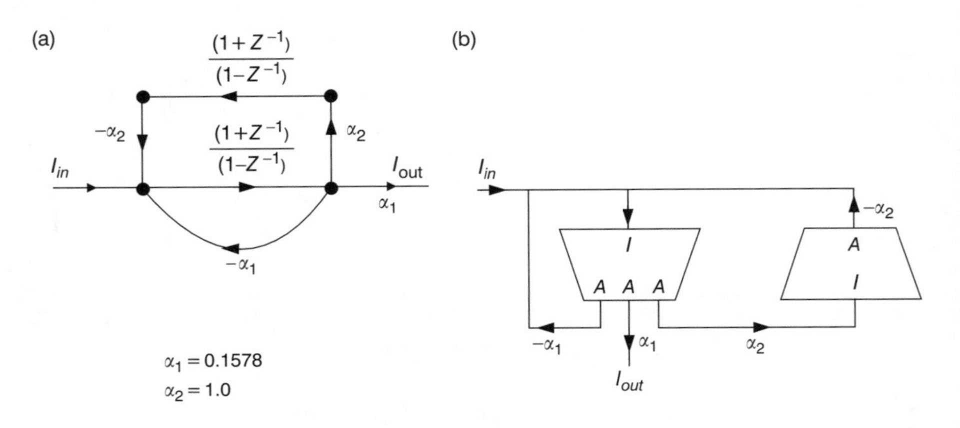

Figure 5.20 Second-order bandpass filter: (a) z-domain signal flow graph; (b) SI filter architecture

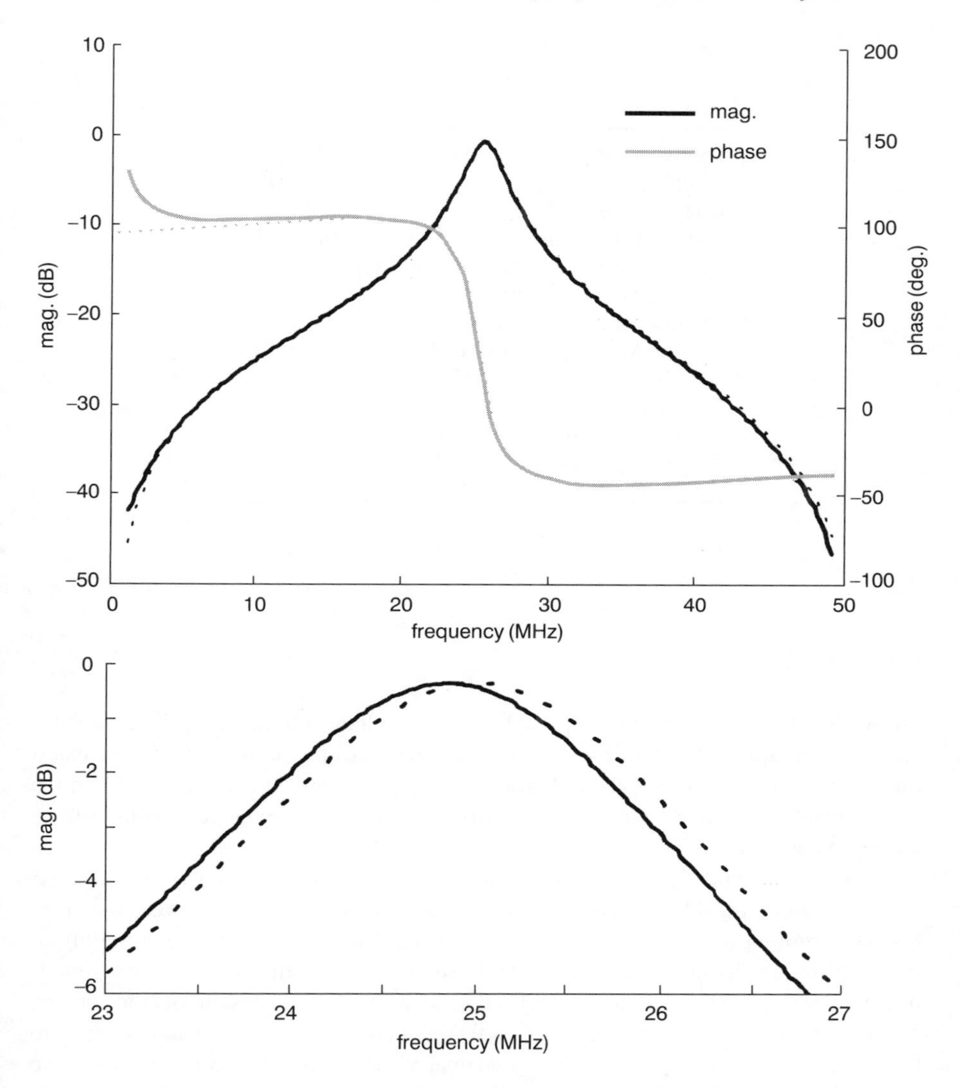

Figure 5.21 Simulated (solid) against ideal (dashed) amplitude and phase responses of the SI filter with the passband detail

The out-of-band second-order intermodulation ($f_1 = 15$ MHz, $f_2 = 40$ MHz) and third-order intermodulation ($f_1 = 15$ MHz, $f_2 = 20$ MHz) products were −68 dB and −61 dB, respectively. For a 1 MHz-spaced two-tone signal centred at 25 MHz, the simulated in-band third-order intermodulation distortion was less than −40 dB. The common-mode rejection was better than 83 dB in the passband frequency range (assuming no device mismatch) and the filter's power dissipation was 1.04 mW. Table 5.4 summarises the filter performance.

Table 5.4 Simulated filter performance summary

Centre frequency	24.8 MHz
Sampling frequency	100 MHz
Quality factor	10.1
Signal-to-noise ratio	48.5 dB
Second-order intermodulation	
$f_1 = 15\,\text{MHz},\ f_2 = 40\,\text{MHz}$	−68 dB
Third-order intermodulation	
$f_1 = 15\,\text{MHz},\ f_2 = 20\,\text{MHz}$	−61 dB
$f_1 = 24.5\,\text{MHz},\ f_2 = 25.5\,\text{MHz}$	−40 dB
Common-mode rejection ratio	83 dB
Power dissipation	1.04 mW
Power supply voltage	1.6 V

5.7 Conclusions

We have established a viable approach for low voltage filters using the switched-current technique. First, we showed that switched currents, unlike switched capacitors, suffer no performance degradation as the supply voltage is lowered. When the supply reaches about 1V, class AB switched currents can be expected to outperform switched capacitors.

Both class A and class AB designs for low voltage have been analysed. Class AB has about an eightfold advantage in terms of signal-to-noise ratio, power consumption or sampling frequency. It also has better precision and linearity due mainly to its signal-independent transconductance. In single-ended memories this gives signal-independent settling and charge-injection errors as well as lower conductance errors. The use of balanced configurations gives further benefits to precision and linearity to both class A and class AB cells. They also enable the use of cross-coupled capacitive feedback to neutralise the memory cell output conductance to give the integrator a theoretically infinite Q-factor. The balanced memory cell proposed accommodates the precision enhancing features without penalty to the other performance vectors.

The low voltage technique has been employed in various switched-current cells. These include a current memory, a bilinear z-transform current integrator and a second-order bandpass filter. Each circuit was designed in a 3.3 V digital CMOS process for 1.6 V operation and 100 MHz double sampling. Simulations were performed on all cells and the bandpass filter gave low power dissipation (1 mW) and high signal frequency (25 MHz) with good signal-to-noise ratio, precision and linearity.

5.8 References

1 VALLANCOURT, D. and TSIVIDIS, Y. P.: 'Sampled Current Circuits', *IEEE International Symposium on Circuits and Systems*, May 1989, pp. 1592–1595.
2 GROENEVELD, W., SCHOUWENAARS, H. and TERMEER, H.: 'A Self-Calibration Technique for High-Resolution D/A Converters', *IEEE International Solid-State Circuits Conference*, 1989, pp. 22–23.
3 NAIRN, D. G. and SALAMA, C. A. T.: 'Current Mode Analogue-to-Digital Converters', *IEEE International Symposium on Circuits and Systems*, May 1989, pp. 1588–1591.
4 WEGMANN, G. and VITTOZ, E. A.: 'Very Accurate Dynamic Current Mirrors', *Electronics Letters*, 11th May 1989, Vol. 25, No. 10, pp. 644–646.
5 HUGHES, J. B., BIRD, N. C. and MACBETH, I. C.: 'Switched-Currents, A New Technique for Analogue Sampled-Data Signal Processing', *IEEE International Symposium on Circuits and Systems*, May 1989, pp. 1584–1587.
6 HUGHES, J. B. and MOULDING, K.: 'Switched-Current Signal Processing for Video Frequencies and Beyond', *IEEE Journal of Solid-State Circuits*, Vol. SC-28, March 1993, pp. 314–322.
7 OLIAEI, O. and LOUMEAU, P.: 'A Low Supply SI Cell Using Threshold Control', *International Conference on Electronic Circuits and Systems*, 1998, pp. 341–344.
8 WORAPISHET, A., HUGHES, J. B. and TOUMAZOU, C.: 'Class AB Technique for High Performance Switched-Current Memory Cells', *International Symposium on Circuits and Systems*, 1999, pp. 456–459.
9 WORAPISHET, A., HUGHES, J. B. and TOUMAZOU, C.: 'Low Voltage Class AB Two-Step Sampling Switched-Currents', *International Symposium on Circuits and Systems*, 2000, pp. 413–416.
10 HUGHES, J. B., MOULDING, K., RICHARDSON, J. R., BENNETT, J., REDMAN-WHITE, W., BRACEY, M. and SOIN, R. S.: 'Automated Design of Switched-Current Filters', *IEEE Journal of Solid-State Circuits*, Vol. 31, No. 7, July 1996, pp. 898–907.
11 MOENECLAEY, N. and KAISER, A.: 'Design Techniques for High-Resolution Current-Mode Sigma-Delta Modulators', *IEEE Journal of Solid-State Circuits*, Vol. 32, No. 7, July 1997, pp. 953–958.
12 BRACEY, M., REDMAN-WHITE, W., RICHARDSON, J. and HUGHES, J. B.: 'A Full Nyquist 15MS/s 8-b Differential Switched-Current A/D Converter', *IEEE Journal of Solid-State Circuits*, Vol. 31, No. 7, July 1996, pp. 945–951.
13 HUGHES, J. B., WORAPISHET, A. and TOUMAZOU, C.: 'Switched-Capacitors versus Switched-Currents, A Theoretical Comparison', *IEEE International Symposium on Circuits and Systems*, May 2000, Vol. II, pp. 409–412.
14 HUGHES, J. B. and WORAPISHET, A.: 'Switched-Capacitors or Switched-Currents, Which Will Succeed?', Chapter 11, in TOUMAZOU, C. and MOSCHYTZ, G. (Eds.): 'Understanding Trade-Offs in Analog Design', Kluwer. In press.

15 TOUMAZOU, C., HUGHES, J. B. and BATTERSBY, N. C. (Eds.): *Switched-Currents: An Analogue Technique for Digital Technology*, Peter Peregrinus Ltd, 1993.
16 HUGHES, J. B. and MOULDING, K. W.: 'A Switched-Current Double Sampling Bilinear Z-Transform Filter Technique', *International Symposium on Circuits and Systems*, 1994, pp. 5.293–296.
17 HUGHES, J. B. and MOULDING, K. W.: 'Error Neutralisation in Switched-Current Memory Cells', *International Symposium on Circuits and Systems*, 1999, pp. 460–463.
18 SITDHIKORN, R., WORAPISHET, A. and HUGHES, J. B.: 'Low-Voltage Class AB Switched-Current Technique', *Electronics Letters*, Vol. 36, 17th August 2000, pp. 1449–1450.
19 HUGHES, J. B.: 'Top-Down Design of a Switched-Current Video Filter', *IEE Proc., Circuits Devices Syst.*, Vol. 147, No. 1, February 2000, pp. 73–81.

Chapter 6

Analogue adaptive filters

Anthony Carusone and David A. Johns

6.1 Introduction

Filters are general signal processing blocks used in virtually every modern electronic system. Whenever a filter's parameters must track poorly controlled or time-varying conditions, adaptive filters are an attractive option. Currently, the vast majority of adaptive filters are implemented digitally and a wealth of literature has been published on the topic [1]. This chapter provides an overview of analogue adaptive filters which represent an important niche in adaptive filter theory and practice.[1]

At low speeds, adaptive filtering is easily and efficiently performed using digital circuits. On the other hand, analogue filters are preferable at high speeds when low power consumption, small integrated area and moderate linearity are required. As digital logic continues to shrink and increase in speed, the minimum speed at which analogue signal processing becomes beneficial increases. Therefore, our focus will be on analogue adaptive filters suitable for high-speed applications, where they will continue to be an important part of systems for years to come. At the same time, considerable attention is paid to historical developments. Although the developments themselves may no longer be of practical use, they provided a foundation for the development of modern filters in use today.

The designer of a modern analogue adaptive filter is required to simultaneously consider both system- and circuit-level issues. However, to provide this chapter with some logical organisation, a top-down approach is used. Specifically, the reader is first introduced to the algorithms (Section 6.2) and filter structures (Section 6.3) used in analogue adaptive filters. Next, the circuit techniques required to implement these algorithms and structures in VLSI hardware are discussed in Section 6.4. Section 6.5

[1] This chapter is based on the February 2000 *IEE Proceedings: Circuits, Devices and Systems* paper, 'Analogue Adaptive Filters – Past and Present' by A. Carusone and D. A. Johns.

concerns itself with specific applications where analogue adaptive filters are currently used. Finally, future directions are surmised.

6.2 Adaptive algorithms

This section will focus upon the mechanisms by which analogue filters can adapt to optimise their performance in an unknown and possibly time-varying environment. Specifically, consider a completely general analogue filter (continuous-time or discrete-time) with N variable parameters $\boldsymbol{p} = [p_1 p_2 \dots p_N]^T$. Let ε denote an error function used to quantify the filter's performance. That is, the error ε will be smallest when the filter parameters are equal to their optimal values, $\boldsymbol{p}^*$. The problem, then, is to search $\Re^N$-space over all possible combinations of filter parameters $\boldsymbol{p}$ to find the minimum value of $\varepsilon(\boldsymbol{p})$. In this general formulation, filter adaptation is seen to be nothing more than a classical optimisation problem for which a myriad of possible algorithms have been developed. However, at present few of these algorithms can be implemented in VLSI and applied to the real-time adaptation of analogue filters. Therefore, this section will focus primarily upon algorithms which are already used or which show potential for use in analogue adaptive systems. The strengths and weaknesses of several algorithms are discussed, and appropriate references are provided for further reading.

The least mean-square algorithm and variants thereof are described in Section 6.2.1 and Section 6.2.2, respectively. These are by far the most popular algorithms in use today. Section 6.2.3 discusses DC offset effects in analogue implementations of the LMS algorithm which represent a major performance limitation in many modern analogue adaptive systems. Heuristic algorithms based, for example, upon the rise-fall times and zero-crossings of certain analogue waveforms have also been used to optimise analogue filters; examples of these approaches are presented in Section 6.2.4. In Section 6.2.5, a simple algorithm based upon a random search of the filter's parameter space is described. Finally, more complicated algorithms are briefly mentioned in Section 6.2.6, including the recursive least squares algorithm and algorithms based upon hyperstability, simulated annealing and genetic optimisation. Currently, these have only been applied to digital adaptive filters. However, as our ability to integrate digital and analogue circuits improves, they may find use in analogue adaptive filters too.

6.2.1 *Least mean-square (LMS) algorithm*

A general LMS adaptive filter is shown in Figure 6.1. It has two inputs (filter input, u, and desired output, d) and two outputs (filter output, y, and error signal, $e = d - y$). The specific configuration of these inputs and outputs will depend upon the application. For instance, in an adaptive echo canceller u is the transmit waveform, d is the received signal plus echo and e is the echo-cancelled received signal. In an adaptive equaliser u is the received signal, d is a training sequence and y is the equaliser output.

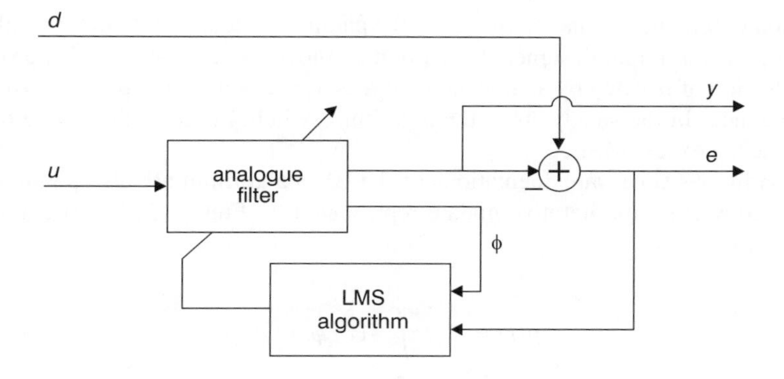

Figure 6.1 An LMS analogue adaptive filter as a 2-input, 2-output system

The performance criterion used for LMS adaptation is the mean-squared error (MSE) of the filter output, y, with respect to the desired output, d. That is, the error function to be minimised is

$$\varepsilon(\boldsymbol{p}) = \mathrm{E}[(d - y)^2] = \mathrm{E}[e^2] \tag{6.1}$$

where the operator $\mathrm{E}[\cdot]$ denotes expectation. The LMS algorithm is a gradient search optimiser which means that adaptation proceeds by updating the filter's parameters iteratively in a direction opposite the gradient $\nabla_{\boldsymbol{p}}\varepsilon$. In discrete time, the update rule is

$$\boldsymbol{p}(k+1) = \boldsymbol{p}(k) - \mu\nabla_{\boldsymbol{p}}\varepsilon \tag{6.2}$$

where μ is a constant which determines the rate of adaptation.

The simple yet brilliant idea put forward by Widrow and Hoff in Reference 2 was to drop the expectation operator when substituting Eqn. (6.1) into Eqn. (6.2). In doing so, they are taking the instantaneous value of the squared-error $e^2 = (d - y)^2$ to be a noisy estimate of its expected value. The resulting update rule is

$$\boldsymbol{p}(k+1) = \boldsymbol{p}(k) + 2\mu\, e(k)\, \boldsymbol{\phi}(k) \tag{6.3}$$

where $\boldsymbol{\phi}(k)$ is the gradient vector $\nabla_{\boldsymbol{p}} y(k)$. This, then, is the LMS algorithm for adaptive filters. The method used to obtain the gradient signals $\boldsymbol{\phi}(k)$ will depend upon the filter structure and is discussed for several different cases in Section 6.3.

The choice of the constant μ is critical. To ensure stability of the algorithm, μ is restricted to positive values smaller than some fixed value, μ_{MAX}, which depends upon the filter structure and the statistics of the filter's input. Eventually the filter parameters, $\boldsymbol{p}$, will converge to a local minimum in the performance surface defined by $\varepsilon(\boldsymbol{p})$. If the MSE at this point is nonzero, the actual MSE will exceed the minimum MSE due to random fluctuations in $\boldsymbol{p}$ caused by the noisy gradient estimates in Eqn. (6.3). These fluctuations, and hence the excess MSE, are decreased by taking a small value for μ. Unfortunately, decreasing μ also decreases the rate of convergence since each iteration will induce a smaller change in the parameter values. With these conflicting requirements in mind, designers have often employed a "gear-shifting" approach to the selection of μ whereby a large value of μ is used at start-up to provide fast initial convergence. In the steady state, the algorithm switches to a smaller value of μ to decrease the excess MSE.

Continuous-time implementations of the LMS algorithm are also possible simply by converting the iterative update represented by Eqn. (6.3) into a continuous integral [3]:

$$\boldsymbol{p}(t) = 2\,\mu \int_{-\infty}^{t} e(\tau)\,\boldsymbol{\phi}(\tau)\,d\tau \tag{6.4}$$

Like all gradient search techniques the LMS algorithm descends the performance surface to a local minimum of the error function [1]. Since it is possible for recursive filters with adapted poles to have multimodal performance surfaces, the LMS algorithm may converge to sub-optimal filter parameter values [4]. However, when the adapted filter is non-recursive or of high enough order the performance surface can be guaranteed unimodal [5] and, hence, the LMS algorithm provides convergence to the globally optimum parameter values, $\boldsymbol{p}^*$.

6.2.2 *Variants of the LMS algorithm*

It was known early on that the product $e \cdot \boldsymbol{\phi}$ in Eqns. (6.3) and (6.4) can be generalised to include any nonideal multiplier of the form

$$e \times \boldsymbol{\phi} = e \cdot f(\boldsymbol{\phi}) \text{ or } f(e) \cdot \boldsymbol{\phi} \tag{6.5}$$

so long as f is a monotonically increasing function [6]. As early as 1970, this fact was being used to simplify the implementation of the LMS algorithm by simply taking the sign of the gradient signal. This has been referred to as the "clipped-data LMS" or "sign-data LMS" (SD-LMS) algorithm in the literature. In discrete time, the update rule is

$$\boldsymbol{p}(k+1) = \boldsymbol{p}(k) - 2\,\mu\, e(k)\, \mathrm{sgn}(\boldsymbol{\phi}(k)) \tag{6.6}$$

The "sign-error LMS" (SE-LMS) and "sign-sign LMS" (SS-LMS) algorithms are similarly defined by taking the sign of the error and/or gradient signals in Eqn. (6.3) [7]:

$$\boldsymbol{p}(k+1) = \boldsymbol{p}(k) - 2\mu\, \mathrm{sgn}(e(k))\, \boldsymbol{\phi}(k) \tag{6.7}$$

$$\boldsymbol{p}(k+1) = \boldsymbol{p}(k) - 2\mu\, \mathrm{sgn}(e(k))\, \mathrm{sgn}(\boldsymbol{\phi}(k)) \tag{6.8}$$

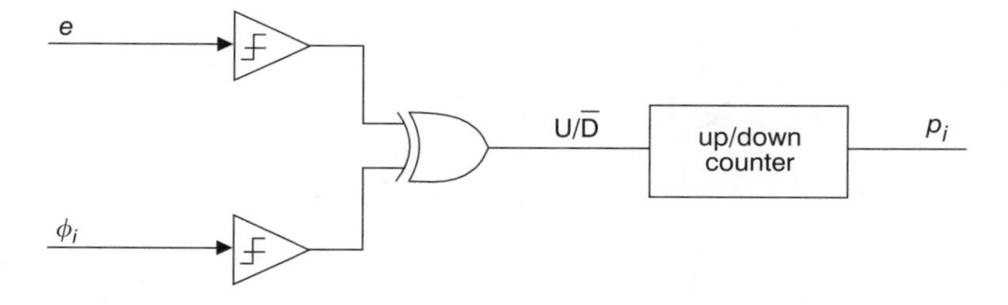

Figure 6.2 Implementation of SS-LMS algorithm for one parameter

These simplifications can also be applied to the continuous-time LMS algorithm by making straightforward modifications to Eqn. (6.4):

$$\boldsymbol{p}(t) = 2\mu \int_{-\infty}^{t} \operatorname{sgn}(e(\tau))\, \boldsymbol{\phi}(\tau)\, d\tau \tag{6.9}$$

$$\boldsymbol{p}(t) = 2\mu \int_{-\infty}^{t} e(\tau) \operatorname{sgn}(\boldsymbol{\phi}(\tau))\, d\tau \tag{6.10}$$

$$\boldsymbol{p}(t) = 2\mu \int_{-\infty}^{t} \operatorname{sgn}(e(\tau)) \operatorname{sgn}(\boldsymbol{\phi}(\tau))\, d\tau \tag{6.11}$$

The hardware advantages of these algorithms over conventional LMS are great. If the adaptation is to be performed using analogue circuitry, these simplifications eliminate the need for linear multipliers which are particularly challenging in pure-CMOS VLSI. If digital circuits are used for the adaptation, a full multiplier is reduced to a (trivial) one-bit multiplier. In the case of SS-LMS the multiplier becomes a single exclusive-OR gate and the integrator becomes an up/down counter as shown in Figure 6.2.

On the other hand, these algorithms converge more slowly than the LMS algorithm and with greater excess MSE in the steady state [8, 9]. Fortunately, these shortcomings can be compensated for by careful selection of the constant μ and, perhaps, by using gear-shifting. A more serious problem for the SD-LMS and SS-LMS algorithms is that instability has been demonstrated due to gradient misalignment [10]. That is, the vector term $\operatorname{sgn}(\boldsymbol{\phi}(k))$ in Eqns. (6.6) and (6.8) is not necessarily parallel to $\boldsymbol{\phi}(k)$, so it is possible for the filter parameters $\boldsymbol{p}$ to "climb" the performance surface, thereby increasing the output's MSE until the algorithm diverges. Nevertheless, these algorithms are often used in practice and have proven useful in many applications.

6.2.3 DC offset effects in analogue LMS

A block diagram of the LMS parameter update rule appears in Figure 6.3. If the parameter updates are being performed using analogue circuitry, DC offsets will

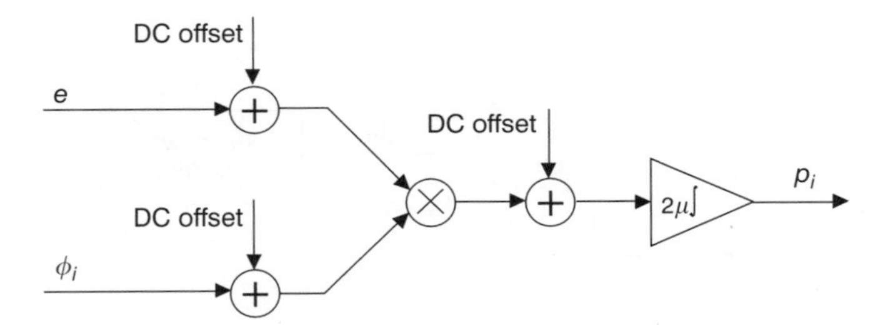

Figure 6.3 Analogue implementation of the LMS parameter update equation

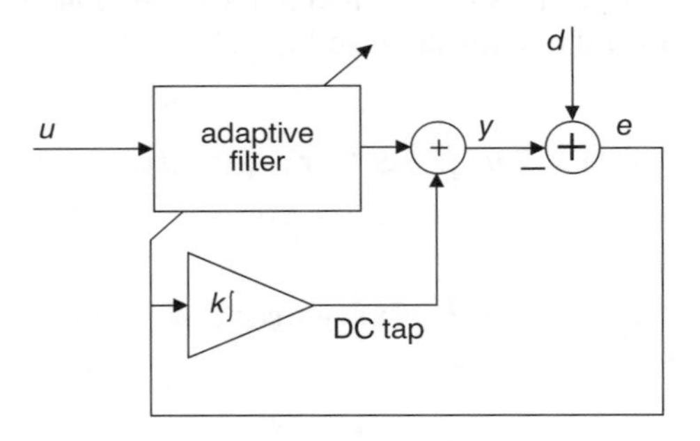

Figure 6.4 DC tap for adaptive offset cancellation

appear at the inputs to the multiplier and integrator blocks. It has been shown that these offsets prevent the LMS algorithm from adapting to the optimal filter parameter values $\boldsymbol{p}^*$ [11–13]. In fact, DC offsets represent a significant performance limitation in many analogue adaptive filters [14–16]. Even if the updates are to be performed using digital circuitry, there may still be some offset on the sampled signals e and ϕ_i which result in excess MSE at the filter output in the steady state.

It was shown in Reference 17 that the SE-LMS and SS-LMS algorithms are somewhat more robust than full-LMS with respect to DC offsets. An algorithmic approach to combating DC offset effects in transversal filters was proposed in Reference 11, requiring another set of N adapted coefficients. Circuit-level techniques for offset-compensation in analogue adaptive filters have also been used with varying degrees of success in, for instance, References 13, 18 and 19. A relatively simple way to eliminate any DC offset on the error signal e is to add a DC offset cancellation tap to the filter output. Shown in Figure 6.4, this tap can be included in any filter structure and essentially forces the error signal, e, to have zero DC content. Unfortunately, it does not eliminate excess MSE entirely. DC offsets introduced between the multiplier and the parameter update integrator persist. However, by applying a high gain to the (offset-free) error signal, excess MSE is minimised. Combining a DC offset cancellation tap with the SS-LMS algorithm results in a median-based offset compensation scheme [20]. The scheme is said to be "median-based" because it forces the median of the error signal to be zero rather than the mean. Fortunately, in many applications

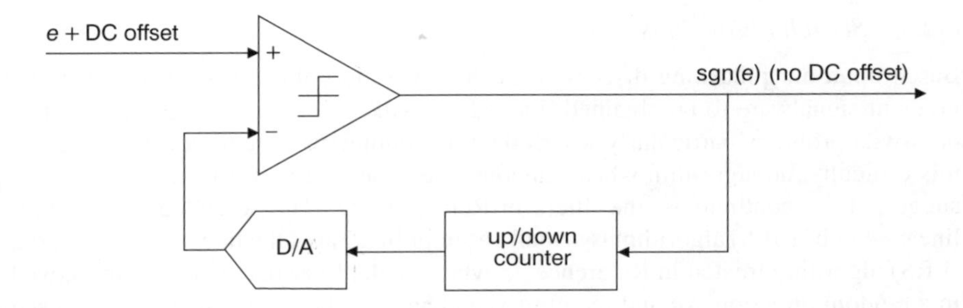

Figure 6.5 Median-based DC offset compensation scheme

the median and mean are identical. The hardware realisation shown in Figure 6.5 provides a relatively low-complexity technique for combating DC offset effects in many analogue adaptive filters.

6.2.4 Heuristic algorithms

A number of heuristic algorithms have been developed for specific applications based on adjusting filter parameters to satisfy some desirable and easily observed condition. For instance, in Reference 15 the resonant frequency of a continuous-time biquad is adapted to track a sinusoidal input by forcing zero correlation between the biquad's lowpass and notch outputs. Since the biquad structure used in Reference 15 already has lowpass and notch signals available at internal nodes, little additional circuitry was required to implement the adaptive algorithm.

In Reference 16 both the resonant frequency and Q-factor of a lowpass continuous-time biquad were adapted for baseband pulse shaping. The resonant frequency was adjusted to force the delay between zero-crossings at the filter input and output to equal some optimum value. Likewise, the Q-factor was adjusted to ensure that the maximum slope of the output waveform was equal to some prescribed value. Again, all of the signals required for adaptation (including the derivative of the output waveform) were available in the biquad, so little additional analogue signal processing was required.

Finally, in Reference 21 the high frequency gain of a receive filter for baseband pulse amplitude modulation was adjusted to eliminate any overshoot or undershoot in the received waveform. The overshoot/undershoot was easily measured by observing the output waveform after each level transition. In combination with adaptive gain compensation, this simple scheme provided enough equalisation for that particular application.

All of these algorithms share a few things in common. First, they are relatively simple, both conceptually and in terms of their hardware implementation. Second, each is specifically tailored to a specific application and a specific analogue filter structure. Third, they are generally limited to situations where adapting one or two parameters is sufficient. So although these heuristic approaches are often efficient, they are not easily generalised to new applications, particularly those requiring high order filters with several adapted parameters.

6.2.5 *Search algorithms*

Based upon our preceding discussion of the LMS algorithm, it is unclear how the gradient signals are to be obtained. Indeed we will see in Section 6.3 that this is a nontrivial problem, particularly when adapting continuous-time filters. To overcome this difficulty, an algorithm which randomly searches the performance surface was suggested for continuous-time filters in Reference 22. The so-called "sequential-linear-search" (SLS) algorithm is actually a simplification of the linear random search (LRS) algorithm treated in Reference 23 whereby the filter parameters $\boldsymbol{p}$ are moved in a random direction $\Delta\boldsymbol{p}$ and the observed change in the error function $\Delta\varepsilon$ is used to estimate the gradient $\nabla_{\boldsymbol{p}}\varepsilon$ according to

$$\nabla_{\boldsymbol{p}}\varepsilon \propto \frac{\Delta\varepsilon \cdot \Delta\boldsymbol{p}}{|\Delta\boldsymbol{p}|^2} \tag{6.12}$$

The advantage of this approach over LMS adaptation is that no gradient signals ϕ are required. Unfortunately, as with all gradient-based algorithms, convergence to sub-optimum filter parameters is possible in recursive filters. The SLS algorithm was applied to an adaptive recursive continuous-time filter implemented with analogue active-RC circuits in Reference 24. Although successful adaptation was demonstrated, it should be noted that a model-matching configuration was used in which the MSE of the filter output decays towards zero in steady state. In an adaptive echo canceller or equaliser, some finite MSE will always remain in steady state. Hence, an echo canceller or equaliser using SLS or RLS would have to be capable of accurately measuring changes in the MSE which are much smaller than its absolute magnitude.

6.2.6 *More complex algorithms*

There are a plethora of other algorithms originally developed for digital adaptive filters which can be applied to analogue adaptive filters – so many, in fact, that only a brief catalogue will be presented here. Although all of these were developed as discrete-time algorithms, continuous-time parameter update equations exist for many which are well suited to analogue circuit implementation. Of course, if a digitally programmable analogue filter is used, digital circuits can be used to implement virtually any adaptive algorithm.

Adaptive lattice algorithms for digital filters were introduced in the early 1980s [25, 26]. A continuous-time algorithm suitable for analogue circuit implementation followed in 1986 [27]. A significant advantage of this approach is that the order of the analogue filter and the number of adapted parameters can be varied to minimise power consumption.

The recursive least-squares (RLS) algorithm [28] can provide faster convergence than the LMS algorithm. A continuous-time implementation of the RLS algorithm was proposed in Reference 29.

Algorithms based upon hyperstability exist in both discrete time [30] and continuous time [31–33] for recursive filters. Other optimisation algorithms, such as those based upon simulated annealing [34] and genetic optimisation [35, 36] are still an active area of research in digital adaptive filters. Unlike the LMS and SLS algorithms, it has been proven that convergence to optimal parameter values can be guaranteed using these algorithms, even when the filter's poles are being adapted.

Although all of these algorithms are somewhat more complicated than the LMS algorithm, each offers distinct advantages. At present, few have been actually applied to analogue adaptive filters. However, as our ability to integrate digital circuits alongside analogue circuits improves so will our ability to apply these algorithms to analogue adaptive filters. Hence, the potential for future research in this area is significant.

6.2.7 Analogue versus digital adaptation

Although analogue filters can offer lower power consumption and smaller integrated area than digital filters at high speeds, the implementation of the adaptation algorithm is a separate consideration. There are two options: continuous-time or switched-capacitor analogue circuits, or digital circuits.

Digital implementations of the adaptation algorithm offer several advantages over analogue implementations. Digital circuitry is generally easier to design, test and scale for new process technologies. The filter parameters can be easily initialised at start-up from on-chip RAM or ROM to speed up initial convergence of the adaptation algorithm. This is particularly significant in magnetic storage applications where faster convergence translates into shorter training sequences and, hence, greater storage densities. Furthermore, there is evidence that digital adaptation circuitry can be robust with respect to DC offsets – a major limitation in analogue adaptive filters [37]. The filter parameters can be 'frozen' during adaptation, if desired; if analogue circuitry is used, the integrators will generally have some leakage which makes it difficult to pause the adaptation. Digital adaptation circuitry also makes possible some of the more complex algorithms mentioned above. Finally, in mixed-signal systems there is usually considerable on-chip DSP which is not being fully utilised at start-up. This DSP hardware can be reused for the adaptation algorithm so no additional digital circuitry is required.

Of course, one must also consider how the adaptation circuitry will interface to the filter. Depending on which algorithm is chosen, the adaptation circuitry may require access to any or all of the following signals:

- filter input
- filter output
- output error with respect to a reference signal
- internal state signals
- control signals for determining the filter's parameters.

In some systems, all of the required signals will be available in digital form. Otherwise it may be necessary to include extra D/A and A/D converters in the design in order to use digital adaptation circuitry. Often, this will not be as difficult as it sounds since the resolution of the data converters can be quite low. For instance, if the sign-sign LMS algorithm is used, only 1-bit versions of the error and gradient signals are required. Furthermore, if fast convergence of the adaptation algorithm is not required, it may be possible to reduce the sampling rate of the data converters.

6.3 Filter structures

It should be clear from the preceding discussion that the best adaptive algorithm for a particular situation is often partly dictated by the filter's structure. In fact, many of the algorithms described in Section 6.2 are only applicable to one particular filter structure. Even the LMS algorithm, which appears to be very general, is not well suited to recursive filters in which the poles are adapted. Furthermore, the gradient signals, $\boldsymbol{\phi}$, in Eqns. (6.3)–(6.11) are obtained very differently in different filter structures. Clearly, then, an important criterion in selecting a filter structure is that a suitable adaptation algorithm exists, preferably with a straightforward and robust hardware implementation. Also, the filter structure must be realisable using analogue circuits and should not go unstable during adaptation. The filter structures in this section are discussed with these criteria in mind. Since both discrete-time and continuous-time signal processing functions are possible using analogue circuits, filter structures falling into both categories will be presented.

Transversal filters are perhaps the simplest and most popular adaptive filters. As such, they are discussed first in Section 6.3.1. Lattice filters, unlike transversal filters, are capable of implementing infinite impulse responses and are briefly described in Section 6.3.2. Continuous-time biquad, Laguerre and orthonormal ladder filters are then discussed in Sections 6.3.3, 6.3.4 and 6.3.5, respectively, with emphasis placed upon the particular advantages of each for adaptive signal processing.

6.3.1 Transversal filters

As shown in Figure 6.6, the input to a transversal filter is applied to an equally spaced tapped delay line. Each tap is then multiplied by a variable tap gain, and the results are summed together to form the filter output. The adapted filter parameters, $\boldsymbol{p}$, are the

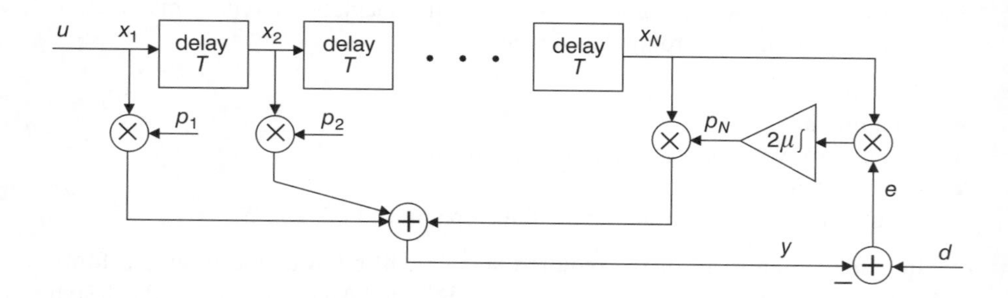

Figure 6.6 Block diagram of an adaptive transversal filter showing the adaptation algorithm for one filter parameter

tap gains. If we denote the filter input as u and the filter output as y, the input–output relationship with N taps spaced time T apart is

$$y(t) = \boldsymbol{p}(t)^T \boldsymbol{x}(t) \tag{6.13}$$

where

$$\boldsymbol{x}(t) = [u(t)\ u(t-T) \cdots u(t-(N-1)T)]^T$$

This structure is sometimes implemented with continuous-time inputs, outputs and delay elements, particularly in adaptive antenna applications [3]. However, it is common for the input and/or output signal to be sampled at a fixed rate $1/T$, in which case a discrete-time description of the input–output relationship is more common. Time indices $k = 0, 1, 2, \ldots$ are used instead of $t = 0, T, 2T, \ldots$:

$$y(k) = \boldsymbol{p}(k)^T \boldsymbol{x}(k) \tag{6.14}$$

where

$$\boldsymbol{x}(k) = [u(k)\ u(k-1) \ldots u(k-N+1)]^T$$

In either case, only the zeros are adapted; all poles are fixed and stable. By virtue of this fact, the transversal filter is unconditionally stable. Another advantage of transversal filters is that any stable transfer function can be approximated with arbitrary accuracy by a transversal filter of sufficient length. The performance surface of a transversal filter using the minimum-MSE criteria has been shown to be an N-dimensional paraboloid. As a result, convergence to the global minimum is guaranteed by using a gradient search technique, such as the LMS algorithm. Furthermore, the gradient signals required for LMS adaptation are easily obtained at the tap outputs. That is, in discrete time,

$$\boldsymbol{\phi}(k) = \boldsymbol{x}(k) \tag{6.15}$$

Substituting Eqn. (6.15) into the discrete-time LMS update rule in Eqn. (6.3) yields

$$\begin{aligned} \boldsymbol{p}(k+1) &= \boldsymbol{p}(k) - 2\mu\, e(k)\, \boldsymbol{x}(k) \\ &= \boldsymbol{p}(k) - 2\,\mu\, e(k)\ [u(k)u(k-1) \ldots u(k-N+1)]^T \end{aligned} \tag{6.16}$$

This expression has a straightforward hardware implementation which is detailed in Figure 6.6 for one adapted parameter. If Eqn. (6.16) is to be implemented digitally, only one A/D converter is required at the filter input along with a shift register to store the sequence $[u(k)u(k-1)\cdots]$ as shown in Figure 6.7. Furthermore, if the sign-data or sign-sign algorithm is used, the A/D converter becomes a single comparator [57, 59].

With all of these advantages, it is no surprise that the transversal filter with LMS adaptation is the most popular structure for adaptive signal processing today. Of course, transversal filters are limited to finite impulse responses. Although stable IIR filters can be well approximated by a transversal filter of sufficient length, recursive structures can often provide the desired frequency response using less hardware.

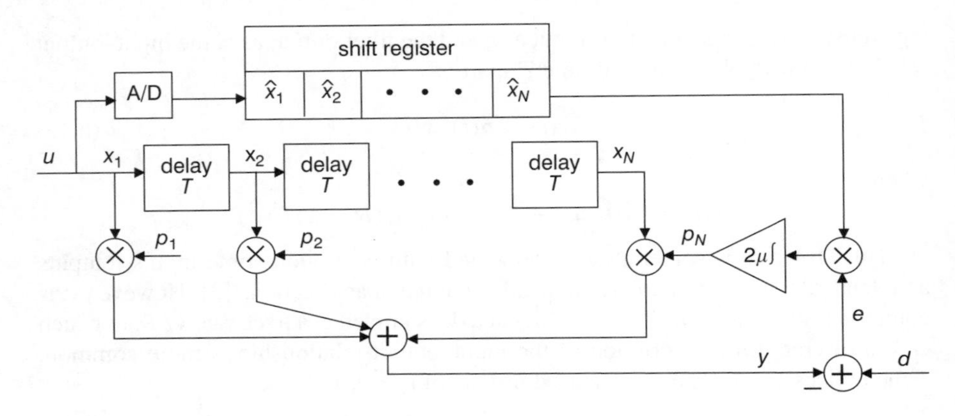

Figure 6.7 Using a shift register to store the gradient signals in a transversal adaptive filter

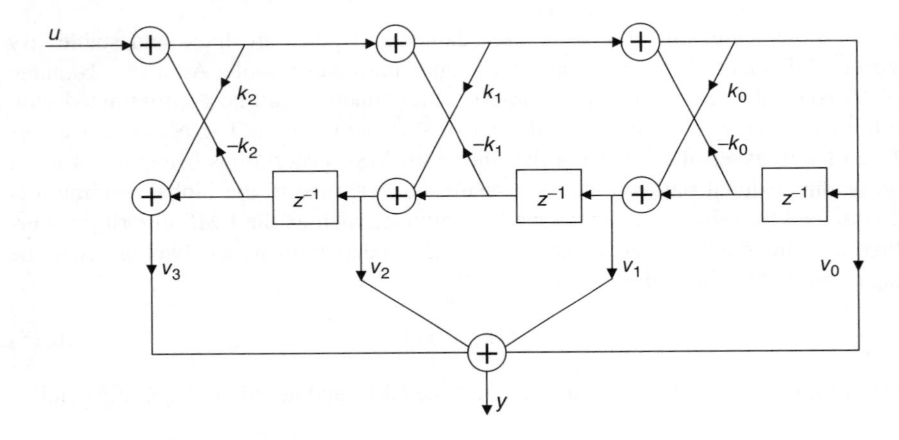

Figure 6.8 A third-order lattice filter for discrete-time signal processing

6.3.2 Lattice filters

Unlike transversal filters, any discrete-time transfer function (FIR or IIR) can be realised using a lattice structure of sufficient order. A third-order lattice filter is shown in Figure 6.8. Notice that it has a regular and hardware-efficient structure, making it suitable for VLSI implementation. As mentioned earlier, adaptive algorithms for lattice filters have been researched and reported extensively [25–27]. An interesting feature of lattice filters is that their stability is guaranteed as long as all parameters k_i in Figure 6.8 satisfy the following condition:

$$|k_i| < 1 \tag{6.17}$$

Therefore, the stability of a lattice filter can be maintained while adapting its poles simply by restricting the gain parameters, k_i, to satisfy Eqn. (6.17). Although the implementation of fixed lattice filters using analogue circuits has been researched [38], analogue adaptive lattice filters are not yet common.

6.3.3 Biquads

Although a biquad is really just any second-order transfer function, it possesses several features which make it a popular choice for analogue adaptive filters. Let us begin by examining a general continuous-time biquadratic transfer function,

$$H(s) = \frac{a_2 s^2 + a_1 s + a_0}{s^2 + b_1 s + b_0} = \frac{a_2 s^2 + a_1 s + a_0}{s^2 + \frac{\omega_0}{Q} s + \omega_0^2} \tag{6.18}$$

Since transfer functions of any order with real-valued coefficients have poles and zeros in complex-conjugate pairs, they can be factored into a series of purely real second-order transfer functions and, therefore, implemented by a cascade of biquads. This has been such a popular technique for the realisation of high order fixed analogue filters that there now exists a wealth of theoretical and experimental information on the design of analogue biquads. Countless analogue biquad circuits have been catalogued [39] and implemented in every process technology. However, unlike fixed analogue filters, biquads are rarely cascaded in adaptive analogue filters. In a fixed cascade of biquads, the gain of each biquad is selected to ensure that all of the filter's internal nodes have the same dynamic range. This type of "dynamic range scaling" is not possible in an adaptive filter where the gains are constantly changing. Therefore, it is difficult to guarantee the dynamic range of an adaptive cascade of biquads since it will depend greatly upon the transfer function implemented.

A useful feature of many biquads is that the lowpass, highpass, bandpass and/or notch outputs are simultaneously available at various nodes in the circuit. These additional outputs can be used as gradient signals for the adaptation process, thereby simplifying the adaptation hardware [15, 16]. Also, it is often easy to control the resonant frequency ω_0 and Q-factor of an analogue biquad without approaching instability. For these reasons, an adaptive analogue biquad is useful for many applications, particularly when accompanied by additional analogue and/or digital adaptive signal processing for higher accuracy.

6.3.4 Laguerre filters

In the early 1930s Norbert Wiener recognised that a set of orthogonal functions named after the French mathematician E. Laguerre could be useful in the design of analogue filters. Although they are complicated functions in the time-domain Eqn. (6.19), they have simple rational Laplace transforms Eqn. (6.20), which Wiener felt could be

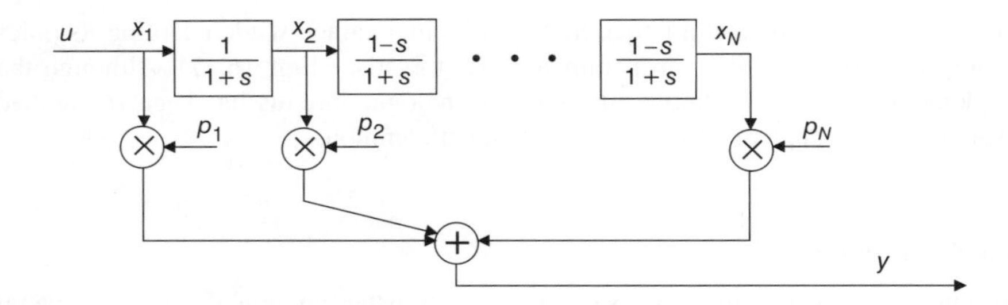

Figure 6.9 Laguerre filter structure

easily implemented using lumped passive elements [40]:

$$\begin{cases} L_0(s) = \dfrac{1}{\sqrt{\pi}}\dfrac{1}{1+s} \\ L_n(s) = \left(\dfrac{1-s}{1+s}\right)^n L_0(s), \quad \text{for } n > 0 \end{cases} \tag{6.19}$$

$$\begin{cases} L_n(t) = e^{-t}\displaystyle\sum_{k=0}^{n}(-1)^{n-k}2^{k+1/2}\frac{n!}{(k!)^2(n-k)!}t^k, \quad \text{for } t \geq 0 \\ L_n(t) = 0, \quad \text{for } t < 0 \end{cases} \tag{6.20}$$

The Laguerre filter structure, shown in Figure 6.9, is based upon Wiener's observations and is still in use today. It realises any linear combination of the transfer functions $L_0(s), L_1(s), \ldots L_N(s)$. That is,

$$H_{Laguerre}(s) = [L_0(s)\ L_1(s) \ldots L_n(s)]\boldsymbol{p} \tag{6.21}$$

Since the Laguerre functions are an infinite set spanning all finite-energy time signals, any linear transfer function can be accurately approximated using a Laguerre filter of sufficient order with the appropriate parameter values, $\boldsymbol{p}$.

In practice, the transfer functions in Eqn. (6.20) would be denormalised to move the pole frequency above the frequency range of interest. In that case, each allpass filter section $(1-s)/(1+s)$ will have a nearly constant group delay. Hence, the Laguerre filter is similar to a continuous-time transversal filter. Like transversal filters, Laguerre filters are well suited to LMS adaptation because the gradient signals ϕ_i are equal to the internal state signals x_i (i.e. Eqns. (6.15) and (6.16) also apply to LMS adaptation of Laguerre filters). However, only the filter zeros are adapted. Although stability of the filter is guaranteed by this, there are situations where the ability to adapt a pole can significantly reduce overall circuit size and/or power consumption. Laguerre filters are further limited by the fact that all poles must be coincident and on the real axis.

6.3.5 Orthonormal ladder filters

The orthonormal ladder filter is a relatively new structure which offers several advantages for adaptive analogue signal processing. It is, essentially, a straightforward

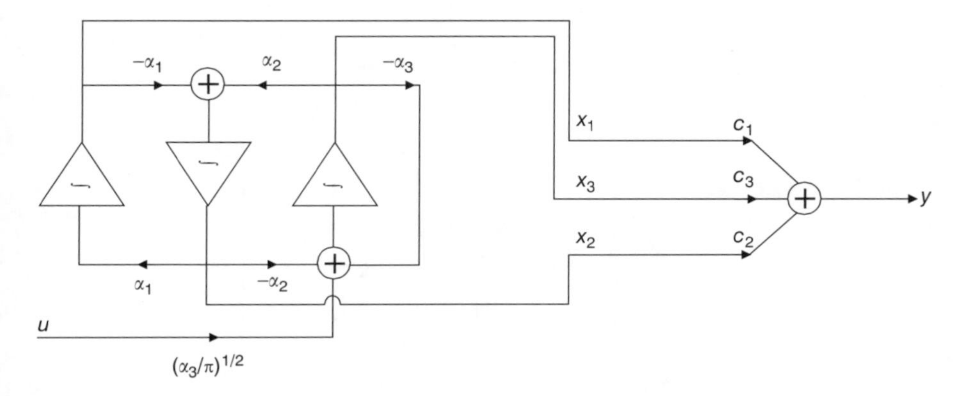

Figure 6.10 *A third-order orthonormal ladder filter. The number of inputs to the output summing node is equal to the filter's order*

implementation of the state-space equations which come about in a singly terminated LC ladder filter:

$$\dot{\boldsymbol{x}}(t) = \boldsymbol{A}\boldsymbol{x}(t) + \boldsymbol{b}u(t) \tag{6.22}$$

$$y(t) = \boldsymbol{c}^T \boldsymbol{x}(t) \tag{6.23}$$

$$\boldsymbol{A} = \begin{bmatrix} 0 & \alpha_1 & & & \boldsymbol{0} \\ -\alpha_1 & 0 & \alpha_2 & & \\ & -\alpha_2 & \cdot & \cdot & \\ & & \cdot & 0 & \alpha_{N-1} \\ \boldsymbol{0} & & & -\alpha_{N-1} & -\alpha_N \end{bmatrix}, \quad \boldsymbol{b} = \begin{bmatrix} 0 \\ \cdot \\ 0 \\ \sqrt{\alpha_N/\pi} \end{bmatrix}, \quad \boldsymbol{c} = \begin{bmatrix} c_1 \\ c_2 \\ \cdot \\ c_N \end{bmatrix}$$

The structure for a third-order orthonormal ladder is shown in Figure 6.10. Continuous-time integrators, summers and gain stages are used as fundamental building blocks. Note that, as the order of the filter increases, so does the number of inputs required at the output summing node. To prevent this from limiting the bandwidth of a circuit implementation, an alternative structure which makes use of multiple feed-ins is presented in Figure 6.11. For a complete overview of orthonormal ladder filters, the reader is referred to Reference 41.

By suitably selecting the filter parameters α_i and c_i or β_i, any rational transfer function can be realised by an orthonormal ladder. If only the transfer function zeros are to be adapted, only the parameters c_i or β_i are updated and the orthonormal ladder is used as an adaptive linear combiner. In this case, with a white input signal, the integrator outputs are orthogonal (i.e. uncorrelated). This ensures fast and robust convergence using the LMS algorithm [1].

Alternatively, the poles of an orthonormal ladder filter can be adapted via the parameters α_i while maintaining near-optimum dynamic range scaling. The gradient signals required to iteratively update each parameter α_i are obtained by applying the corresponding state signals, x_i, to duplicates of the ladder filter. The added complexity and power consumption which this implies is severely limiting in practice.

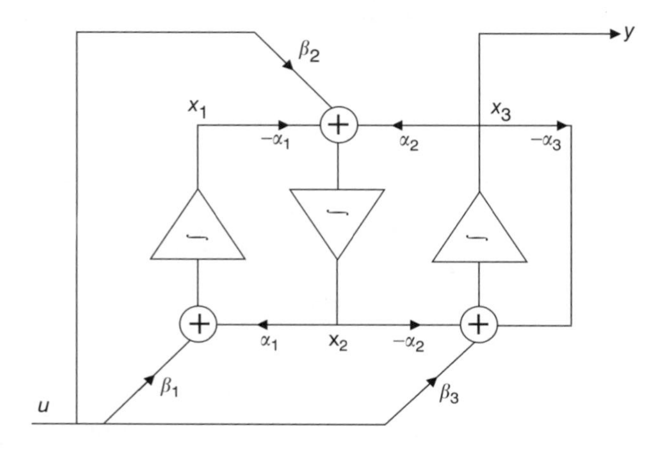

Figure 6.11 *A third-order orthonormal ladder filter using multiple feed-ins of the input signal instead of an output summing node*

However, the additional circuitry can be shared by multiple coefficient update blocks as described in Reference 14. Of course, whenever poles are being adapted care must be taken to ensure that both the filter and the adaptation algorithm remain stable.

6.4 Circuit implementations

Until now, our discussion has made reference to several basic analogue building blocks such as delay elements, summers, multipliers and integrators. Obviously, the implementation of these black boxes is an important consideration. Implementation issues are the subject of this section. Although all of the circuit techniques described here were originally applied to the design of fixed analogue filters, the design of an adaptive analogue filter presents certain unique challenges. For instance, matching, process and temperature variations, which can make the design of accurate fixed analogue filters extremely difficult, are often not critical issues in adaptive analogue filters because the filter is automatically optimised.[2] However, in order for the filter's transfer characteristic to be adapted, it must be possible to change the filter's parameters via some control signals. Finally, since many modern applications require an analogue adaptive filter to be integrated alongside digital circuitry, circuits and techniques amenable to pure CMOS implementation are now preferred.

Discrete-time circuits based upon charge-coupled devices were the subject of much research in the 1970s and are discussed first in Section 6.4.1. Modern discrete-time systems make use of integrated capacitors in sample-and-hold stages for charge storage and are discussed in Section 6.4.2. Switched-capacitor building blocks for

[2] In fact, adaptive techniques are often employed in fixed analogue filters to improve their accuracy [14].

adaptive filters are the subject of Section 6.4.3. Then, continuous-time analogue filter circuits are discussed including MOSFET-C and transconductance-C circuits in Section 6.4.4. The implementation of common building blocks, their limitations and, where possible, techniques for overcoming those limitations are presented in each section.

6.4.1 Filters based on charge-coupled devices (CCDs)

Before the advent of CCDs, acoustic surface wave devices (SWDs) were the most common means of implementing fixed analogue delays and, hence, of realising discrete-time signal processing [42]. However, the maximum delay times which can be implemented by SWDs are severely limited by size. Therefore, CCDs were an active area of research throughout the 1970s. They are now rarely, if ever, applied to adaptive signal processing and are discussed here primarily for completeness and historical interest.

CCDs store a charge proportional to the input voltage in isolated wells of a semiconductor material. Clock signals applied to electrodes above the semiconductor are then used to transfer the charge packets along a linear path. At each stage, it is possible to measure the stored charge nondestructively. Transversal filters are implemented using CCDs by forming weighted sums of these measured charges. For a more detailed description of the operation of CCDs, the reader is referred to Reference 43.

Unfortunately, the transfer of charge along a CCD delay line is not perfect. Some fraction of the charge, ϵ, is left behind at each stage. Finite charge-transfer inefficiency (CTI) in a CCD tapped delay line will cause the impulse response of a transversal filter, $h_k{}'$, to differ from its programmed value, h_k, according to

$$h_k{}' = (1-\epsilon)^k h_k + \sum_{l=0}^{k-1} h_l \frac{k!}{l!(k-l)!} \epsilon^{k-l}(1-\epsilon)^l \tag{6.24}$$

It was shown in Reference 44 that, as long as the CTI is known *a priori*, the distortion term can be compensated for when calculating the programmed parameter values, h_k. Fortunately, in adaptive filters this occurs automatically in the course of adaptation. However, for very long transversal filters the cumulative effects of CTI will still be limiting. In Reference 45, rather than sequentially passing the samples along a long chain of cells, each successive input sample is routed to a separate CCD cell where the sample value was measured by “sloshing” the current back and forth rather than passing it along a delay line. This eliminated the cumulative effects of CTI.

Another property of CCDs is that the stored charge stored must be unipolar. Therefore, to handle bipolar signals, a DC offset, u_{dc}, must be introduced at the input. The result is a DC offset at the output, y_{dc}, which is dependent upon the filter's tap weights:

$$y_{dc} = u_{dc} \sum_{l=0}^{N-1} h_l \tag{6.25}$$

Since the tap weights in an adaptive filter are constantly changing, this offset term may limit the output's dynamic range. The offset can be cancelled by operating two filters in parallel – one with input $u(k)$ and the other with input $-u(k)$ – then taking the output differentially between the two filters [46].

Of course, adaptive CCD filters require a mechanism by which the tap weighting coefficients can be changed. There are two basic techniques by which the measured charge can be programmably weighted: using variable conductances, or using binary-weighted elements and digital control. In Reference 47, the design of an adaptive transversal filter using CCDs with variable conductances is described. The analogue gate voltage on a MOSFET in the triode region is varied to realise a variable conductance and, hence, a variable gain. However, the MOSFET conductance is somewhat nonlinear. Techniques for combating this nonlinearity are presented in Reference 48. Also, variations in the MOSFETs' threshold voltages cause a "fixed pattern" noise at f_S/N and harmonics thereof which can limit the filter's dynamic range. The advantage of this approach is that it provides a power- and area-efficient hardware implementation.

Alternatively, it is possible to operate M binary-weighted CCD tapped delay lines in parallel, as shown in Figure 6.12, and use an M-bit binary control word to weight the contribution of each tap to the output summer [49]. Using this technique, the nonlinearities introduced by MOSFET conductances are avoided. Furthermore, fixed pattern noise is eliminated because transistor matching is not relied upon to accurately determine the tap weights. Of course, area and/or power consumption are increased tremendously. An example of this approach is presented in Reference 46.

Finally, it should be noted that, although our discussion has focused upon transversal filters, by feeding the filter output back to its input it is possible to realise recursive filters using CCD tapped delay lines. Fixed filters have been realised this way [50], but adaptive recursive filters using CCDs have not been studied.

6.4.2 *Filters based on sample-and-hold circuits*

Diffusion capacitors were being used in bucket-brigade devices (BBDs) to sample analogue waveforms for discrete-time signal processing long before most of the research

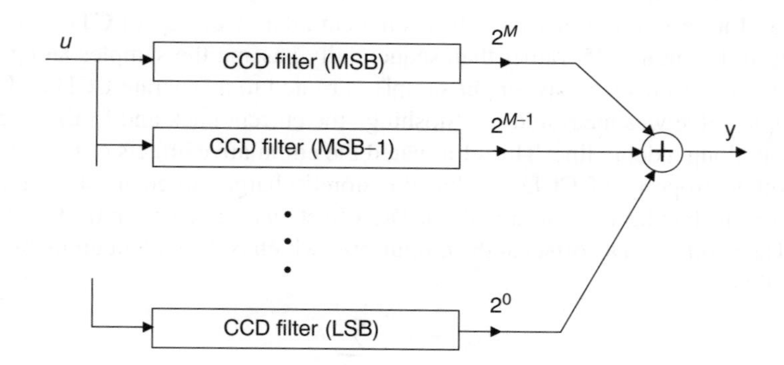

Figure 6.12 Binary-weighted CCD tapped delay lines for a digitally programmable transversal filter

on CCDs described above [51]. CCDs were preferred for a long time because they had a smaller CTI than BBDs. However, capacitors can be realised in a standard MOS process while CCDs cannot, so the use of fixed sample-and-hold stages quickly replaced CCDs in transversal filtering applications.

The simplest approach is to use a chain of sample-and-hold (S/H) circuits as shown in Figure 6.13. There are two main problems with implementing a tapped delay line this way. First, two S/Hs per tap are required because each one must sample during the hold phase of the previous S/H. Second, the DC offsets and noise introduced by each S/H accumulate as the signal propagates along the chain. Both problems can be solved by introducing a "rotating tap weight structure" [52], whereby the analogue input voltage is sampled at fixed sites while the tap gains are digitally shifted along the length of the filter, as shown in Figure 6.14. Since the analogue sample values are

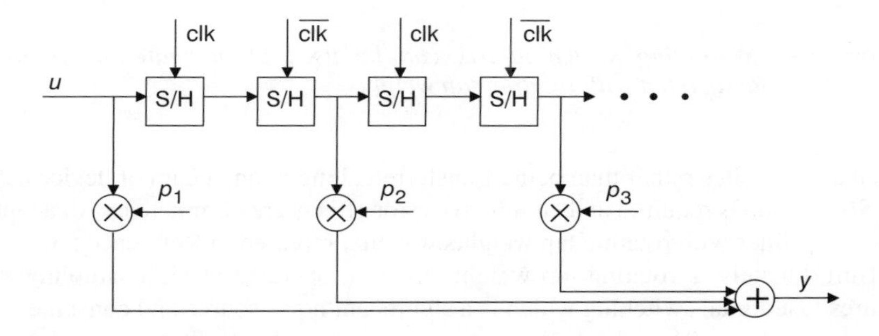

Figure 6.13 A transversal filter implemented with a chain of S/H circuits

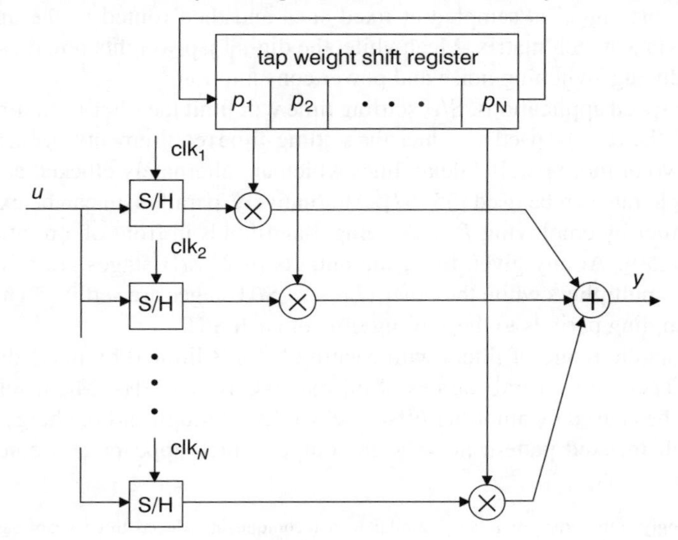

Figure 6.14 A rotating tap weight structure for realising a transversal filter using sample-and-hold stages

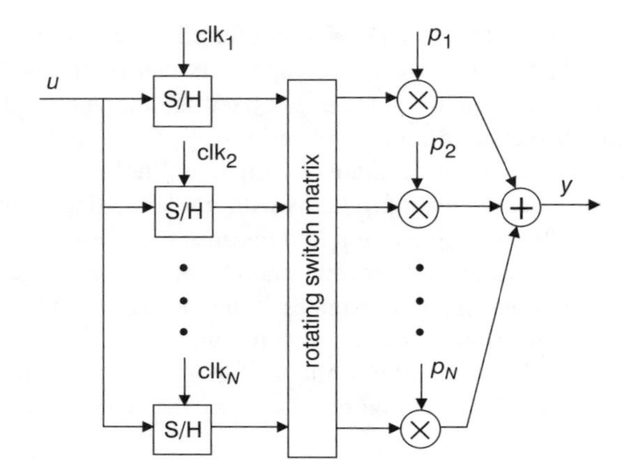

Figure 6.15 A rotating switch matrix can be used to alleviate the problems associated with rotating tap weights

stored at fixed sites rather than being transferred along a long chain of devices, only one S/H per tap is required and cumulative error effects are eliminated.[3] An adaptive transversal filter with rotating tap weights was first reported in Reference 53.

Unfortunately, a rotating tap weight structure operated at high sampling rates requires fast digital switching which is a significant noise source and consumes a lot of dynamic power. These problems are addressed by using a rotating switch matrix between the sample-and-hold stages and the multipliers [54] as in Figure 6.15. The input analogue signal is sampled at fixed sites and then routed to the appropriate multiplier via a switch matrix. Meanwhile, the digital tap weights remain stationary, thereby reducing switching noise and power consumption.

In high speed applications, S/H settling time will limit the filter's maximum clock rate. Parallelism can be used to reduce the settling-time requirements on each S/H. For instance, two or more parallel delay lines which are alternately clocked at a fraction of the sample rate can be used [55–57]. Alternatively, parallelism can be exploited in an N-tap filter by employing $L > N$ sample-and-holds in front of a rotating switch matrix [58, 59]. At any given time, the outputs of N S/H stages are routed to the appropriate multipliers while the other $(L - N)$ S/H stages are settling. This adds an extra L sampling periods to the settling time of each S/H.

The dynamic range of filters with rotating S/Hs is limited by noise due to mismatches.[4] There are several sources of mismatch between S/Hs. Mismatches in DC offsets can be caused by amplifier offsets, clock feedthrough and/or charge injection. These result in fixed pattern noise at the output which appears as a constant tone

[3] Interestingly, this structure is very similar to a technique introduced three years earlier for the elimination of CTI in a CCD transversal filter [45].

[4] If a delay chain of S/Hs is used instead of rotating S/Hs, some sources of mismatch can be compensated for by adapting the tap weights and a DC offset tap.

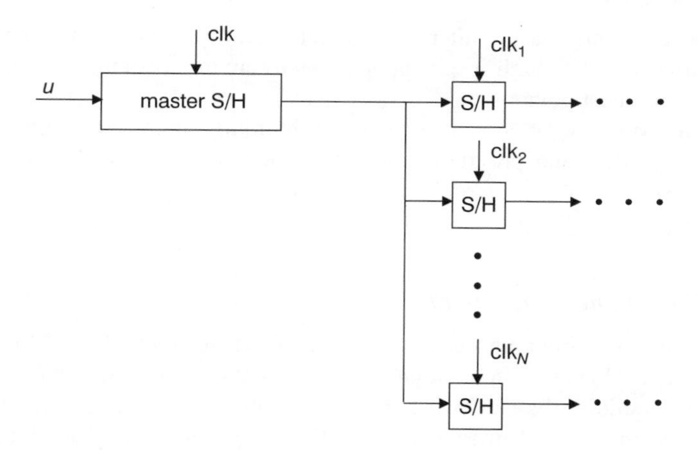

Figure 6.16 A master S/H preceding all others eliminates sampling-phase mismatches but must be clocked at the Nyquist rate

in the output spectrum at f_S/L and its harmonics, where L is the number of S/H stages. Mismatches in the gain and sampling phase of each S/H cause sideband noise around f_S/L and its harmonics. Fully differential signalling can be used to reduce the effects of clock feedthrough and charge injection [60]. Sometimes, a master S/H can be added to the design preceding all other S/Hs as in Figure 6.16. The master S/H eliminates sampling-phase mismatches [59], but it must be clocked at the signal's Nyquist rate. Together, these considerations restrict discrete-time analogue filters to applications where high speed operation and only moderate performance is required.

6.4.3 Switched-capacitor building blocks

As stated earlier, M-bit digitally programmable tap weights can be implemented using M binary-weighted copies of the filter, as shown in Figure 6.12 for a CCD filter. However, this technique consumes a lot of integrated circuit area. The advent of switched-capacitor multiplying digital-to-analogue converters (MDACs) in the mid-1970s [61, 62] made digitally programmable tap weights more practical. MDACs have been used in transversal filters with BBD [63], CCD [64] and sample-and-hold [55] tapped delay lines.

Around the same time, switched-capacitor filters became an active area of research. The zeros and/or poles of a switched-capacitor filter can be made programmable using a number of parallel binary-weighted capacitors [65, 66]. A simple adaptive equaliser was realised using switched-capacitor circuits in Reference [66]. Unfortunately, switched-capacitor circuits are unsuitable for very high speed applications since switching noise becomes more significant as the clock frequency is increased. Furthermore, to ensure quick settling times the unity-gain frequency of the op-amps in a switched capacitor circuit should generally be at least five times greater than the signal bandwidth. Finally, all discrete-time filters must be preceded by a continuous-time anti-aliasing filter. This is a strong motivation to look at continuous-time adaptive filters for high speed applications where the advantages of an analogue signal path are most obvious.

For an overview of switched-capacitor building blocks including MDACs, discrete-time filters and programmable capacitor arrays, the reader is referred to Reference 67.

6.4.4 Continuous-time circuitry

In the discrete-time filters discussed so far, the fundamental building block was the delay element. However, in analogue continuous-time filters, the integrator is the fundamental building block. Two popular ways of implementing continuous-time integrators are illustrated in Figure 6.17. They represent two broad categories of analogue continuous-time filters: MOSFET-C filters and transconductance-C (g_m-C) filters. Each of these categories is discussed in detail in Chapters 2 and 1, respectively. In both cases, the circuit bandwidth is limited by parasitic poles in the active circuitry. These should be a factor of 10 or more beyond the filter's cutoff frequency to ensure that they do not affect the overall transfer function [68]. Compared to operational amplifiers, transconductors have simpler circuitry with few, if any, parasitic internal nodes, resulting in superior high frequency performance. Therefore, they are usually preferred in high frequency analogue adaptive filters.

Fixed continuous-time filters generally require a tuning mechanism to guarantee the accuracy of on-chip time constants over process, voltage and temperature variations. These tuning mechanisms are the subject of Chapter 7. However, if the on-chip time constants are included in the adaptation, they will be automatically tuned by the adaptation algorithm, thus obviating the need for additional tuning circuitry.

The time constant for a MOSFET-C integrator (Figure 6.17(a)) is given by $R_m C$, where R_m is the triode MOSFET drain-source resistance. In g_m-C integrators (Figure 6.17(b)), the time constant is C/g_m. In either case, there are two ways to adjust the integrator time constant: (i) vary the capacitance, C; or (ii) maintain a constant capacitance and vary g_m or R_m. It has been shown that programmable filters based upon constant capacitances are optimal with respect to noise and dynamic

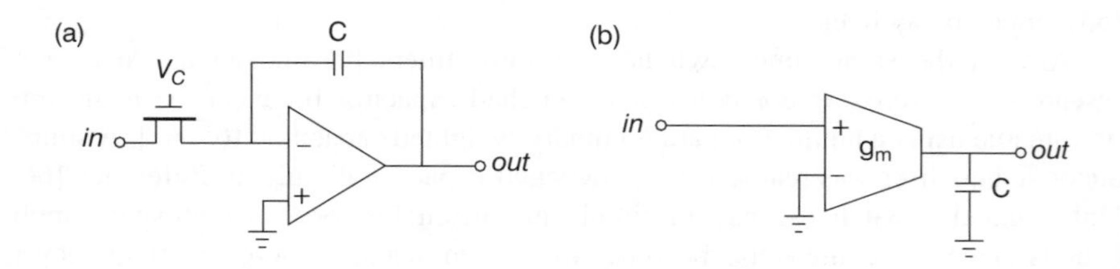

Figure 6.17 Continuous-time (single-ended) integrators suitable for integrated analogue filters: (a) MOSFET-C integrator; (b) g_m-C integrator

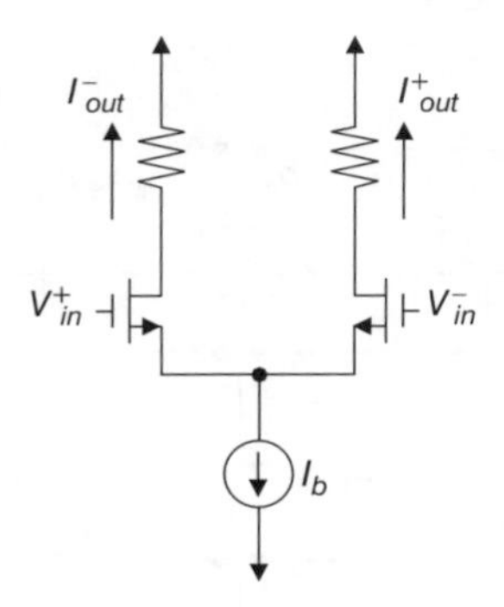

Figure 6.18 Differential pair CMOS transconductor

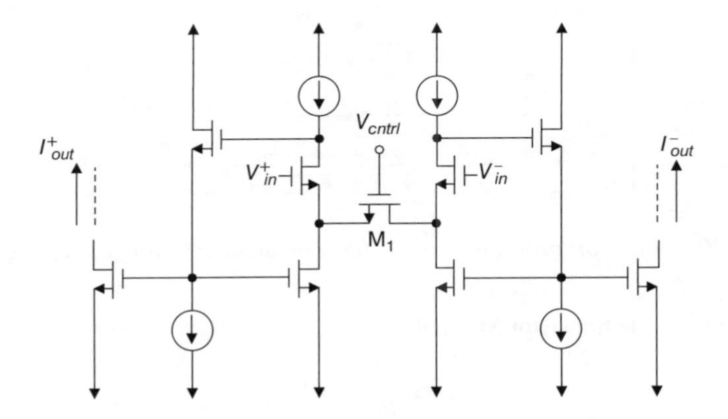

Figure 6.19 Differential CMOS transconductor based on triode transistor [70]

range [69]. Therefore, the designer of an analogue adaptive filter should use programmable conductances rather than programmable capacitors. Of course, even when the filter time constants are not adapted, programmable conductances are often used as programmable gains in continuous-time and discrete-time analogue adaptive filters. Therefore, the remainder of this section will discuss circuit techniques for implementing integrated programmable conductances.

In low to moderate linearity, high speed applications, simple differential pair transconductors can be used (Figure 6.18). The transconductance can be controlled by varying the bias current, I_b. Of course, this may also influence the output common mode voltage which can be a problem.

If higher linearity is desired in a CMOS process, a transconductor circuit based upon a fixed resistor or triode transistor is used. Figure 6.19 shows an example of a fully differential circuit whose transconductance is proportional to the drain-source

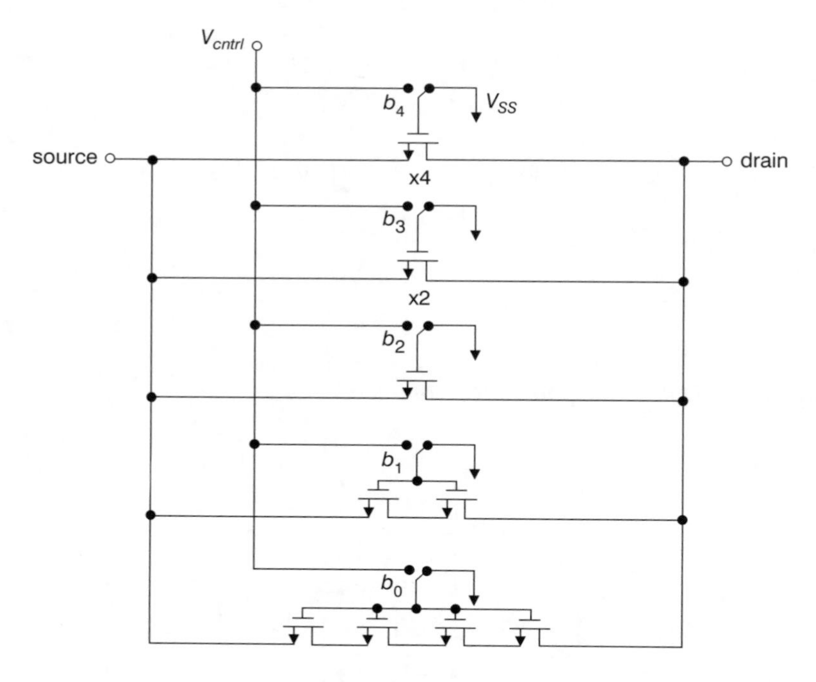

Figure 6.20 A five-bit programmable triode conductance with gate voltage V_{cntrl}

resistance of triode transistor M_1 [70],

$$g_m \propto \frac{1}{R_{ds1}} \propto V_{gs1} \tag{6.26}$$

By varying the gate voltage V_{cntrl} the circuit's transconductance is controlled. Alternatively, M_1 can be replaced by a binary-weighted array of triode transistors as shown in Figure 6.20. In this case, the gate voltages can be switched on and off to provide digital control. A problem with these approaches is that V_{cntrl} must be referenced to the common mode input voltage, which is not always accurately known. Instead, an array of integrated resistors can be combined with MOS switches, perhaps at the expense of circuit area and/or speed.

Regardless of the transconductor employed, it is also possible to make use of a digitally programmable current mirror at the output, as shown in Figure 6.21, to change the overall transconductance. Programmable transconductors are discussed further in Chapter 1.

6.5 Applications

This chapter provides an overview of several application areas for analogue adaptive filters. These applications are discussed for two reasons. First, they provide

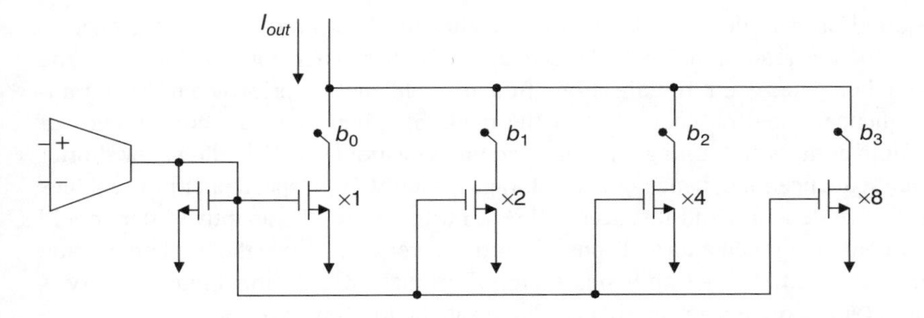

Figure 6.21 *A digitally programmable CMOS current mirror can be used to change the gain of a transconductor*

excellent examples of the circuits and systems described thus far. Second, specific applications have always profoundly influenced research directions in this field and will undoubtedly continue to do so in the future.

6.5.1 Early applications

Analogue adaptive filters were in use long before the advent of integrated circuits. Inspired by Norbert Wiener's work in the early 1930s, Y. W. Lee reported an efficient cascade of passive elements implementing a Laguerre filter structure in 1932 [71, 72]. The so-called *Lee–Wiener network* was used by Wiener in the early 1940s to perform flight-trajectory prediction for use in World War II anti-aircraft fire control. An important contribution was made by Julian Bigelow at this time whose high input impedance amplifiers were necessary for the accurate realisation of the tap gains, p_i, in Figure 6.9. Their work was declassified at the end of the war and the mathematics was reported by Wiener in 1949 [73].

Adaptive antenna systems became an important application for adaptive filters in the late 1960s. An adaptive interference cancellation system for beamforming in antenna arrays was desired because the location of interfering sources could not be known *a priori*. The LMS algorithm was employed. At the time, digital electronics was not fast enough to perform the signal-path filtering. Therefore, analogue systems were often used. A review of adaptive antenna systems in the 1960s is provided by Reference 3. The LMS algorithm was also used in the 1960s to solve optimisation problems using analogue computers [74]. Finally, a summary of medical applications pursued in the 1960s and 1970s is provided in Reference 75.

6.5.2 Digital magnetic storage

Digital magnetic storage channels emerged as the primary application area for analogue adaptive filters in the 1990s. The signals received from the read head in a magnetic storage channel are baseband pulses for which some forward equalisation is required prior to detection. Adaptive equalisation is desired because the characteristics of the read signal will depend upon the particular zone of the magnetic medium being accessed. Initialisation of the read channel is performed using a training sequence stored on each zone of the medium. Therefore, fast convergence of the adaptive filters and timing recovery circuits is desirable since it allows for shorter training sequences and, hence, greater storage capacity. High speed operation enables high storage densities and fast access times. Furthermore, the adaptive filter should have a small integrated area and consume little power to facilitate the implementation of an entire read channel on a single chip. Fortunately, only moderate linearity is required (approximately > 40 dB) to obtain satisfactory bit error rates.

Figure 6.22 shows three possible architectures for a digital magnetic storage read channel. The adaptive equalisation can be implemented using either digital (Figure 6.22(a)), analogue discrete-time (Figure 6.22(b)) or analogue continuous-time (Figure 6.22(c)) filters. In all-digital systems (Figure 6.22(a)), some partial equalisation is still performed by a fixed lowpass anti-aliasing filter [76]. This is done to reduce the dynamic range and resolution required in the A/D converter and to shorten the length of the digital equaliser required. By making the analogue filter adaptive the digital circuitry and A/D converter complexity are reduced. In some systems, this approach is combined with an analogue Viterbi detector to eliminate the need for A/D conversion altogether [77].

Analogue discrete-time transversal filters with five to ten taps are common in current commercial systems. The delay lines are generally implemented using the S/H techniques described in Section 6.4.2. However, in Reference 78 a cascade of continuous-time Bessell allpass filters implemented with MOSFET-C circuits was used. An indirect tuning scheme locked the delay time of each MOSFET-C circuit to the system's sampling frequency.

Tap weighting in the transversal filters can be implemented using either switched-capacitor MDACs [56], programmable transconductors [58] or, in a BiCMOS process, Gilbert multipliers [59]. If either of the latter two techniques is used, a current output is obtained and the output summer is easily realised by connecting together all output nodes. In general, digitally programmable tap weights are useful since the optimal filter coefficients for each zone of the magnetic medium can be stored in a digital RAM. The stored values are then used to initialise the adaptation algorithm at the start of each read operation, thereby ensuring rapid convergence. In this case, the adaptation algorithm must be implemented digitally. The SS-LMS [59, 79] or SD-LMS [55] algorithms are popular because they are easily realised with digital circuitry.

Recently, continuous-time adaptive analogue equalisers have been examined for the magnetic storage channel. These offer several distinct advantages over both digital and discrete-time analogue adaptive equalisers. First, since a continuous-time lowpass anti-aliasing filter is required prior to sampling when digital or discrete-time filters are used, it would save power and area if the adaptive equaliser could be combined with the lowpass filter and implemented in a single circuit. Second, a continuous-time IIR filter with just a few adapted parameters can provide performance comparable to that of a higher order FIR filter. Third, when the equaliser is inside the system's timing recovery loop, as is the case with a discrete-time or digital equaliser, the delay around the loop can cause slow convergence at start-up. A seventh-order adaptive continuous-time equaliser with four adapted zeros is described in Reference 80. Its performance is better than a fixed seventh-order continuous-time filter combined with a 9-tap adaptive FIR filter [81]. In Reference 82, a continuous-time seventh-order orthonormal ladder filter implemented in a CMOS process using g_m-C cells is used as both a lowpass anti-aliasing filter and an adaptive equaliser. Two parameters are adapted using simple digital SD-LMS circuitry. Although only one zero is adapted, the performance is comparable to systems with 5-tap adaptive FIR filters. These results are very encouraging and certainly warrant further research.

6.5.3 Wired digital communications

An adaptive equaliser for digital communications was first proposed by Lucky in 1965 [83]. Since then, data rates over wired communication channels have increased by several orders of magnitude. Modern applications often require multiple adaptive filters for echo cancellation and crosstalk cancellation as well as equalisation [84]. Although many of the adaptive functions in wired data communications can be efficiently performed digitally, analogue adaptive filters play a critical role when high speed and low power are required.

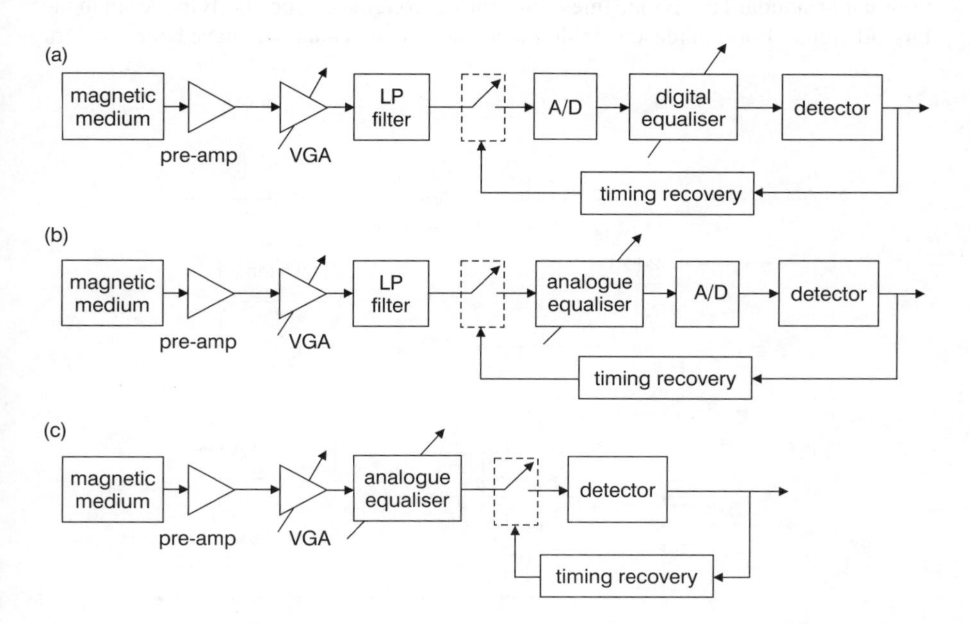

Figure 6.22 System architectures for digital magnetic recording read channels. The adaptive feedforward equaliser can be realised using (a) digital, (b) discrete-time analogue or (c) continuous-time analogue circuitry

Analogue adaptive equalisers offer essentially the same advantages in wired digital communications as they do in magnetic storage applications (Figure 6.22): smaller circuit area and power consumption at high speeds, largely due to the reduced A/D converter specifications. Again, analogue continuous-time equalisation can eliminate the start-up problems associated with having an adaptive equaliser inside a timing recovery loop. However, a key difference between digital communication and magnetic storage applications arises when the wired channel's transfer characteristic has a pole frequency far below the information signal's bandwidth. For example, transformer coupling can introduce a pole at a few kilohertz in multimegabit-per-second channels. As a result, the channel's impulse responses may be very long and a transversal filter of great length would be required to perform equalisation. Therefore, IIR adaptive filters are often preferred over FIR transversal structures.

Analogue adaptive filters are also used for echo cancellation in full-duplex communication systems where they are required to eliminate any vestiges of the transmit signal from the receive path. Figure 6.23 shows a full-duplex system with digital (a) and analogue (b) adaptive echo cancellation. Note that the entire echo path in Figure 6.23(a), including transmit D/A, line driver, receive filter and A/D, must be highly linear to allow for linear echo cancellation in the digital domain. Therefore, an analogue adaptive echo canceller will often ease the D/A and line driver specifications, as well as the A/D specifications. These savings are particularly significant in applications using standard telephone lines since the echo signal can be 30 dB louder than the far-end signal. For this reason, analogue adaptive echo cancellers have been used for

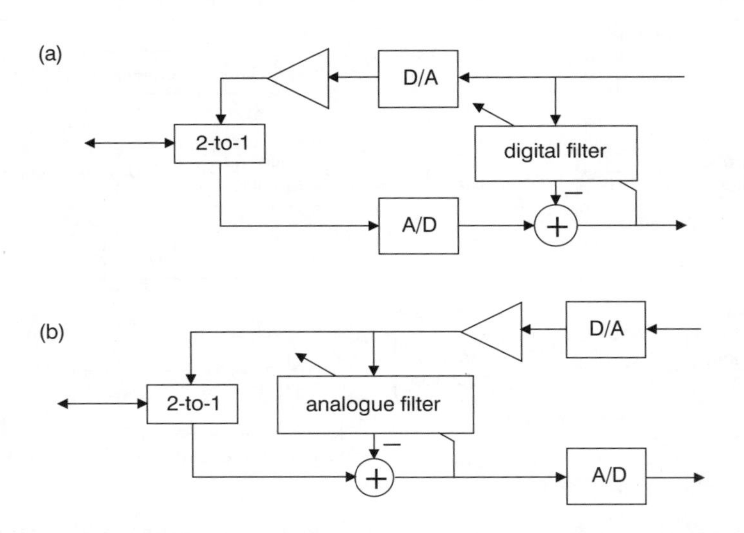

Figure 6.23 Adaptive echo cancellation in a full-duplex wired digital communication transceiver: (a) digital; (b) analogue

voiceband modems [85], ISDN [86, 87] and, more recently, asymmetric digital subscriber lines (ADSL) [88]. Again, due to low frequency poles in the echo-path transfer characteristic, the impulse responses required are often very long, so continuous-time IIR filters are common [85, 88].

For very high speed applications, an A/D and digital equaliser running in the GHz range may be impractical, so analogue continuous-time equalisation is the only option. Baseband communication over high quality coaxial cable is an example of this [89, 90]. In Reference 90 a bipolar analogue circuit is used for adaptive cable equalisation up to 2.5 Gb/s. Although new digital process technologies will certainly reach these speeds, there will still be a frequency limit beyond which analogue adaptive signal processing remains more efficient.

6.6 Conclusions

In the future, it is clear that, to improve the reliability and cost of signal processing electronics, entire systems must be integrated on to a single chip. In many applications, an all-digital or all-analogue system is impossible because both analogue and digital external interfaces are required. So we must assume that both analogue and digital circuits will have to coexist on the same IC. The advance of digital process technologies will enable faster digital signal processing with a smaller integrated circuit area and lower power consumption. At the same time it will become increasingly difficult to implement accurate and linear analogue circuits. Therefore, analogue adaptive filters are likely to continue to be targeted at applications where they can simplify or reduce the analogue circuitry required elsewhere on the chip.

An important trend in many digital communication applications is that signal bandwidths are constantly increasing while the required adaptation rates remain fixed. Therefore, longer training sequences and slower adaptation algorithms may be used to simplify the hardware in future adaptive filters.

DC offset effects introduced by using an analogue LMS algorithm were also discussed. They significantly limit the performance of analogue adaptive filters. Digitally programmable analogue filters provide an analogue signal path, and there are indications that they may be more robust in the presence of DC offsets [37]. Future research may consider hardware-efficient methods for digitally adapting general analogue filter structures.

A review of analogue adaptive filters has been provided including adaptive algorithms, filter structures and circuit techniques. A few key application areas for analogue adaptive filters were identified. In spite of the wealth of theoretical and experimental work on analogue adaptive filters which has been described, currently there are still many applications where they are not used to full advantage. It is hoped that the summary provided here will educate engineers in both academia and industry and motivate future work in the area.

6.7 References

1 WIDROW, B. and STEARNS, S. D.: 'Adaptive Signal Processing', Prentice Hall, New Jersey, 1985.

2 WIDROW, B. and HOFF, M. E.: 'Adaptive Switching Circuits', *IRE 1960 WESCON Conv. Rec.*, pp 96–104, 1960.

3 WIDROW, B., MANTEY, P. E., GRIFFITHS, L. J. and GOODE, B. B.: 'Adaptive Antenna Systems', *Proc. IEEE*, Vol. 55, pp. 2143–2159, December 1967.

4 JOHNSON, Jr. C. R. and LARIMORE, M. G.: 'Comments on and additions to "An Adaptive Recursive LMS Filter"', *Proc. IEEE*, Vol. 75, pp. 1399–1401, September 1977.

5 STEARNS, S. D.: 'Error Surfaces of Recursive Adaptive Filters', *IEEE Trans. Circuits and Systems*, Vol. CAS-28, pp. 603–606, June 1981.

6 MITRA, D. and SONDHI, M. M.: 'Adaptive Filtering with Nonideal Multipliers – Applications to Echo Cancellation', *Int. Conf. on Commun. Conf. Rec.*, Vol. 2, p. 30, 1975.

7 HIRSCH, D. and WOLF, W.: 'A Simple Adaptive Equalizer for Efficient Data Transmission', *IEEE Trans. Communications*, Vol. COM-18, p. 5, 1970.

8 CLAASEN, T. A. C. M. and MECKLENBRAUKER, W. F. G.: 'Comparison of the Convergence of Two Algorithms for Adaptive FIR Digital Filters', *IEEE Trans. Acoustics, Speech and Signal Processing*, Vol. ASSP-29, pp. 670–678, June 1981.

9 DUTTWEILER, D. L.: 'Adaptive Filter Performance with Nonlinearities in the Correlation Multiplier', *IEEE Trans. Acoustics, Speech and Signal Processing*, Vol. ASSP-30, pp. 578–586, August 1982.

10 ROHRS, C. R., JOHNSON, C. R. and MILLS, J. D.: 'A Stability Problem in Sign-Sign Adaptive Algorithms', *IEEE Int. Conf. Acoustics, Speech and Signal Processing*, Vol. 4, pp. 2999–3001, April 1986.

11 TZENG, C.-P. J.: 'An Adaptive Offset Cancellation Technique for Adaptive Filters', *IEEE Trans. on Acoustics, Speech and Signal Processing*, Vol. 38, pp. 799–803, May 1990.

12 JOHNS, D. A., SNELGROVE, W. M. and SEDRA, A. S.: 'Continuous-Time LMS Adaptive Recursive Filters', *IEEE Trans. Circuits and Systems*, Vol. 38, pp. 769–778, July 1991.

13 MENZI, U. and MOSCHYTZ, G. S.: 'Adaptive Switched-Capacitor Filters Based on the LMS Algorithm', *IEEE Trans. Circuits and Systems*, Vol. 40, pp. 929–942, December 1993.

14 KOZMA, K. A., JOHNS, D. A. and SEDRA, A. S.: 'Automatic Tuning of Continuous-Time Integrated Filters Using an Adaptive Filter Technique', *IEEE Trans. Circuits and Systems*, Vol. 38, pp. 1241–1248, November 1991.

15 KWAN, T. and MARTIN, K.: 'An Adaptive Analog Continuous-Time CMOS Biquadratic Filter', *IEEE J. Solid-State Circuits*, Vol. 26, pp. 859–867, June 1991.

16 SHOVAL, A., SNELGROVE, W. M. and JOHNS, D. A.: 'A 100 Mb/s BiCMOS Adaptive Pulse-Shaping Filter', *IEEE J. Selected Areas in Communications*, Vol. 13, pp. 1692–1702, December 1995.

17 SHOVAL, A., JOHNS, D. A. and SNELGROVE, W. M.: 'Comparison of DC Offset Effects in Four LMS Adaptive Algorithms', *IEEE Trans. Circuits and Systems, II*, Vol. 42, pp. 176–185, March 1995.

18 QIUTING, H.: 'Offset Compensation Scheme for Analogue LMS Adaptive FIR Filters', *Electronics Letters*, Vol. 38, pp. 1203–1205, June 1992.

19 KUB, F. J. and JUSTH, E. W.: 'Analog CMOS Implementation of High Frequency Least-Mean Square Error Learning Circuit', *Proc. 1995 IEEE Int. Symp. Circuits and Systems*, pp. 74–75, February 1995.

20 SHOVAL, A., JOHNS, D. A. and SNELGROVE, W. M.: 'Median-Based Offset Cancellation Circuit Technique', *IEEE Int. Symp. Circuits and Systems*, pp. 2033–2036, May 1992.

21 EVERITT, J., PARKER, J. F., HURST, P., NACK, D. and KONDA, K. R.: 'A CMOS Transceiver for 10-Mb/s and 100-Mb/s Ethernet', *IEEE J. Solid-State Circuits*, Vol. 33, pp. 2169–2177, December 1998.

22 MIKHAEL, W. B. and YASSA-GREISS, F. F.: 'A Frequency-Domain Adaptive Echo-Cancellation Algorithm', *IEEE Int. Symp. Circuits and Systems*, pp. 720–723, April 1981.

23 WIDROW, B. and McCOOL, J. M.: 'A Comparison of Adaptive Algorithms Based on the Methods of Steepest Descent and Random Search', *IEEE Trans. Antennas and Propagation*, Vol. AP-24, p. 615, September 1976.

24 MIKHAEL, W. B. and YASSA, F. F.: 'An Implementation of a Continuous-Time Frequency-Domain Adaptive System', *IEEE Int. Symp. Circuits and Systems*, pp. 585–588, May 1982.

25 PARIKH, D., AHMED, N. and STEARNS, S. D.: 'An Adaptive Lattice Algorithm for Recursive Filters', *IEEE Trans. on Acoustics, Speech and Signal Processing*, Vol. ASSP-28, pp. 110–111, February 1980.

26 AYALA, I. L.: 'On a New Adaptive Lattice Algorithm for Recursive Filters', *IEEE Trans. on Acoustics, Speech and Signal Processing*, Vol. ASSP-30, pp. 316–319, April 1982.

27 LEV-ARI, H.: 'Continuous-Time, Discrete-Order Lattice Filters', *Proc. 1986 IEEE Int. Conf. Acoustics Speech and Signal Processing*, pp. 2947–2950, April 1986.

28 HAYKIN, S.: *Adaptive Filter Theory, Third Edition*, Prentice Hall, New Jersey, 1996.

29 LEV-ARI, H., CIOFFI, J. M. and KAILATH, T.: 'Continuous-Time Least-Squares Fast Transversal Filters', *Proc. 1987 IEEE Int. Conf. Acoustics, Speech and Signal Processing*, pp. 415–418, April 1987.

30 LARIMORE, M. G., TREICHLER, J. R. and JOHNSON, C. R.: 'SHARF: An Algorithm for Adapting IIR Digital Filters', *IEEE Trans. on Acoustics, Speech and Signal Processing*, Vol. ASSP-28, pp. 428–440, August 1980.

31 JOHNSON, C. R. and TAYLOR, T.: 'CHARF Convergence Studies', *13th Asilomar Conf. Circ. Syst. and Comp.*, pp. 403–407, 1979.

32 CORNETT, F.: 'Continuous-Time Adaptive Recursive Signal Processing', *Proc. IEEE Southeastcon '92*, Vol. 1, pp. 223–225, April 1992.

33 AIELLA, S. R. and CORNETT, F.: 'Continuous-Time Adaptive Recursive and Nonrecursive Filters', *Proc. IEEE Southeastcon '93*, April 1993.
34 EDMONSON, W., PRINCIPE, J., SRINIVASAN, K. and WANG, C.: 'A Global Least Mean Square Algorithm for Adaptive IIR Filtering', *IEEE Trans. Circuits and Systems, II*, Vol. 45, pp. 379–384, March 1998.
35 NAMBIAR, R., TANG, C. K. K. and MARS, P.: 'Genetic and Learning Automata Algorithms for Adaptive Digital Filters', *IEEE Int. Conf. Acoustics, Speech and Signal Processing*, pp. IV41–IV44, April 1992.
36 NG, S. C., LEUNG, S. H., CHUNG, C. Y., LUB, A. and LAU, W. H.: 'The Genetic Search Approach – A New Learning Algorithm for Adaptive IIR Filtering', *IEEE Signal Processing Magazine*, pp. 38–46, November 1996.
37 CARUSONE, A. and JOHNS, D. A.: 'Obtaining Digital Gradient Signals for Analog Adaptive Filters', *Proc. 1999 IEEE Int. Symp. Circuits and Systems*, Vol. 3, pp. 54–57, May 1999.
38 NOWROUZIAN, B.: 'Theory and Design of LDI Lattice Digital and Switched-capacitor Filters', *IEEE Proceedings*, Vol. 139, pp. 517–526, August 1992.
39 SU, K. L.: *Analog Filters*, Chapman & Hall, London, 1996.
40 MASANI, P. R.: *Norbert Wiener 1894–1964*, Birkhauser, Basel, Boston, and Berlin, 1990.
41 JOHNS, D. A., SNELGROVE, W. M. and SEDRA, A. S.: 'Orthonormal Ladder Filters', *IEEE Trans. Circuits and Systems*, Vol. 36, pp. 337–343, March 1989.
42 HARTMANN, C. S., BELL, D. T. and ROSENFELD, R. C.: 'Impulse Model Design of Acoustic Surface-wave Filters', *IEEE Trans. Microwave Theory and Techniques*, Vol. MTT-21, pp. 162–175, April 1973.
43 TOMPSETT, M. F.: 'Charge Transfer Devices', *J. Vacuum Science and Technology*, Vol. 9, pp. 1166–1181, July/August 1972.
44 BUSS, D. D. and BAILEY, W. H.: 'Applications of Charge Transfer Devices to Communication', *Proc. Int. Conf. CCD Applications*, pp. 83–93, 1973.
45 TIEMANN, J. J., ENGELER, W. E., and BAERTSCH, R. D.: 'A Surface-Charge Correlator', *J. Solid-State Circuits*, Vol. SC-9, pp. 403–410, December 1974.
46 HAKEN, R. A., PETTENGILL, R. C. and HITE, L. R.: 'A General Purpose 1024-Stage Electronically Programmable Transversal Filter', *IEEE J. Solid-State Circuits*, Vol. SC-15, December 1980.
47 WHITE, M. H., MACK, I. A., BORSUK, G. M., LAMPE, D. R. and KUB, F. J.: 'Charge-Coupled Device (CCD) Adaptive Discrete Analog Signal Processing', *IEEE J. Solid-State Circuits*, Vol. SC-14, pp. 132–147, February 1979.
48 TSIVIDIS, Y., BANU, M. and KHOURY, J.: 'Continuous-Time MOSFET-C Filters in VLSI', *IEEE Trans. Circuits and Systems*, Vol. CAS-33, pp. 125–140, February 1986.
49 BUSS, D. D., HEWES, C. R., de WIT, D. E. and BRODERSEN, R. W.: 'Charge Couple Devices for Analog Signal Processing', *IEEE Int. Symp. Circuits and Systems*, pp. 708–720, 1977.
50 BOUNDEN, J. E., EAMES, R. and ROBERS, J. B. G.: 'MTI Filtering for Radar with Charge Transfer Devices', *Proc. CCD Technology and Applications Conf.*, pp. 206–213, September 1974.

51 SANGSTER, F. L. J. and TEER, K.: 'Bucket Brigade Electronics – New Possibilities for Delay, Time-Axis Conversion, and Scanning', *IEEE J. Solid-State Circuits*, Vol. SC-4, pp. 131–136, June 1969.
52 HAQUE, Y. A. and COPELAND, M. A.: 'Design and Characterization of a Real-Time Correlator', *IEEE J. Solid-State Circuits*, Vol. SC-12, pp. 642–649, December 1977.
53 AHUJA, B. K., COPELAND, M. A. and CHAN, C. H.: 'A Sampled Analog MOS LSI Adaptive Filter', *IEEE J. Solid-State Circuits*, Vol. SC-14, pp. 148–154, February 1979.
54 SONNTAG, J., AGAZZI, O., AZIZ, P., BURGER, H., COMINO, V., HEIMANN, M. *et al.*: 'A High Speed, Low Power PRML Read Channel Device', *IEEE Trans. Magnetics*, Vol. 31, pp. 1186–1195, March 1995.
55 GOMEZ, R., ROFOUGARAN, M. and ABIDI, A. A.: 'A Discrete-Time Analog Signal Processor for Disk Read Channels', *IEEE Int. Solid-State Circuits Conf. Dig. Tech. Papers*, pp. 212–213, February 1993.
56 UEHARA, G. T. and GRAY, P. R.: 'Parallelism in Analog and Digital PRML Magnetic Disk Read Channel Equalizers', *IEEE Trans. on Magnetics*, Vol. 31, pp. 1174–1179, March 1995.
57 YAMASAKI, R. G., PAN, T., PALMER, M. and BROWNING, D.: 'A 72 Mb/s PRML Disk-Drive Channel Chip with an Analog Sampled-Data Signal Processor', *IEEE Int. Solid-State Circuits Conf. Dig. Tech. Papers*, pp. 278–279, February 1994.
58 XU, D., SONG, Y. and UEHARA, G. T.: 'A 200 MHz 9-Tap Analog Equalizer for Magnetic Disk Read Channels in 0.6 μm CMOS', *IEEE Int. Solid-State Circuits Conf. Dig. Tech. Papers*, pp. 74–75, February 1996.
59 KIRIAKI, S., VISWANATHAN, T. L., FEYGIN, G., STASZEWSKI, B., PIERSON, R., KRENIK, B., de WIT, M. and NAGARAJ, K.: 'A 160-MHz Analog Equalizer for Magnetic Disk Read Channels', *IEEE J. Solid-State Circuits*, Vol. 32, pp. 1839–1850, November 1997.
60 SUNTER, S. K., GIRCZYC, E. F. and CHOWANIEC, A.: 'A Programmable Transversal Filter for Voice-Frequency Applications', *IEEE J. Solid-State Circuits*, Vol. SC-16, pp. 367–372, August 1981.
61 MCCREARY, J. and GRAY, P. R.: 'All-MOS Charge Redistribuition Analog-to-digital Conversion Techniques – Part I', *IEEE J. Solid-State Circuits*, Vol. SC-10, pp. 371–379, December 1975.
62 ALBARRAN, J. F. and HODGES, D. A.: 'A Charge-transfer Multiplying Digital-to-analog Converter', *IEEE J. Solid-State Circuits*, Vol. SC-11, pp. 772–779, December 1976.
63 TANAKA, S. C., TSENG, H.-F., LN, L. T. and CHEN, P.-J.: 'An Integrated Real-Time Programmable Transversal Filter', *IEEE J. Solid-State Circuits*, Vol. SC-15, pp. 978–983, December 1980.
64 ENOMOTO, T., YASUMOTO, M., ISHIHARA, T. and WATANABE, K.: 'Data Communication ICs', *IEEE Int. Solid-State Circuits Conf. Dig. Tech. Papers*, pp. 150–151, February 1982.

65 ALLSTOT, D., BRODERSEN, R. and GRAY, P.: 'An Electrically Programmable Analog NMOS Second Order Filter', *IEEE Int. Solid-State Circuits Conf. Dig. Tech. Papers*, pp. 76–88, February 1979.
66 MARTIN, K. and SEDRA, A. S.: 'Switched-Capacitor Building Blocks for Adaptive Systems', *IEEE Trans. Circuits and Systems*, Vol. CAS-28, pp. 576–584, June 1981.
67 GREGORIAN, R., MARTIN, K. W. and TEMES, G. C.: 'Switched-Capacitor Circuit Design', *Proc. IEEE*, Vol. 71, pp. 941–966, August 1983.
68 NAUTA, B.: 'A CMOS Transconductance-C Filter Technique for Very High Frequencies', *IEEE J. Solid-State Circuits*, Vol. 27, pp. 142–153, February 1992.
69 PAVAN, S. and TSIVIDIS, Y.: 'Time-Scaled Electrical Networks – Properties and Applications in the Design of Programmable Analog Filters', *IEEE Trans. Circuits and Systems, II*, Vol. 47, pp. 161–165, February 2000.
70 WELLAND, D. R. *et al.*: 'A Digital Read/Write Channel with EEPR4 Detection', *IEEE Int. Solid-State Circuits Conf.*, pp. 276–277, San Francisco, February 1994.
71 LEE, Y. W.: 'Synthesis of Electric Networks by Means of Fourier Transforms of Laguerre's Functions', *J. Mathematical Physics*, pp. 83–113, 1932.
72 LEE, Y. W.: *Statistical Theory of Communication*, Wiley, New York, 1960.
73 WIENER, N.: *Extrapolation, Interpolation, and Smoothing of Stationary Time Series with Engineering Applications*, The MIT Press, Cambridge; Wiley, New York; Chapman & Hall, London, 1949.
74 BEKEY, G. A. and KARPLUS, W. J.: *Hybrid Computation*, New York, Wiley, 1968.
75 WIDROW, B., GLOVER, Jr. J. R., MCCOOL, J. M., KAUNITZ, J., WILLIAMS, C. S., HEARN, R. H. *et al.*: 'Adaptive Noise Cancelling: Principles and Applications', *Proc. IEEE*, Vol. 63, pp. 1692–1716, December 1975.
76 CIDECIYAN, R. D., DOLIVO, F., HERMANN, R., HIRT, W. and SCHOTT, W.: 'A PRML System for Digital Magnetic Recording', *IEEE J. Selected Areas in Communications*, Vol. 10, pp. 38–56, January 1992.
77 SHAKIBA, M. H., JOHNS, D. A. and MARTIN, K. W.: 'BiCMOS Circuits for Analog Viterbi Decoders', *IEEE Trans. Circuits and Systems, II*, Vol. 45, pp. 1527–1537, December 1998.
78 SANDS, N. P., HAUSE, M. W., LIANG, G., GROENEWOLD, G., LAM, S., LIN, C.-H. *et al.*: 'A 200 Mb/s Analog DFE Read Channel', *IEEE Int. Solid-State Circuits Conf. Dig. Tech. Papers*, pp. 72–73, February 1996.
79 PARSI, N., RAO, N., BURNS, R., CHAIKEN, A., CHAMBERS, M., CHEUNG, R. *et al.*: 'A 200 Mb/s PRML Read/Write Channel IC', *IEEE Int. Solid-State Circuits Conf. Dig. Tech. Papers*, pp. 66–67, February 1996.
80 PAI, P. K. D., BREWSTER, A. D. and ABIDI, A.: 'Analog Front-End Architectures for High-Speed PRML Magnetic Read Channels', *IEEE Trans. on Magnetics*, Vol. 31, pp. 1103–1108, March 1995.
81 PAI, P. K. D., BREWSTER, A. D. and ABIDI, A.: 'A 160-MHz Analog Front-End IC for EPR-IV PRML Magnetic Storage Read Channels', *IEEE J. Solid-State Circuits*, Vol. 31, pp. 1803–1816, November 1996.

82 BROWN, J. E. C., HURST, P. J., ROTHENBERG, B. C. and LEWIS, S. H.: 'A CMOS Adaptive Continuous-Time Forward Equalizer, LPF, and RAM-DFE for Magnetic Recording', *IEEE J. Solid-State Circuits*, Vol. 34, pp. 162–169, February 1999.
83 LUCKY, R.: 'Automatic Equalization for Digital Communication', *Bell Systems Technical Journal*, Vol. 44, pp. 547–588, April 1965.
84 HATAMIAN, M., AGAZZI, O. E., CREIGH, J., SAMUELI, H., CASTELLANO, A. J., KRUSE, D. *et al.*: 'Design Considerations for Gigabit Ethernet 1000Base-T Twisted Pair Transceivers', *IEEE Custom Integrated Circuits Conf.*, pp. 335–342, June 1998.
85 ROESGEN, J. P. and WARREN, G. H.: 'An Analog Front End Chip for V.32 Modems', *IEEE Custom Integrated Circuits Conf.*, pp. 16.1/1-16.1/5, May 1989.
86 SMOLKA, G. J.: 'Analog CMOS Circuits for ISDN', *Proc. 1988 IEEE Int. Symp. Circuits and Systems*, Vol. 2, pp. 1927–1930, June 1988.
87 AGAZZI, O., HODGES, D. A., MESSERSCHMIT, D. G. and LATTIN, W.: 'Echo Canceller for a 80 kbs Base-band Modem', *IEEE Int. Solid-State Circuits Conf. Dig. Tech. Papers*, pp. 144–145, February 1982.
88 PECOURT, F., HAUPTMANN, J. and TENEN, A.: 'An Integrated Adaptive Analog Balancing Hybrid for Use in (A)DSL Modems', *IEEE Int. Solid-State Circuits Conf. Dig. Tech. Papers*, pp. 252–253, February 1999.
89 HUSS, S. D., CRANFORD, H. C. and GYURCSIK, R. S.: 'An Integrated Two-Parameter Adaptive Cable Equalizer Using Continuous-Time Filters', *Proc. 1997 IEEE Int. Symp. Circuits and Systems*, Vol. 2, pp. 969–972, June 1997.
90 SHAKIBA, M. H.: 'A 2.5 Gb/s Adaptive Cable Equalizer', *IEEE Int. Solid-State Circuits Conf. Dig. Tech. Papers*, pp. 396–397, February 1999.

Chapter 7
On-chip automatic tuning of filters

Rolf Schaumann and Aydin I. Karsilayan

7.1 Introduction

We have learnt from the material in the previous chapters (refer also to the references in Reference 1) that a variety of well understood methods is available for designing continuous-time, or analogue, filters in fully integrated form for operating frequencies ranging from the audio range to RF. The synthesis strategies were seen to be the same, in principle, as the ones employed for any analogue active filter; they make use of capacitors, resistors and gain, that is, active devices, in a feedback configuration. Integrated filters can be implemented in any integrated-circuit (IC) technology, such as CMOS, bipolar or GaAs. The preferred technology is CMOS because analogue filters will normally reside on predominantly digital ICs and for economic reasons must be compatible with digital IC technology.

As all analogue filters, integrated filters require accurate component values to realise the prescribed frequency, quality factor and gain parameters. These are normally not available in an IC because of process tolerances and varying operating conditions, and changing the components on an IC may prove difficult or impossible. To provide more room for accommodating parameter deviations, the designer may be tempted to solve tolerance problems by designing a filter with tighter passband and steeper roll-off constraints than required. This solution can as a rule not be recommended because it leads to a higher-order filter, uses more power and area, and increases costs. In addition, the higher filter order and steeper roll-off characteristics result in larger sensitivities [2], which are likely to make matters worse and negate the advantages sought in an over-designed filter. Consequently, a better approach for fighting the effect of tolerances is automatic electronic tuning.

Quality factors and gain constants are dimensionless quantities and, as such, their implementation depends on accurate *ratios of like components*, such as resistor ratios or capacitor ratios. Implementing accurate ratios of like components is a forte of IC technology: resistor ratios of better than 1 percent accuracy can readily be obtained and capacitor ratios with tolerances below 0.1 percent are achievable. With careful layout and close proximity of the circuits on the IC chip, even ratios of entire circuit modules, such as transconductances, g_{mi}/g_{mj}, can be obtained quite accurately as long as the ratios are moderate, preferably of integer value and near unity. On the other hand, frequency parameters are set by time constants, that is, by RC or LC products, or by C/g_m ratios. These require the realisation of precise *absolute values of electrically different devices*. Absolute values depend on process parameters, aging, and environmental operating conditions, such as temperature or bias, which are difficult to control during processing and operation of the IC. Consequently, as is well known, absolute values of IC components cannot be expected to be accurate or even repeatable: errors of 20–50 percent or even larger are not uncommon. A further complication arises because filter parameters are affected by parasitic components. This is true especially as operating frequencies approach hundreds of megahertz or even gigahertz where the nominal circuit capacitors may be as small as a fraction of a picofarad and, therefore, in the range of device or layout and routing parasitics on the IC chip.

Although we might expect that on an IC the dimensionless quality factor, Q, could be realised precisely via ratios of like components, in practice this is found not to be correct. The effect of parasitic components, device losses and, especially, small phase shifts in the feedback loops of the active circuits are known to result in large and unpredictable errors in quality factors (see, for example, Eqn. (7.13)). For large values of Q, lagging loop-phase errors, often referred to as *excess phase*, of even a small fraction of a degree can result in large Q enhancement and even oscillations.

Although in some analogue filter applications[1] large tolerances can be accepted, this discussion demonstrates that after design and processing an analogue filter generally must be adjusted to meet specifications. The filter's frequency parameters, and even quality factors and gain need to be tuned. *Tuning* of a filter on an IC in the traditional sense is often not possible or too difficult and expensive because the circuit components are usually not accessible. In some circumstances, we may attempt to change elements digitally, e.g. by opening or closing MOS switches in a capacitor array, or we may adjust the filter components by cutting or fusing electrical links. We must bear in mind, though, that switches in the signal path may introduce parasitic resistors and capacitors that contribute errors themselves and call for further tuning. Occasionally, laser trimming might be used to achieve (irreversible) fine-tuning. Or, we may be able to change ('trim') external resistors to vary a bias current or voltage and thereby change the DC gain of a transconductor, the time constant of an integrator or the value of a junction capacitor. The filter is thereby tuned by varying the components electronically. All these methods have been proposed and used.

The mentioned additional design or processing steps can be employed to combat processing tolerances, but they are normally not adequate when environmental conditions, such as power supply or temperature, must be expected to change during filter

[1] For example, in lowpass antialiasing filters where the ratio of clock frequency to signal frequency is large.

operation. Then, filter components must be adjusted, preferably continuously, during operation, that is, the filter must be *tuned automatically* by some on-chip electronic control scheme. Evidently, for this plan to work we must make certain at the outset of a design that the chosen filter topology is electronically adjustable, i.e. that pertinent filter components or parameters are variable electronically, for example, by changing the bias conditions.

For the discussion in this chapter, we consider the tuning circuitry as auxiliary to the filters. It does not aid the analogue signal-processing task, but does add to its cost. Because the proposed tuning circuitry will reside on the IC together with the filters and any other working analogue or digital systems, it must satisfy certain constraints to ensure that any overhead expense connected with automatic tuning is minimal. The more important ones are:

- The silicon area of the control circuitry must be small when compared to that of the filter, i.e. the circuitry must be relatively simple so that Si real estate is conserved.
- The tuning circuitry must be 'quiet', that is, it must not generate excessive thermal noise or 'signal noise', such as cross-talk or modulation, that can reduce the dynamic range of the filter.
- Particularly important for battery-operated systems, power consumption of the tuning circuitry should be small so that the filter's power budget is minimised.

To help us develop the automatic tuning circuitry, we need to understand the principle of the operation it must perform. Let us assume a filter was designed with the transfer function $H(s, \mathbf{k})$, where $\mathbf{k}$ is the vector of the circuit components consisting of the variable, or *tuned*, elements $\mathbf{t}$ and the fixed, or *untuned*, elements $\mathbf{u}$, that is $\mathbf{k} = \mathbf{t} + \mathbf{u}$. $\mathbf{t}$ and $\mathbf{u}$ consist of the nominal values $\mathbf{t}_0$ and $\mathbf{u}_0$ and the deviations $\Delta\mathbf{t}$ and $\Delta\mathbf{u}$: $\mathbf{t} = \mathbf{t}_0 + \Delta\mathbf{t}$, $\mathbf{u} = \mathbf{u}_0 + \Delta\mathbf{u}$. Further, the deviations $\Delta\mathbf{t}$ consist of the changes $\Delta\mathbf{t}_t$ effected by tuning and any component errors $\Delta\mathbf{t}_\varepsilon$, resulting from processing or other causes: $\Delta\mathbf{t} = \Delta\mathbf{t}_t + \Delta\mathbf{t}_\varepsilon$. The realised transfer function is then

$$H(s, \mathbf{t}, \mathbf{u}) = \frac{N_m(s, \mathbf{t}, \mathbf{u})}{D_n(s, \mathbf{t}, \mathbf{u})} = H(s, \mathbf{t}_0 + \Delta\mathbf{t}, \mathbf{u}_0 + \Delta\mathbf{u}) \quad (7.1)$$

The tuning circuitry must detect and measure the errors

$$\varepsilon(\Delta\mathbf{t}) = H(s, \mathbf{t}_0 + \Delta\mathbf{t}, \mathbf{u}_0 + \Delta\mathbf{u}) - H(s, \mathbf{t}_0, \mathbf{u}_0) \quad (7.2)$$

in the filter's operation. Then the system must apply appropriate tuning signal(s) to selected component(s), t_i, in $\mathbf{t}$ so that the tuning changes, Δt_i, reduce this function to its nominal value $H(s, \mathbf{t}_0, \mathbf{u}_0)$. Notice that the tuning changes must correct not only the errors in the untuned components, $\mathbf{u}$, but also those of the tuned elements, $\mathbf{t}$.

To be able to detect errors, the tuning circuit needs a *reference* according to which filter performance is judged. Typically, a single sine wave or other periodic signal, such as a clock, at the appropriate frequency is used (but see, for example, [3, 4] for

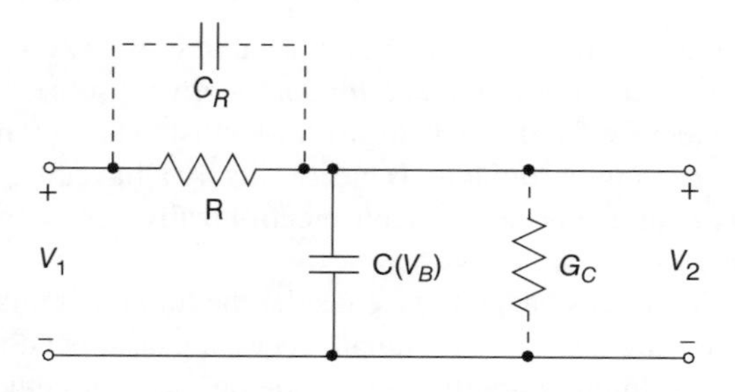

Figure 7.1 *A first-order RC lowpass filter with parasitic components*

multiple-frequency tuning approaches).[2] The tuning circuitry compares the filter's measured response to the reference signal with the known nominal response or with the reference signal itself. It must then identify not only the presence of errors in the operation, but also their signs and sizes. With that information, correction signal(s), usually DC bias voltages or currents, are generated by the tuning circuitry and fed back to the filter in a closed-loop control scheme so that the errors are reduced to zero or at least minimised. All these operations must be performed automatically and, of course, without interfering with the filter's signal processing function.

Let us briefly look at a very simple example to help illustrate the points made. Consider the first-order RC lowpass filter in Figure 7.1. Disregarding for now the parasitic components (shown with dashed connections) and assuming the capacitor is tuned (belonging to set **t**) by an applied bias voltage V_B and the resistor is untuned (belonging to **u**), we find the transfer function

$$V_2/V_1 = H(s, C(V_B), R) = 1/[sC(V_B)R + 1] \tag{7.3}$$

Assuming that in processing R and C the errors ΔR_ε and ΔC_ε are acquired from their nominal values R_0 and C_0, and that the tuning change is $\Delta C_t(V_B)$, the realised time constant equals $[C_0 + \Delta C_\varepsilon + \Delta C_t(V_B)]\,[R_0 + \Delta R_\varepsilon]$. For the filter to have the right cut-off frequency, the RC product must be corrected to $C_0 R_0$, which can be achieved by adjusting the capacitor. A particularly simple magnitude-locked-loop approach was proposed in Reference 5. It equates $\mid \mathrm{H}(j\omega_{ref}) \mid$ of Eqn. (7.3) to an accurately designable voltage divider $R_2/(R_1 + R_2)$ so that the RC product is adjusted to

$$CR = \frac{1}{\omega_{ref}} \sqrt{\left(1 + \frac{R_1}{R_2}\right)^2 - 1} \tag{7.4}$$

[2] The transfer function H in Eqn. (7.1) is expressed as a ratio of two polynomials of orders m and n, with $m \leq n$, and has $m + n - 1$ coefficients. Consequently, for perfect tuning, theoretically $m + n - 1$ pieces of information must be available, for example, by applying n reference signals at n frequencies, each providing information about gain and phase at those frequencies. However, constraints on the complexity of the tuning circuitry will, as a rule, limit the number of reference signals to one.

The equation illustrates a result that is typical of all automatic tuning techniques: *The accuracy and long-term stability of RC time constants, which determine the frequency parameters of a filter, are as good as those of the reference frequency and of a ratio of like components on the IC.*

So far, we have neglected the presence of parasitics in the lowpass filter of Figure 7.1. If the capacitor losses G_C and the capacitor C_R bridging the resistor R are included in the analysis, the transfer function becomes

$$\frac{V_2}{V_1} = H_p(s) = \frac{sC_R + G}{s(C + C_R) + G + G_C} \tag{7.5}$$

A comparison with Eqn. (7.3) shows that $H_p(s)$ has acquired a parasitic zero at $-G/C_R$ and that the pole position has shifted from $1/(CR)$ to

$$\sigma = -\frac{G + G_C}{C + C_R} = -\frac{1}{CR} \times \frac{1 + R/R_C}{1 + C_R/C} \tag{7.6}$$

We see here that parasitic components may have two main effects:

1. Parasitics cause changed values of components.
2. Parasitics may result in additional poles or zeros.

The consequence of the first of these effects is the same as are changes in nominal component values. In the example, G_C and C_R appear just like tolerances $\Delta G = G_C$ and $\Delta C = C_R$ of the elements G and C. They can, in principle, always be corrected by tuning.

The second effect, the generation of new critical frequencies, is much more serious because these parasitic poles and zeros may destroy the transfer function shape and no known method of tuning can reduce the 'expanded' transfer function

$$H_p(s, \mathbf{t}, \mathbf{u}, \mathbf{p}) = \frac{N_r(s, \mathbf{t}, \mathbf{u}, \mathbf{p})}{D_q(s, \mathbf{t}, \mathbf{u}, \mathbf{p})} \tag{7.7}$$

to its nominal form. $H_p(s)$ is expanded from $H(s)$ in Eqn. (7.1) by the presence of parasitics, $\mathbf{p}$, since, as a rule, the degrees r and q of the numerator and denominator polynomials of H_p are higher than m and n of H. If parasitics are of concern, as they always are, particularly as frequencies increase, the tuning operation can generally attempt only to reduce the error defined through a suitable norm, $\| \bullet \|$, to a minimum,

$$\varepsilon(\Delta\mathbf{t}) = \|H(s, \mathbf{t}_0 + \Delta\mathbf{t}, \mathbf{u}_0 + \Delta\mathbf{u}, \mathbf{p}) - H(s, \mathbf{t}_0, \mathbf{u}_0)\| \to \min \tag{7.8}$$

The effort to solve this problem is normally excessive, because we must bear in mind that the operation must be performed on-chip, with simple circuitry, and with little or no computing power available. Nevertheless, an adaptive algorithm has been used in an attempt at removing parasitic effects on filter performance [6, 7]. More often, the complexity of the function $H_p(s, \mathbf{t}, \mathbf{u}, \mathbf{p})$ is reduced by 'partitioning', such as by using a cascade design where all effects are localised to one second-order section; see, for example, Reference 4.

As a rule, filter requirements are specified in the frequency domain, but the frequency-domain parameters, such as quality factor and pole frequency, do of course affect the time-domain response. Consequently, several methods have been proposed, e.g. References 8–10, that apply a pulse- or step-reference input to the filter and deduce filter parameter errors from the transient response. For instance, the natural frequency in the time domain is related to the pole frequency in the frequency domain, and the exponential decay of the natural response is a function of Q.

Conceptually, all these tuning schemes are simple enough, but two practical challenges must be solved. The first one arises because generally the reference signal(s) cannot be applied to the filter simultaneously with the main input signal to be processed. Analogue filters are linear systems so that applying the reference and the main signals simultaneously will cause undue modulation or other interference. This will corrupt the responses to the main signal or to the reference signal, and likely to both of them. In any event, such interference is generally unacceptable because it will change the information content of the main filter signal, and it will make it difficult to identify the reference and interpret the reference response. The solution generally adopted by IC filter designers is a *master-slave* system (Figure 7.2). It will be discussed in Section 7.2. The second difficulty is well known to test engineers. It is concerned with *testing*, that is, measuring the error(s) and identifying the filter components that cause the error(s). Since in the present situation, *tuning*, we must deal with on-chip built-in self-test *and correction* (BISTAC), the whole procedure must be

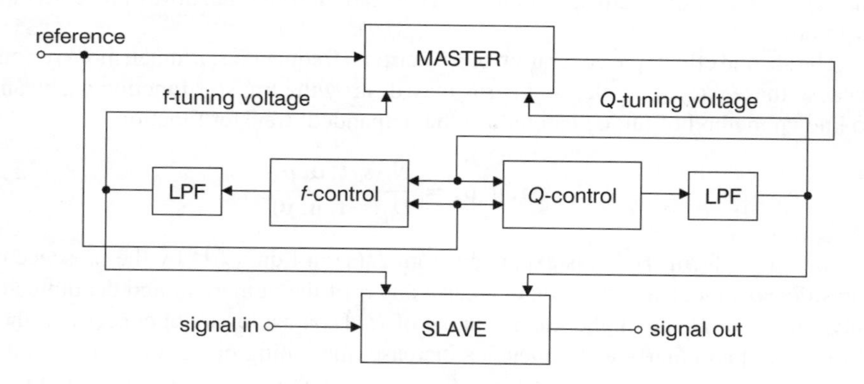

Figure 7.2 Master-slave tuning system. The master circuit models all relevant performance parameters of the main filter, the slave. Since the behaviours of master and slave on the IC match and track, the correction (tuning) signals applied to tune the master are also applied to tune the slave filter. The f (frequency)-control circuitry compares the master filter's response with the reference signal, measures errors in frequency parameters and generates appropriate tuning signals. The Q-control circuitry, if used, performs the analogous function for errors in quality factor(s). The low-pass filters (LPF) are used to clean the control voltages of any remaining ripple

accomplished with moderate on-chip circuitry, and little or no computing power to solve the system-identification question. Typically, the problem is handled by selecting suitable filter circuits (see Section 7.4) and carefully modelling the dependence of critical performance parameters on filter components.

Even if an easily tunable filter structure is found, and a suitable tuning scheme and circuitry are identified, it is still highly advisable to design the filter so that it meets the *nominal* specifications as closely as possible. Electronic tuning is normally effected by changing the bias conditions, and the tuning task will be easier if only moderate errors need to be corrected. In addition, if very wide tuning ranges are required, other important filter parameters that depend on bias conditions, such as linearity or dynamic range, are likely to suffer. Careful modelling, predistortion methods to account for parasitics, and possibly design centring and statistical methods, should be employed in the design process.

7.2 On-line tuning

We pointed out in Section 7.1 that any errors in filter performance are identified by applying a known reference signal to the filter and comparing the filter's response with the reference. This reference can generally not be applied to the filter while the filter is in operation (*on-line*) because of unacceptable interference between main and reference signals. The solution commonly adopted is to construct on the IC chip a model of the filter and apply the reference signal to this model. The model circuit, labelled the *master*, should be simple but must be of sufficient complexity to represent (model) accurately all relevant performance criteria of the main filter, labelled the *slave*.

The master may be as simple as an integrator[3] [8] or as complex as a duplicate [4] of the slave filter. The on-chip control system then *tunes the master* by applying the generated correction or tuning signal(s), typically DC voltages or currents, in a closed-loop feedback arrangement to the master until the errors are zero or small enough to be acceptable. Assuming that master and slave on the same IC chip match and track in their behaviour, the errors of master and slave are expected to match and track as well. Under these conditions, the correction signal(s) generated by the tuning circuitry for the master can also be applied to the slave to correct its errors: the slave is tuned by following the master. Figure 7.2 shows a block diagram of the system.

It is evident that the master circuit should be relatively simple to keep Si area and power consumption manageable. Although first-order blocks have been used with some success, the most commonly used master circuits are second-order filters or resonators built to match the slave's performance. A potential difficulty arises when

[3] It has also been proposed (see, for example, Reference 11) to choose simply a transconductor, g_m, as the master and electronically equate it to the reciprocal of an external reference resistor, $g_m = 1/R_{\text{ext}}$, by setting the bias. The bias of all filter transconductors is then set equal to that of the master transconductor. This method is generally not adequate because critical to filter operation are frequency parameters, g_m/C, whose errors cannot be discovered by a simple DC measurement.

the required filter is of order higher than two, requiring, for example, a cascade of several second-order sections that usually have different pole frequencies and quality factors. Although in principle these differences are set by capacitor or other component ratios, in practice, at least at higher frequencies and quality factors, the presence and effects of parasitics must be accounted for. Thus, matching and tracking difficulties must be expected when designing a single second-order master to match different slave sections. Several different masters could be used [4], but the cost of the overhead will normally preclude such solutions. Fortunately, methods are available that permit a high order filter to be designed with only identical sections or resonators, where better matching and tracking across the chip can be expected. Most of these methods use multiple-feedback techniques [2], such as the primary resonator block (PRB) method [12]. See Reference 13 for an in-depth discussion of the problem. Although some mismatch problems are alleviated, multiple-loop feedback techniques have a serious drawback in that their global feedback loops introduce hard-to-control phase errors, which severely affect the transfer function shape [14–16].

It has been pointed out throughout this volume that signals in analogue integrated filters should be applied differentially for improved noise reduction, linearity and dynamic range. This fact opens a possible solution [17] for applying a reference signal to the main filter while it is in operation and largely avoiding master-slave matching and tracking problems. The reference can be applied as a common-mode signal that will be eliminated in the fully differential signal path. Hence, it will not interfere with the main signal. At the same time, the undisturbed common-mode reference response can be recovered by adding the differential filter outputs. The approach requires single-ended amplifiers or OTAs and the construction of pseudo-differential circuitry. It has been shown to compare favourably with the master-slave approach [18]. A similar solution, referred to orthogonal reference tuning, has been proposed in Reference 19. In this approach, a known reference signal is present in the filter passband together with the RF signal being processed. If reference signal and processed signal are (nearly) orthogonal, they can be separated at the filter output.

The reader will appreciate that the master-slave method really tunes only the master filter; slave-filter tuning can only be as good as the achievable matching on the IC. Great care must be taken in laying out the model (the master) circuit in reference to the main filter (the slave).[4] The master should be in close proximity to the slave to minimise the effects of process nonuniformity across the chip. On the other hand, the proximity of master and slave increases the potential signal noise and intermodulation because of parasitic feed-through of reference signals to the main filter. In any event, the master must be a good model of the slave's behaviour. For example, when process tolerances cause the frequency parameters of the slave to decrease, the frequency parameters of the master ought to be affected in the same way. Or, when parasitic phase lags in the slave's feedback loops cause the quality factor to increase, the same phase shifts ought to be seen in the master and should cause the

[4] Even the common-mode-reference approach [17, 18] requires good matching because the two single-ended circuits used to construct the pseudo-differential filter, of course, must match. Matching problems are mitigated, though, by requiring matching only between two identical circuit blocks.

same Q enhancement. Master-slave matching has been observed to work well for frequencies up to several megahertz, with only a few percent matching error, e.g. References 4 and 20, but the last requirement, *parasitics matching*, can be expected to be increasingly problematic as operating frequencies increase. Indeed, for communication circuits at several hundred megahertz or even gigahertz, filter capacitors become a small fraction of a picofarad and the effects of parasitic components can no longer be neglected or even be assumed small. Filter operation at a few hundred megahertz is fundamentally determined by parasitic components so that, even with very careful layout of a master-slave system, matching at frequencies above 100 MHz is bound to be poor and unreliable [21].

7.3 Off-line tuning

Because of the unavoidable errors resulting from matching problems in the master-slave approach, it has been proposed to tune the main filter itself while its signal is not connected: the filter is *off-line*. The discussion can be brief, because the tuning principle is rather similar to the master-slave approach, except that no master is used. Figure 7.3 illustrates the concept. The reference signal is applied to the filter while the main signal is absent, causing no interference, and the necessary tuning voltage is stored in a hold circuit when the main signal is reapplied. Tuning takes a few milliseconds and there are many signal-processing applications where the filter has suitable inactive or 'quiet' periods that can be used for tuning. Examples are the field fly-back time of a video signal, time intervals outside the receive periods in cell

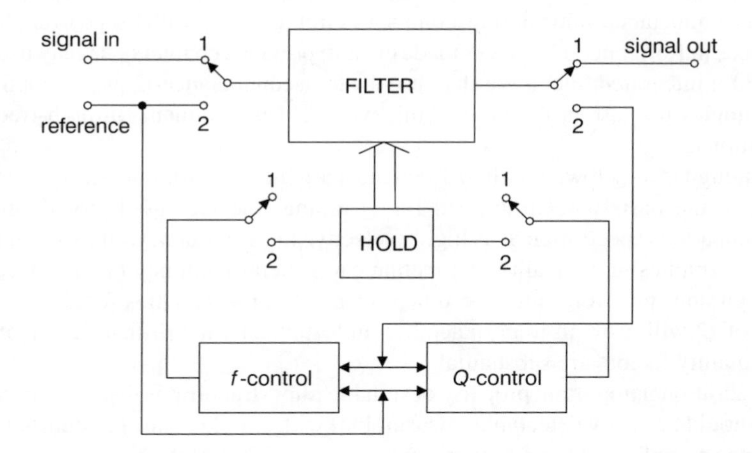

Figure 7.3 Off-line tuning system. The switches are in positions 1 for filtering and in positions 2 for tuning. The reference signal is applied to the filter when the main signal is absent. When filtering must resume, the required values of the tuning bias voltages are stored on a hold circuit (on capacitors or in digital memory) until the next update period

phones using time-division multiple access systems [10], or generally the stand-by period of a cell phone when a ringing signal can alert the system that tuning should be activated. The same principle can be used in situations where the main signal cannot be switched off or interrupted. As proposed in Reference 22, in that case duplicate filters are implemented of which one is tuned while the second one performs the signal processing operation. Using appropriate care to avoid switching transients, the two filters are interchanged periodically so that the signal-processing path is always tuned correctly.

7.4 Suitable filter structures

To help minimise the effects of component tolerances, any topology considered for an integrated filter should obviously have low sensitivities to the elements. Further, since the main electronic parameters that must be tuned in a g_m-C filter are the transconductances, it is helpful to use a design with only identical transconductors because it leads to simpler tuning algorithms and better matching across the chip. Fortunately, several biquads and all ladder simulations fall into this category [2]. With these elementary precautions met, it is clear that any topology considered for integrated analogue filters must be tunable by electronic means. Since the pole frequencies are most critical to a filter's performance [2], this means mainly that the frequency parameters, which are determined by RC products or g_m/C ratios,[5] must be adjustable by changing a bias voltage or current. Because the commonly used MOS capacitors are fixed,[6] resistors or transconductors are employed as variable components. MOS transistors biased in the triode region offer a resistance tunable by the applied gate bias voltage [25], and the transconductance value of transconductor circuits varies with bias current (see the references in Reference 1). As was made clear in previous chapters, we may use either method for integrated filters. We shall use the transconductance-C (g_m-C) method for our examples because at the time of this writing it is dominant for high-frequency applications.

Although low-Q lowpass filters filled most urgent needs for integrated filter applications in the past, with little need for Q-tuning, the increased crowding of the communications spectrum makes highly selective high-frequency filters more important. This implies large quality factors that can generally not be obtained accurately by design and, therefore, must be tuned. Consequently, schemes for the automatic tuning of Q will grow in importance but, unfortunately, the difficulties involved in tuning quality factors are substantial.

An additional important property of suitable filter structures is that filter-parameter errors should be easily detectable (measurable) with simple control circuitry. In addition, since complicated tuning algorithms are generally not feasible, any error should

[5] If spiral inductors in LC filters are used at gigahertz frequencies, some scheme must be employed to make the inductors electronically variable [23, 24]. See Chapter 3.

[6] We do not consider capacitor arrays where the switch parasitics may cause additional difficulties at high frequencies.

be correctable by varying as few components as possible. In most filter topologies, this condition is satisfied for the frequency parameters because they are proportional to a basic g_m/C ratio to which all others are related by accurate capacitor ratios. Thus, frequency tuning is accomplished by applying the same adjustment to *all* transconductors;[7] this adjustment will not alter any other filter parameters because they must necessarily depend on transconductance or capacitor *ratios*, which are not changed. As mentioned, tuning of quality factors is considerably more complicated and a contemplated filter topology should be very carefully analysed at the outset of any design to assess its suitability for Q-tuning. For instance, attempting to identify Q errors by measuring the differences of (at least) two frequencies and two gain levels would be an intricate and error-prone undertaking. It is much easier to observe the simpler relationship (proportionality) between Q and gain at a circuit's pole frequency that is present in many filter topologies.

For the design of high-order filters, simulated transconductance-C ladders should be used if possible. It makes no difference whether the element replacement method (replacing inductors by gyrator-C combinations) or operational simulation (using signal-flow graphs) is employed. The final circuits are the same [26]. The advantages of ladder structures are their low sensitivities to component tolerances and that, as suggested earlier, they can be built with all-identical transconductors [2]. It is also important that all internal nodes of a ladder are loaded by a circuit capacitor that can absorb any capacitive parasitics. This feature ensures that the filter retains its degree so that additional parasitic poles or zeros are avoided. A second-order example is shown in the biquad in Figure 7.4. If ladders are not appropriate, for example for the design of equalisers, multiple-loop feedback configurations[8] [2] or cascade connections of biquads are used.

A second-order circuit that was found suitable in practice is the two-integrator loop biquad in Figure 7.4, with a bandpass output at V_2 and a lowpass output at V_3. It will become evident in the next two sections that in popular tuning schemes it is convenient to have filters with both lowpass and bandpass outputs. As explained in the heading of Figure 7.4, the circuit may also be interpreted as an active simulation of a parallel RLC resonator. Before analysing the biquad, we first note that at node 2 we have the parasitic capacitance $C_{p1} = 2C_{in} + 3C_{out}$, consisting of the input capacitances $C_{2in} + C_{3in} = 2C_{in}$ and the output capacitances $C_{1out} + C_{2out} + C_{4out} = 3C_{out}$. As shown in dashed form, C_{p1} is in parallel with the conductance $G_{p1} = 3g_{out} = g_{1out} + g_{2out} + g_{4out}$. At node 3 is the parasitic capacitance[9] $C_{p2} = 2C_{in} + C_{out} = C_{4in} + C_{Load} + C_{3out}$ in parallel with the conductance $G_{p2} = g_{3out} = g_{out}$. Labelling the *total* capacitors, including parasitics, at the corresponding nodes $C_i = \hat{C}_i + C_{pi}$, $i = 1, 2$, and assuming for simplicity all transconductors to be identical, $g_{mi} = g_m, i = 1, \ldots, 4$,

[7] For this reason it is preferable to build a filter with only identical transconductors.

[8] It is noted, though, that multiple-loop feedback topologies suffer from severe deviations caused by the phase errors in the global feedback loops [14–16].

[9] We assumed here that the lowpass output is loaded by a similar biquad whose input can be modelled as a capacitance $C_{Load} = C_{in}$.

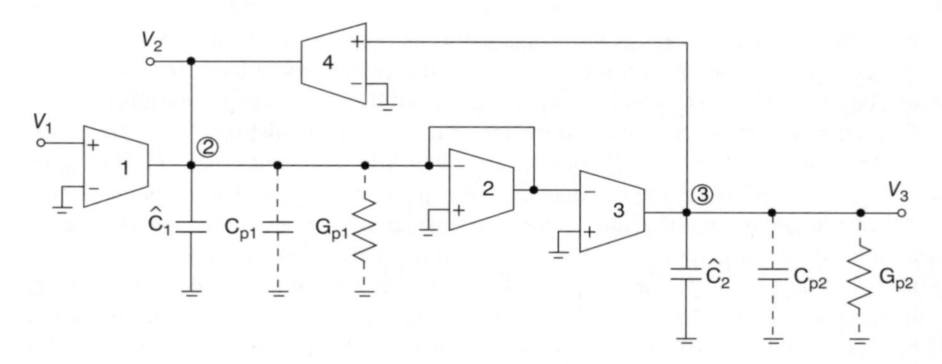

Figure 7.4 *A practical two-integrator-loop second-order section consisting of a lossy noninverting integrator ($g_{m1} - C_1 - g_{m2}$) and a lossless inverting integrator ($g_{m3} - C_2$), with the feedback loop closed by g_{m4}. Parasitic capacitors and loss resistors are shown dashed. V_2 and V_3, respectively, are bandpass and lowpass outputs; other transfer functions can be obtained by lifting fractions f_1C_1 and f_2C_2, $0 \le f_i \le 1$, of the capacitors C_1 and C_2 off ground and connecting these fractions to the input. Note that g_{m3} and g_{m4} form a gyrator and $C_2/(g_{m3}g_{m4})$ represents an inductor. Thus, the circuit can also be understood as arising from an active simulation of a passive RCL $[1/g_{m2} - C_1 - C_2/(g_{m3}g_{m4})]$ resonator. An attractive feature of this circuit is that all internal nodes are loaded by circuit capacitors so that parasitic capacitors can be absorbed and do not increase the transfer-function order with additional poles and zeros*

the two node equations at nodes 2 and 3 yield the bandpass transfer function

$$\frac{V_2}{V_1} = s\frac{g_m}{C_1}\left(1 + \frac{G_{p2}}{sC_2}\right) \Big/ \left[s^2 + s\left(\frac{G_{p1}}{C_1} + \frac{G_{p2}}{C_2} + \frac{g_m}{C_1}\right) + \frac{g_m^2 + G_{p2}(g_m + G_{p2})}{C_1C_2}\right] \tag{7.9}$$

In the ideal case with no parasitics ($C_{in} = 0$, $g_{out} = 0$), Eqn. (7.9) becomes

$$\frac{V_2}{V_1} \approx \frac{sg_m/\hat{C}_1}{s^2 + sg_m/\hat{C}_1 + g_m^2/(\hat{C}_1\hat{C}_2)} \tag{7.10}$$

It shows that the nominal (ideal) midband gain H_0, quality factor Q_0 and centre-frequency ω_0 are given by

$$H_0 = 1, \quad Q_0 = \omega_0\hat{C}_1/g_m = \sqrt{\hat{C}_1/\hat{C}_2} \quad \text{and} \quad \omega_0 = g_m/\sqrt{\hat{C}_1\hat{C}_2} \tag{7.11}$$

Similarly, the lowpass function equals

$$\frac{V_3}{V_1} = -\frac{g_m^2}{C_1C_2} \Bigg/ \left[s^2 + s\left(\frac{G_{p1}}{C_1} + \frac{G_{p2}}{C_2} + \frac{g_m}{C_1}\right) + \frac{g_m^2 + G_{p2}(g_m + G_{p2})}{C_1C_2} \right] \Bigg|_{G_p, C_p = 0}$$

$$\approx -\frac{g_m^2/(\hat{C}_1\hat{C}_2)}{s^2 + sg_m/\hat{C}_1 + g_m^2/(\hat{C}_1\hat{C}_2)} \quad (7.12)$$

Several observations are in order. Although the bandpass function acquired a parasitic negative real zero at $-G_{p2}/C_2$, similar to the earlier example, Eqn. (7.5), notice that, as predicted, the second-order functions retained their degrees in spite of the numerous parasitics. We pointed out earlier that this is an important property of any filter topology to be used for full integration because parasitic poles and zeros tend to destroy the desired behaviour, but are almost impossible to remove by automatic tuning. On the other hand, as we saw, changes in component values due to parasitics, such as in C_1 and C_2 in Eqn. (7.9), can readily be corrected by tuning. We notice that the centre frequency has shifted by a small amount and, especially, that the quality factor is decreased because of the losses and parasitic capacitors. Since $g_m \gg G_p$, the frequency shift is normally negligible but will be corrected in the tuning schemes explained below. Looking more carefully at Q, the fact that transconductors have a small parasitic phase lag, $-\omega\tau$, must be taken into account. Modelling it as a right half-plane zero, $g_m = g_{m0}\exp(-s\tau) \approx g_{m0}(1 - s\tau)$, an expression for the realised quality factor, Q_R, can be derived as

$$Q_R \approx \frac{Q_0}{1 + Q_0\left[G_{p1}/(\omega_0C_1) + G_{p2}/(\omega_0C_2)\right] - 2Q_0\omega_0\tau} \quad (7.13)$$

This equation verifies the well known fact that quality factors, Q, depend quite critically on parasitic losses and phase shifts. In particular, Q is reduced by losses (and leading, $\tau < 0$, phase errors) and is enhanced by the transconductors' lagging ($\tau > 0$) excess phase shift or delay. The effect is quite strong[10] and, indeed, becomes critical for large values of Q because both contributions are multiplied by Q_0. We also point out that the maximum value of Q that a given technology can realise is obtained from Eqn. (7.13) by setting $R = 1/g_{m2} = \infty$, i.e. $Q_0 = \infty$. It is determined by the transconductors' output conductance g_{out} and excess phase $\omega_0\tau$ at the pole frequency:

$$Q_{R,max} \approx \frac{1}{3g_{out}/(\omega_0C_1) + g_{out}/(\omega_0C_2) - 2\omega_0\tau} \quad (7.14)$$

Although both excess phase and losses are always present in any real filter, their effects on Q cannot be assumed to cancel completely since they are caused by different parasitic phenomena. Consequently, as stated before, the values of Q are not

[10] In addition to the evident change of Q in a second-order section as shown in Eqn. (7.13), the parasitic effects in a transconductor, modelled for simplicity as a right half-plane zero, result in an overall change of the transfer function. The passband gain rises toward the upper passband edge and, in a bandpass filter, falls toward the lower passband corner. The effect is that the nominally flat passband is skewed [14, 16].

predictable with any accuracy at the design stage and some means of automatic tuning is called for, in particular when large values of Q are prescribed.

For the circuit in Figure 7.4, a simple relationship exists between gain and Q that makes it easy to measure Q. Neglecting the distant parasitic zero at $-G_{p2}/C_2$, but including the effect of the phase lag $-\omega_0\tau$ it follows from Eqn. (7.9) with Eqns. (7.11) and (7.13) that the realised gain at $\omega = \omega_0$ equals

$$\begin{aligned} H_R(j\omega_0) &\approx \frac{g_m/C_1}{(g_m/C_1) + (G_{p1}/C_1) + (G_{p2}/C_2) - 2\omega_0^2\tau} \\ &= \frac{1}{1 + Q_0\left[G_{p1}/(\omega_0 C_1) + G_{p2}/(\omega_0 C_2)\right] - 2Q_0\omega_0\tau} = \frac{Q_R}{Q_0} \end{aligned} \tag{7.15}$$

that is, the realised gain at the centre frequency is directly proportional to the realised quality factor, Q_R. Thus, when H_R is adjusted in a tuning scheme to its ideal value, unity in this case, the realised value of Q will equal the design value, $Q_R = Q_0$. This easy-to-perform magnitude detection is used almost universally in practical Q-control loops. Refer also to Section 7.6.

Understanding the causes of Q errors makes clear how to approach tuning. Typically, quality factors are adjusted by employing phase correction to counter the effect of $\omega_0\tau$ and/or loss compensation to eliminate the limits imposed by g_{out}. One adjustment is normally adequate to make the needed changes. They are implemented most easily by placing a negative resistor, $-R_N$, in parallel with the transconductor output [15, 27, 28]. $-R_N$ reduces the effective output conductance of the transconductor, g_m, and affects the phase of the integrator built with g_m. As the following equations show, both the DC gain of the integrator transfer function, T_I, and its phase, φ_I are changed simultaneously by varying $R_N = 1/G_N$,

$$T_I = \frac{V_o}{V_i} = \frac{g_m}{sC + g_{out} - G_N}, \qquad \varphi_I = -\tan^{-1}\frac{\omega C}{g_{out} - G_N} \tag{7.16}$$

Evidently, the DC gain can become infinite and the phase can be adjusted to $-90°$. Leading or lagging excess phase can be realised to decrease or increase Q, as is evident from Eqn. (7.13). Observe that the adjustment becomes very sensitive when $g_{out} \approx G_N$.

7.5 Principle of tuning frequency parameters

The universally adopted method of tuning frequency parameters compares a filter's performance to an accurate reference signal that sets either a frequency or a time constant. The most widely used method by far performs the comparison by a phase detector that looks at the reference input and the output of a master lowpass filter, such as the structure in Figure 7.4 with output at V_3, see Eqn. (7.12). The reference frequency should be close to the filter's operating frequency, such as the passband corner. The measurement becomes very prone to errors when the reference frequency is not near the frequency of operation: phase is a nonlinear function of frequency and is not easy to measure accurately; any measurement errors in the phase detector will result directly in frequency-tuning errors. Further, should Q-tuning be performed by phase correction, large tuning errors will result if excess phase is not measured with sufficient accuracy.

The output of the phase detector is a DC voltage that is applied as a bias to the transconductors of the master until the phase is exactly 90°. Evidently, then, from Eqn. (7.12), the pole frequency, $\omega_0 = g_m/\sqrt{C_1C_2}$, is equal to the applied reference frequency ω_R, and the filter is tuned. The phase detector can be an analogue multiplier or the signals can be converted with hard limiters to square waves and the comparison can be accomplished digitally by an EXOR or EXNOR gate. This latter approach avoids some dependence on gain and amplitude information [29] and, as a rule, results in simpler circuitry. The method is referred to as *frequency-locked loop* (FLL) and is illustrated in Figure 7.5(a). The figure assumed a master-slave, *on-line*, tuning system; if *off-line* tuning is permitted, master and slave are the same filter and, as shown in Figure 7.3, the reference signal is applied to the filter when the main signal is off.

For the approach to work as sketched in Figure 7.5(a), a sine wave[11] must be available so that clean phase information can be extracted. If a sine wave with accurate frequency is not available on-chip or in the system, it can be generated by a voltage-controlled oscillator (VCO). [12] In that case, the VCO is used directly as the master, constructed such that the oscillation frequency depends on the same components as the filter's pole frequency. Figure 7.5(b) illustrates this *phase-locked loop* (PLL) method using a VCO. The figure also suggests a method for constructing the VCO that may result in better tracking with the to-be-tuned filter. Evidently, the two methods are very similar in approach and performance (see also the discussion in Reference 20). They both depend on a comparison of the phases along two different signal paths, and careful attention must be paid to account for any parasitic phase shifts that affect the comparison. Any parasitic phase errors in the two signal paths, and also any offset in the phase detector, result in frequency tuning errors [20, 30]. Naturally, this problem

[11] Square or triangular waves, or other periodic signals, may be used if the master is a bandpass filter of fairly high Q so that all interfering harmonics are eliminated. Harmonics in the phase comparison lead to systematic tuning errors [20, 30].

[12] The design of the VCO is not trivial. The ideal $j\omega$-axis pole location ('infinite quality factor') depends critically on parasitics and the oscillator contains a nonlinear element to maintain the oscillation with finite amplitude. Neither of these effects is present in the to-be-tuned filter and are likely to result in systematic frequency errors. As indicated in Figure 7.5(b), an oscillator that avoids the mentioned shortcomings is a positive feedback loop, constructed from a high-Q bandpass *filter* that can duplicate the topology of the to-be-tuned filter, and a hard limiter to generate a square wave [30, 31]. The high-Q, high-gain bandpass filter extracts the fundamental component of the square wave so that the circuit oscillates at the pole frequency of the bandpass. Thus, the oscillation frequency is determined only by the bandpass filter and is independent of the nonlinearity (the limiter). No oscillations occur at any other frequency because the corresponding loop gains are less than unity. A clean sine wave, v_o, and a square wave of the same frequency (controlled by the reference clock v_R) are available in this design. If the achievable matching between the high-Q master bandpass filter and the slave filter(s) is found not to be good enough, the stable oscillator sine wave output, v_o, can be used as a reference signal in an FLL as in Figure 7.5(a), at the cost of additional circuitry.

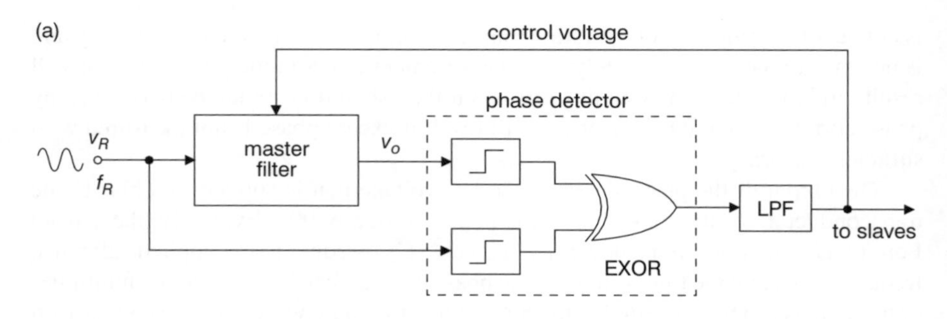

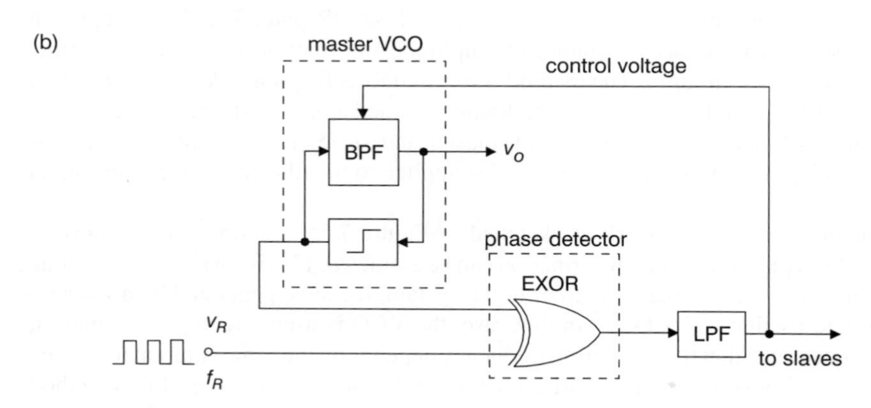

Figure 7.5 *Frequency-control loops. (a) A frequency-locked loop (FLL) using a master lowpass filter. Input reference signal, v_R, and output, v_o, are converted to square waves and processed by an EXOR gate. This phase detector compares v_R and v_o, and adjusts the filter, after cleaning the signal with a lowpass filter (LPF) until the phase shift equals 90°. The pole frequency ω_0 is then tuned to ω_R, and the resulting tuning voltage is sent to the slave filter(s). Any parasitic phase error results in a frequency tuning error, as does a phase detector offset. Also, harmonic distortion in the master filter causes systematic frequency tuning errors [20, 30]. (b) A phase-locked loop (PLL) using a master oscillator (VCO). The oscillator frequency is compared to an accurate system clock and adjusted until the difference is zero (or small enough to be acceptable); the control voltage is then sent to the slave(s). The figure suggests a method for building the oscillator out of a high-Q bandpass filter and a limiter [30, 31]. Better matching between the VCO and the to-be-controlled filter may thereby be obtained*

increases in severity with rising operating frequency and may force the designer to resort to magnitude-locking rather than phase-locking techniques (see below). But, with careful design and layout, the filter's frequency parameter can be tuned to within a fraction of a percent.

The circuit arrangements in Figure 7.5 are possible embodiments of the f-control block in Figure 7.2 when phase detectors are used and frequency-domain information is to be extracted from the master block. If phase information is deemed too inaccurate or hard to measure, peak or envelope detectors (see References 4, 5 and 28), can be used to compare the magnitudes of input and output signals of the master. The method discussed in Reference 32, Figure 7.6(b), illustrates the approach. Analogous

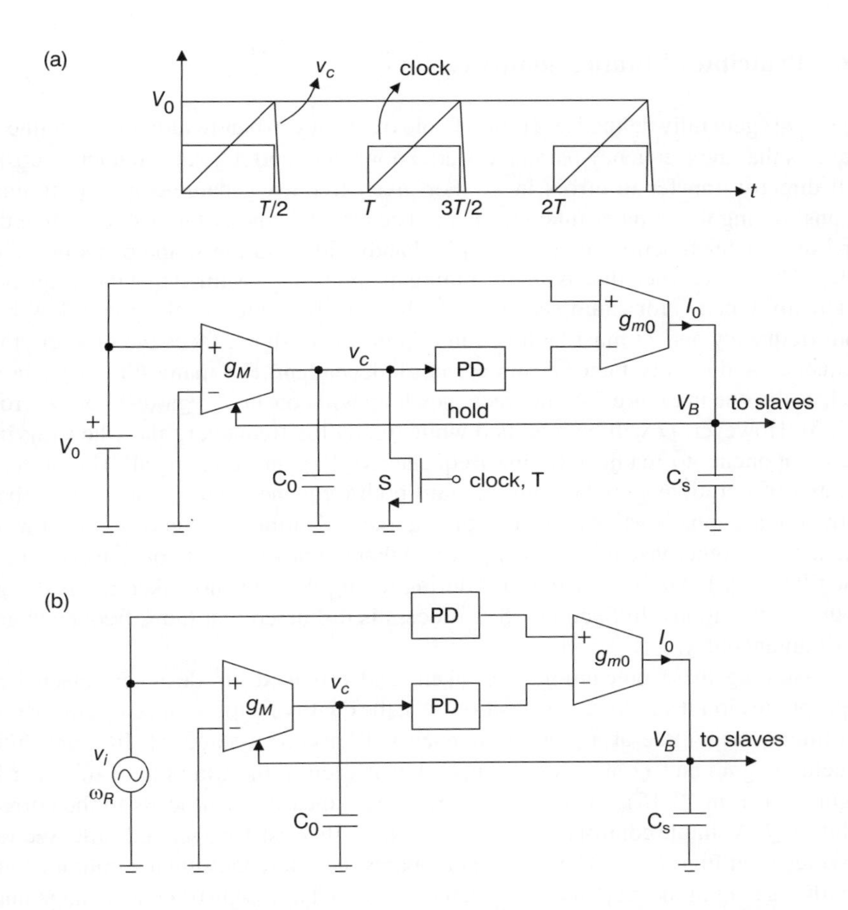

Figure 7.6 *The tuning systems proposed (a) in Reference 8 to equate the integrator time constant to an accurate clock, $g_M/C_0 = T/2$, and (b) in Reference 32 to equate the reciprocal integrator time constant to a reference frequency $g_M/C_0 = \omega_R$. This circuit can operate at very high frequencies; the limit is set by the peak detectors, PD, that do not need to be accurate as long as they are matched. As shown, in both schemes the charge delivered by the current I_0 is stored on C_S and the resulting bias voltage V_B controls the master g_M and all slave transconductors. The value of g_{m0} is not important as long as its DC gain is sufficiently high*

schemes are employed when time-domain information is used for tuning, except that the phase comparator is replaced by suitable digital or analogue electronics that allows the pertinent error information to be deduced. Typically, using a closed-loop control scheme quite similar to the one sketched in Figure 7.5, the frequency parameters are fixed by setting a C/g_m ratio equal to the time-period of a clock signal. A specific approach is discussed in Section 7.7 (see Figure 7.6(a)).

7.6 Principle of tuning quality factors

Since Q is generally defined as a ratio of pole frequency to bandwidth, to determine Q requires that the frequency parameters are known and correct. Any frequency errors will directly transfer to errors in Q. For given frequency parameters, 'Q-tuning' means 'tuning the transfer function shape' because the quality factors determine the type of transfer function, its gain, ripple, bandwidth, and the shape of its phase or delay. Therefore, when discussing Q-tuning methods it is assumed that the frequency is already tuned – for example, by the techniques presented in Section 7.5. When both frequency and Q must be tuned in a filter, the designer faces the problem that frequency and quality-factor errors are not independent. For many filter structures, such as the one in Figure 7.4, the frequency loop will converge regardless of Q errors [4, 33]. However, Q will be modified while tuning the frequency, and when varying the components to readjust Q, the frequency will again be changed. Therefore, a number of iterations may be required until both frequency and Q converge to their correct or acceptable values. The two tuning loops identified in Figure 7.2 must work interactively, otherwise it is entirely possible that the process will not converge at all. The filter may break into oscillations during tuning if care is not taken in the design of the control loops. In Section 7.8, a process is discussed that tunes frequency and Q simultaneously.

Assuming these precautions are taken, and provided the designer selected an appropriate circuit structure (see Section 7.4), the quality factor is tuned by comparing the filter's magnitude at a given frequency with a known standard. In many filter structures, gain and Q are proportional, as was seen in the example of the filter in Figure 7.4, Eqn. (7.15). Therefore, a precise gain measurement leads to the correct value of Q. A simple control method to accomplish this task for a second-order section is sketched in Figure 7.7. The scheme compares the reference signal amplitude with the filter's magnitude response to the reference signal and adjusts a bias voltage until Q is correct.

It is evident from Eqns. (7.13) and (7.14) that the sensitivity of Q adjustments tends to be very high. This is true especially for large values of Q, which are established by loss compensation, i.e. through a difference effect.[13] For example, we may change the polarity of the transconductor g_{m2} in the circuit in Figure 7.4 to realise a negative

[13] As in any parameter established by a difference effect (see, for example, Eqn. (7.16)), the sensitivity is proportional to Q.

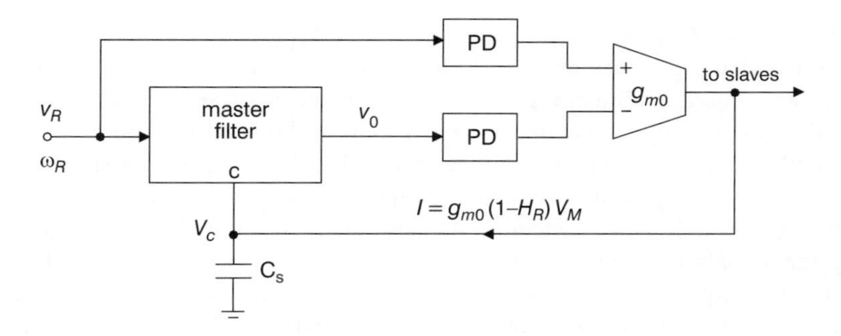

Figure 7.7 *A Q-control scheme. The reference signal $v_R(t) = V_M \sin \omega_R t$ is applied to the master filter with a bandpass output. Let the nominal midband gain of the master filter be $H_0 = 1$ as in Eqn. (7.10). If $H_R = K \neq 1$, a DC amplifier of gain 1/K can be inserted in series with the master to compensate for the effect. Two identical (matched) peak detectors (PD) then find the peak values of the reference signal itself and of the master's response, v_o. The peak values are compared in a transconductor, g_{m0}, that sends the current $I = g_{m0}(1 - H_R)V_M$ to the storage capacitor C_S, where $H_R = Q_R/Q_0$ per Eqn. (7.15). As long as Q is too small, that is, $H_R < 1$, I is positive, C_S is charged and V_C increases. If $H_R > 1$ because Q is too large, I is negative and C_S is discharged for a reduced value of control voltage V_C. V_C is applied to the appropriate control terminal c and the value of H_R is changed (increased or decreased) until the loop stabilises when $H_R = 1$ and the realised value of Q, Q_R, equals the design value Q_0*

resistor $-1/g_{m2}$. The quality factor then becomes, instead of Eqn. (7.13),

$$Q_R \approx \frac{1}{3g_o/(\omega_0 C_1) + g_o/(\omega_0 C_2) - g_{m2}/(\omega_0 C_1) - 2\omega_0 \tau} \tag{7.17}$$

and is adjusted by varying $g_{m2}(V_C)$. Evidently, Q can become infinite and the final fine-tuning must be handled very carefully to avoid gross tuning errors or even oscillations.

7.7 Various tuning methods

In Sections 7.5 and 7.6, the principles of the frequency- and Q-tuning methods were presented. Details depend necessarily on the specific realisations of filter and control circuitry. They are, as a rule, quite involved and cannot be discussed in the space of this short article. To provide a flavour of the range of approaches taken, in this section a number of tuning methods are sketched out that have been proposed in the literature and tried in practice. The reader should consult the cited references for further details of these methods, which can then be adapted to the specific applications.

A successful CMOS chip for a seventh-order 4 MHz elliptic antialiasing lowpass filter for digital video was reported in Reference 29. The circuit uses the 'on-line' master-slave configuration of Figure 7.2 for frequency *and* Q control. The frequency-control block employs a frequency-locked loop as depicted in Figure 7.5(a). The master filter, with transconductors identical to those of the slave, is shown in Figure 7.4. Its lowpass output is compared with the input reference signal of frequency ω_R by an EXNOR phase detector. As described earlier, feedback is applied to bring the signals into quadrature so that the pole frequency equals ω_R. The lowpass corner frequency is controlled to a mean of 4.36 MHz with 0.11 MHz standard deviation. The frequency was found to shift by ± 0.08 MHz over the temperature range $0° \leq T \leq 70°$. The Q-control scheme is that of Figure 7.7 with the peak values of the bandpass output of the master filter (Figure 7.4) and of the input reference signal compared to establish the realised gain. Any deviations of Q are corrected by adjusting the integrator phase, $\omega_{0\tau}$, as suggested in Eqn. (7.15). Q control held the deviations of the passband ripple to 0.26 dB (out of a mean of 1.24 dB) with ± 0.02 dB variation in $0° \leq T \leq 70°$.

A very simple tuning system for an FM audio IC lowpass filter for a VHS video recorder was presented in Reference 8. The authors use the control loop in Figure 7.6(a) to equate the time constant of a master g_M-C_0 integrator to a given clock period. As is shown in the figure, the master transconductor g_M is driven by a constant voltage V_0 and charges the capacitor C_0 with the constant current $g_M V_0$. The switch S, driven by a reference clock with period T, discharges the capacitor every $T/2$ seconds and the capacitor voltage at that time, $v_c(T/2)$, is held and compared by the transconductance g_{m0} with the input voltage V_0. The output current,

$$I_o = g_{m0}(V_0 - V_c) = g_{m0}\left(1 - \frac{g_M}{C_0}\frac{T}{2}\right)V_0 \qquad (7.18)$$

sets a bias voltage or current that tunes g_M until the difference $V_0 - V_c$ is zero. Then, the integrator time constant is fixed at one half the clock period: $C_0/g_M = T/2$.

A similar scheme that uses frequency-domain instead of time-domain information was proposed in Reference 32. Analogous to the system in Figure 7.7 and shown in Figure 7.6(b), a sine wave with accurate reference frequency ω_R is applied to a master g_M-C_0 integrator. The peak values of the output voltage $V_c = V_i g_M/(j\omega_R C_0)$ and the input voltage V_i are obtained from two matched peak detectors, PD, and are compared by the transconductance g_{m0}. The output current

$$I_o = g_{m0}(V_i - V_c) = g_{m0}[1 - g_M/(\omega_R C_0)]V_i \qquad (7.19)$$

generates a bias voltage or current that adjusts g_M until I_o is zero. Then, the integrator unity-gain frequency is determined by ω_R: $g_M/C_0 = \omega_R$. The simple relationship assumes that $\omega_R C_0$ is much larger than any integrator losses, which are neglected.

In these control loops, the time constant is defined only by the accurate reference, the clock T, or the frequency ω_R, respectively, but is independent of the input voltage level. In both methods, the generated bias is sent to all filter integrators. These must be carefully matched to the master integrator, and as in all tuning schemes, care must be taken that the clock or reference frequency and its harmonics do not interfere with the main filter signals. Both schemes rely on a comparison of peak values and avoid using phase information, an advantage for applications at high frequencies. On the other hand, designing slave filter integrators that match the master, including the effects of any parasitics, becomes more difficult as frequencies increase.

The master-slave control scheme of Figure 7.5(a) is generally used to set a frequency parameter ω_0 as $\omega_0 = g_m(V_B)/C$ or $\omega_0 = 1/[R(V_B)C]$, where V_B is a bias-control voltage. Evidently, the method simply controls a time constant so that the same scheme can also be used to set an accurate time delay. Reference 34 discusses the implementation of a well controlled analogue adaptive delay line. The circuit uses a cascade of 26 identical first-order delay slaves and one first-order master. All of them implement the transfer function H with the phase ϕ:

$$H(s) = \frac{1 - sCR}{1 + sCR}, \quad \phi(\omega) = -2\tan^{-1}(\omega RC) \tag{7.20}$$

The group delay is

$$D(\omega) = \left.\frac{2RC}{1 + (\omega RC)^2}\right|_{(\omega RC)^2 \ll 1} \approx 2RC \tag{7.21}$$

and is seen to be approximately constant for small frequencies, $\omega \ll 1/(RC)$. The principle of the operation is as follows. The resistors are realised as MOS devices, tunable by a bias voltage V_B, $R = 1/[K(V_B - V_T)]$, where K is a transconductance parameter and V_T is the threshold voltage. As is shown in Figure 7.8, the control circuit consists of a master allpass filter, a multiplier and a lowpass filter (or integrator) to clean the control voltage V_B. A reference signal v_{ref} with frequency ω_{ref} is applied to

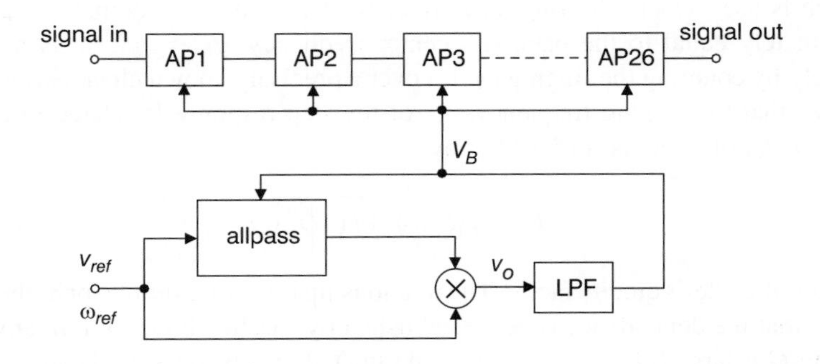

Figure 7.8 *Phase-locked-loop master-slave tuning scheme for a 26-stage analogue CMOS delay line [34]. The control scheme sets the pole frequency of the allpass master filter to the reference frequency and thereby generates a programmable delay equal to 52/ω_{ref}. Further flexibility is attained by using the delay line outputs from intermediate points (taps) in the 26 stages*

the master allpass whose output signal is shifted by the phase ϕ of Eqn. (7.21). This signal is multiplied with v_{ref} to generate the output $v_o(t) = A\cos\phi - A\cos(2\omega_{ref}t + \phi)$, where A is a function of the input amplitude and the multiplier gain. The lowpass filter removes the component at $2\omega_{ref}$ and retains the DC component

$$V_o = A\cos\left[2\tan^{-1}\frac{\omega_{ref}C/K}{V_B - V_T}\right] \qquad (7.22)$$

Assuming the feedback connections have the correct polarity and neglecting offset voltages in the multiplier and integrator, the control system adjusts V_B until the allpass filter has a phase shift of 90°. This happens at the voltage $V_B = V_T + \omega_{ref}C/K$, i.e. when

$$\frac{\omega_{ref}C}{K(V_B - V_T)} = \omega_{ref}CR = 1 \qquad (7.23)$$

so that the allpass pole frequency is tuned to ω_{ref} or, in other words, the delay of a stage is $D = 2/\omega_{ref}$ for a total delay of $D_{total} = 26 \times 2/\omega_{ref}$ if all 26 stages are used.

Applying a reference sine wave to a bandpass filter and judging whether the pole frequency is too high or too low normally requires knowledge of phase behaviour, which is affected by various parasitics and may be inaccurate at high frequencies. An input signal with a rich spectrum, rather than a single frequency, provides more reliable information. Thus, to tune the centre frequency of a 16th-order, 450 kHz, g_m-C bandpass filter for a cell phone, Yamazaki *et al.* [10] used a time-domain approach and applied a step function to the filter input. Because of a very tight power budget, concerns about process uniformity across the chip, and because the time-division multiple-access application makes idle times available for tuning, the authors opted for the direct tuning method (Figure 7.3) shown in Figure 7.9(a). Critical for the system's operation is the precise centre frequency, f_0, of the bandpass filter, which must be accurate to at least 2.2 kHz out of 450 kHz ($< 0.5\%$). A time-domain approach is used where the ringing frequency, f_r, of the time-domain response is approximately equal to the bandpass centre frequency, $f_r \approx f_0$. f_0 is measured accurately by counting the ringing cycles over a precisely known clock period. It can be shown that the ringing frequency, f_r, of the step response is related to the pole frequency, f_0, of a second-order filter by

$$f_r = f_0\sqrt{1 + 1/(4Q^2)} \qquad (7.24)$$

where Q is the filter's quality factor. The relationship is not that simple for higher-order filters so that the dependence is best established by careful simulation. In any event, assuming Q is large ($Q = 9$ in the realised filter), f_r can be measured and is (nearly) equal to the nominal pole frequency when the filter is tuned correctly. To measure the observed ringing frequency, the authors applied the step response to a comparator that converts the ringing response to a square wave (Figure 7.9(b)). Its frequency can readily be counted and compared with a target frequency, f_t, generated by an accurate system clock. A feedback signal is then applied to the filter transconductors to adjust the time constants digitally until the difference $f_r - f_t$ is reduced to a value small

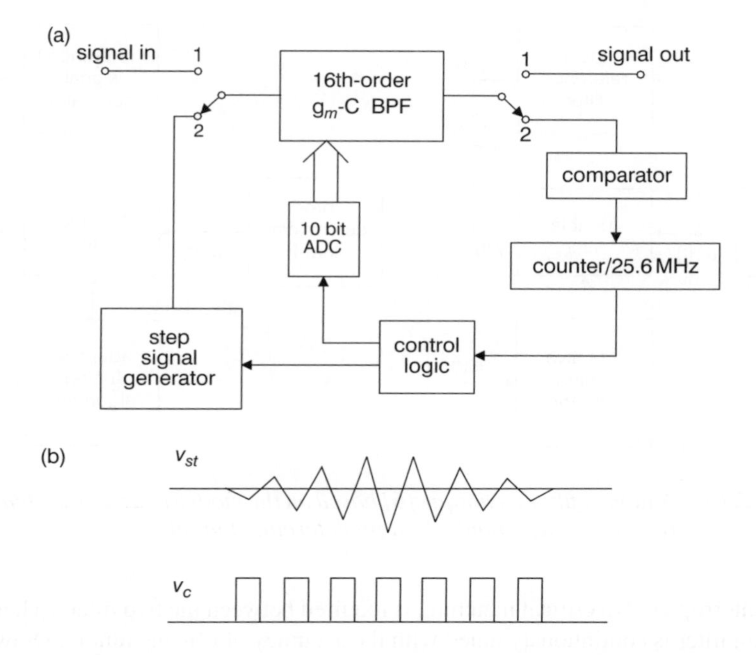

Figure 7.9 *(a) System block diagram for a direct-tuning system for a time-division multiple-access cell-phone system. The switches are in position 1 for signal processing and in position 2 (shown) for tuning. The control-logic block manages the mode control, all switches, and generates the digital word for tuning the transconductors (10 bits; 2 bit for coarse + 8 bit for fine tuning). Tuning accuracy is based on the comparator output v_c, relative to a 25.6 MHz timing clock. (b) The square wave signal v_c is generated from the step response, v_{st}, of the bandpass filter*

enough to be acceptable. Paying careful attention to all error sources and optimising the design, the frequency error of the experimental chip was found to be only 0.1 per cent.

An adaptive filter technique based on the model-matching configuration for continuous-time filters is presented in Reference 6, employing the model-matching system shown in Figure 7.10(a). Applying a spectrally rich input signal, $u_n(t)$, to the filter, an adaptive algorithm (LMS, for this case) is used to adjust the coefficients of the tunable filter so as to minimise the error signal, $e(t)$, which is the difference between the tunable filter output, $y(t)$, and the ideal reference filter output, $\delta_n(t)$. Since the main filter continuously provides the output, the white noise source in Figure 7.10(a) is replaced by the signal to be processed in the actual implementation. Realisation of the ideal reference filter is achieved by employing a second filter that is tuned by an adaptive loop, using a predetermined set of inputs, and the corresponding pre-calculated output, as shown in Figure 7.10(b). The reference filter is tuned periodically using the self-tuning technique [22]; however, the tunable filter output is

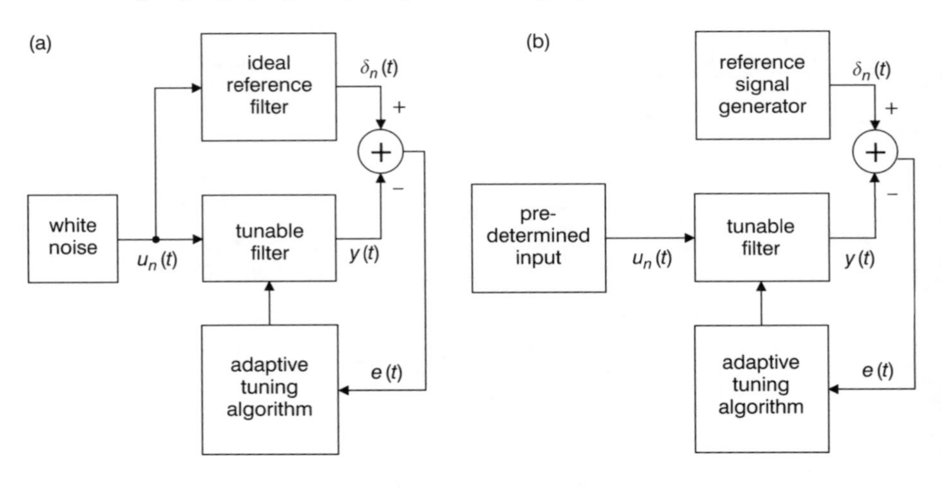

Figure 7.10 *Adaptive filter technique [6] based on the model-matching system using (a) white noise input; (b) a predetermined input*

never interrupted. No critical matching is required between the two filters. Therefore, the main filter is continuously tuned with the accuracy of off-line tuning. Drawbacks of this method are higher circuit complexity and power consumption, and the fact that the input signal is assumed to be spectrally rich. Furthermore, the scheme requires two low-resolution analogue-to-digital converters operating at a frequency greater than the passband of the tunable filter.

In Reference 35, an accurate Q-tuning method for high-Q filters in the IF band is proposed. The method improves on the magnitude-locked loop [5] by utilising the continuous-time LMS algorithm. A block diagram of the proposed adaptive Q-tuning technique is shown in Figure 7.11. The modified LMS equation for this scheme is

$$\dot{V}_Q = \mu(V_{in} - V_{bp})V_{bp} \qquad (7.25)$$

where V_Q is the Q-tuning signal, V_{in} is the input signal, V_{bp} is the bandpass output and μ is a constant. When the filter is tuned, the error signal $[v_{in}(t) - v_{bp}(t)]$ becomes zero. Experimental results showed a 0.75% Q-error for a desired Q-value of 20. One disadvantage of this method is that low-offset multipliers at the frequency of operation are required. Realisation of such blocks at very high frequencies may be challenging.

7.8 Method for tuning high-frequency high-Q bandpass filters

An automatic mixed-mode tuning scheme [21, 28] suitable for analogue filters for very high operating frequencies and large quality factors will be described in some detail in the following. Indeed, a high quality factor is an essential requirement for the method, and increasing Q-values result in lower tuning errors. The tuning principle is based on envelope detection. To obtain centre-frequency and quality-factor information,

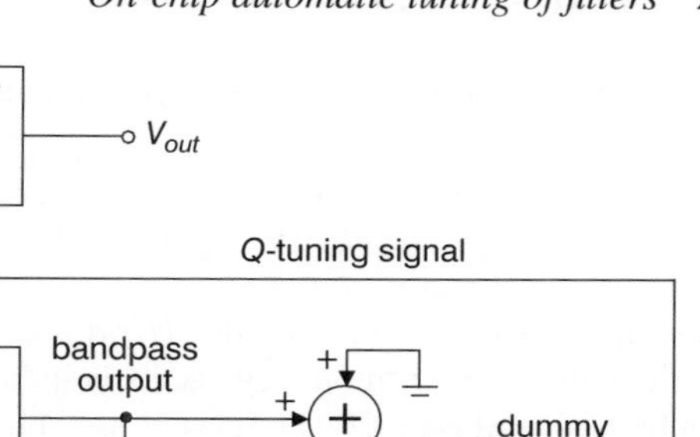

Figure 7.11 Tuning system of Reference 35 using a continuous-time LMS algorithm

the master filter's frequency-control input is swept periodically over its valid range while a fixed reference signal is applied to the filter input. The envelope at the master filter's output is evaluated by frequency- and Q-tuning circuits, and the correct tuning voltages are generated and applied to the slave filter(s).

One advantage of the proposed tuning method is its simplicity. The tuning circuitry is largely digital, with the exception of the envelope detector and the filter itself, which deal with the high-frequency signals. Thus, this method can have a higher frequency limit than other existing techniques, and can be used for very-high-frequency applications, such as in mobile communications. Once the envelope detector is implemented appropriately, the tuning circuit operates on DC voltages and low-frequency logic signals. Another aspect of the proposed scheme is its accuracy and robustness. The operation is insensitive to process tolerance, relying only on matching of components, which even in a standard CMOS process is very good. Unlike in the basic magnitude-locked loop approach (see, for example, References 4 and 5), frequency-tuning errors do not result in Q errors. Frequency- and Q-tuning loops use the filter's output magnitude information rather than the phase.

7.8.1 Time-domain transfer functions

Transfer functions of linear systems, including filters, are normally given in the frequency domain. With $D(s) = s^2 + \omega_x s + \omega_0^2$, where ω_0 is the pole frequency and the parameter ω_x determines the quality factor as $Q = \omega_0/\omega_x$, typical second-order functions for bandpass, lowpass and highpass filters can be expressed as

$$H_{BP}(s) = \omega_0 s \big/ D(s), \quad H_{LP}(s) = \omega_0^2 \big/ D(s), \quad H_{HP}(s) = s^2 \big/ D(s) \qquad (7.26)$$

At a fixed reference frequency $s = j\omega_r$, the magnitude of $H_{BP}(s)$ is

$$\left|H_{BP}(j\omega_r)\right| = \frac{\omega_0\omega_r}{\sqrt{(\omega_0^2 - \omega_r^2)^2 + \omega_x^2\omega_r^2}} = H_{BP}(\omega_0) \tag{7.27}$$

and similarly for $\left|H_{LP}(j\omega_r)\right|$ and $\left|H_{HP}(j\omega_r)\right|$ in Eqn. (7.26). These magnitudes are functions of the pole frequency ω_0, as is indicated by the simplified notation on the right hand side of Eqn. (7.27). If $\omega_0 = \omega_0(t)$ is swept within a certain interval of time, then the output signal envelope can be described by Eqn. (7.27), which we will label the time-domain transfer function. This definition is, of course, not the actual time-domain response; rather, it is a function of time that reflects certain characteristics of the system described by the frequency-domain transfer function.

The filter implementation assumes that ω_0 is controlled by a bias voltage, V_f, such that $\omega_0 = f(V_f)$. V_f is swept and the proposed tuning scheme detects the value of V_f that sets ω_0 to the value for which $H(\omega_0)$ reaches a maximum. The detection process is illustrated in Figure 7.12, which shows the neighbourhood of the peak point of $H(\omega_0)$. Note that for high values of Q, all three types of filters have similar magnitude characteristics near ω_0. After selecting a level KQ, the values V_{f1} and V_{f2} belonging to the corresponding frequencies ω_1 and ω_2 are detected and averaged to approximate V_{fc}. When V_{fc} is applied to the filter, its pole frequency is tuned to $\omega_{0c} = f(V_{fc})$.

Systematic frequency-tuning errors arise due to the following factors: first, $\omega_{0c} = \omega_0(V_{fc}) \neq \omega_r$. This error depends on the desired value Q_D of the quality factor,

$$Q_D = \omega_0/\omega_x\big|_{\omega_0=\omega_r} = \omega_r/\omega_x \tag{7.28}$$

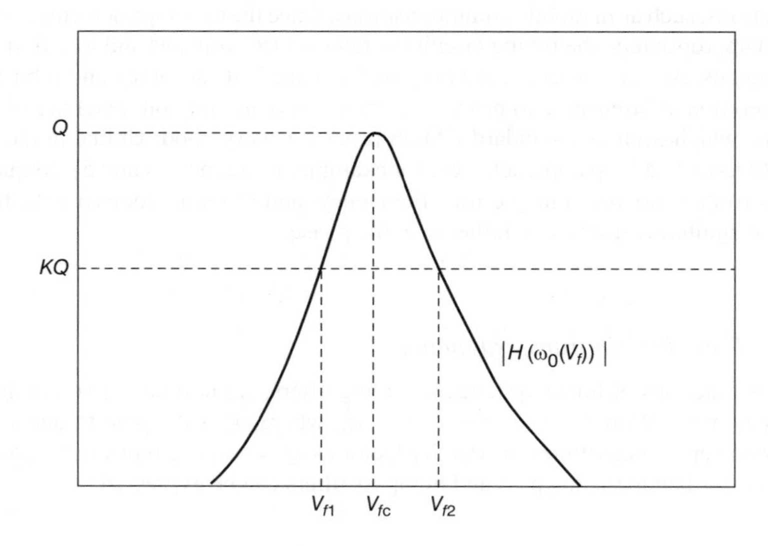

Figure 7.12 Detection of the appropriate tuning voltage

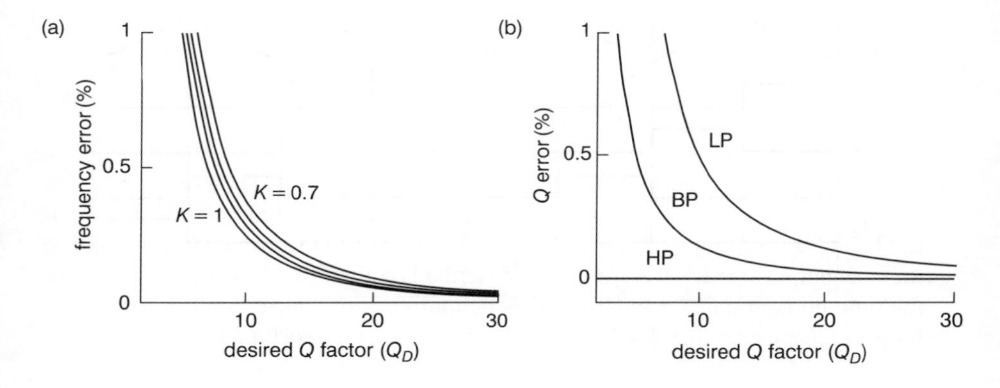

Figure 7.13 *(a) Bandpass frequency error against Q_D for $\omega_0 = k_1 V_f$ and $K = 0.7, 0.8, 0.9, 1$; (b) Q-tuning errors for varying Q_D*

and becomes smaller as Q_D increases. Contributions of the second factor, $V_{fa} = (V_{f1} + V_{f2})/2 \neq V_{fc}$ are caused by the asymmetry of the function $|H(\omega_0)|$ around ω_{0c} and depend on K: the error becomes smaller as K gets closer to 1. The tuning error approaches zero as $K \to 1$ and $Q \to \infty$. Figure 7.13(a) shows the systematic bandpass frequency error as a function of Q_D for several values of K. Similar curves can be obtained for the lowpass and highpass functions [21, 28]. The results show that the worst-case frequency-tuning error is less than 0.5% for $Q > 10$, and less than 0.1% for $Q > 30$. The output level KQ does not affect the error significantly, and the effect becomes negligible at high Q values (less than 0.05% for $Q > 30$). Since the detection of V_{f1} and V_{f2} (refer to Figure 7.12) depends on voltage levels but not on how and when these levels are reached, the result is independent of the time waveform $V_f(t)$, i.e. the sweep function applied at V_f[where $\omega_0 = f(V_f)$] is immaterial and can be any monotonic function of time within the valid range of V_f.

The proposed Q-tuning method sets the peak level of the time-domain transfer function to the desired value, Q_D, of the quality factor. However, as shown in Figure 7.13(b), Q-dependent systematic tuning errors are introduced while tuning lowpass and bandpass biquadratic filters. The worst-case Q-tuning error is less than 0.5% for $Q_D > 10$, and less than 0.1% for $Q > 30$.

Note that the errors discussed so far give only the systematic theoretical limits of the accuracy of the proposed tuning scheme, but they do not include errors arising from the tuning circuit itself, from offsets and other parasitics.

7.8.2 *Frequency tuning*

To detect the control voltage at which the output envelope (the time-domain response) reaches its peak, a reference signal, V_{ref}, at the desired pole frequency ω_r is applied to the master filter and its frequency control voltage, V_{fm}, is swept (cycled) throughout the tuning process as shown in Figure 7.14. At the end of each cycle, the correct

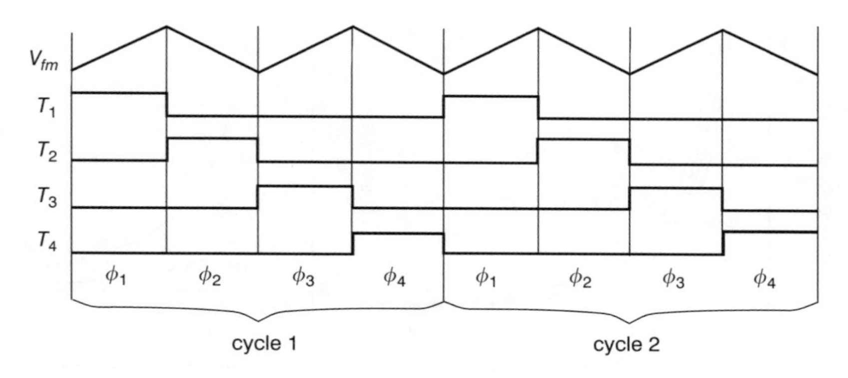

Figure 7.14 Frequency-control voltage, V_{fm} , and clock signals

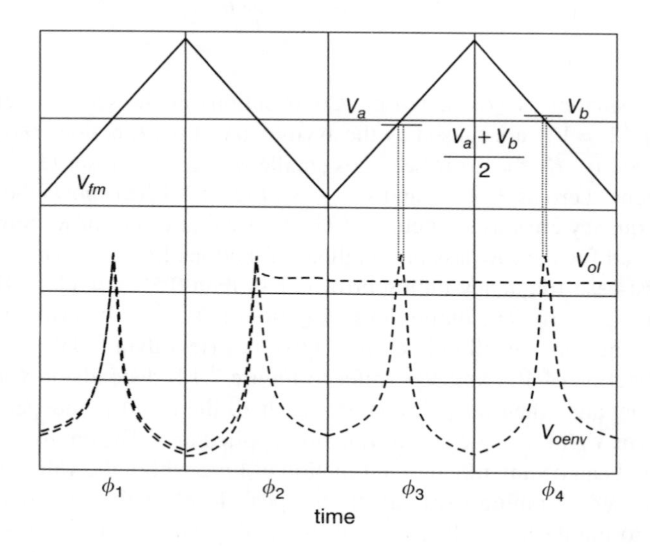

Figure 7.15 Phases of the frequency tuning process

frequency tuning voltage is detected for the current value of Q that is assumed to be constant from ϕ_1 to ϕ_4. Q is updated at the beginning of a new cycle. The frequency tuning phases are sketched in Figure 7.15, where V_{oenv} is the filter output envelope and V_{ol} is the level previously labelled as KQ. Referring back to Figure 7.12,

$$K = \frac{V_{ol} - (V_{oenv})_{min}}{(V_{oenv})_{max} - (V_{oenv})_{min}} \tag{7.29}$$

$$V_a = V_{f1}, \quad V_b = V_{f2} \tag{7.30}$$

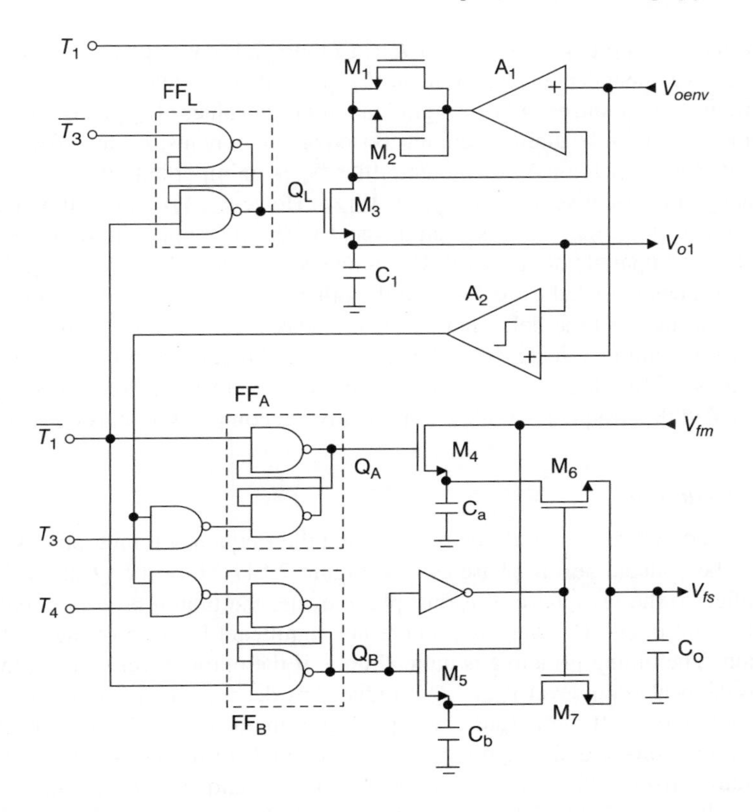

Figure 7.16 Frequency tuning circuit

The predominantly digital frequency tuning circuit is shown in Figure 7.16. The tuning process during $\phi_1 - \phi_4$ is described as follows:

ϕ_1: Q_L, Q_A and Q_B (the outputs of the flip-flops FF_L, FF_A and FF_B, respectively) are set to V_{DD}. Since the op-amp A_1 is in unity-gain configuration (M_1 and M_3 are closed), V_{o1} tracks the filter output envelope V_{oenv}.

ϕ_2: The transistor M_1 is turned off, and the $A_1 - M_2 - M_3 - C_1$ combination operates as a peak detector. At the end of this phase (or just at the beginning of the next phase), Q_L is reset to 0, causing M_3 to turn off. Because of parasitic discharge, the finite gain of A_1, and clock feedthrough from the switch M_3, V_{o1} is set to a voltage level slightly below the peak level of V_{oenv}. C_1 holds this level until the end of phase ϕ_4.

ϕ_3: The output envelope V_{oenv} and V_{o1} are compared in A_2. At the first point in ϕ_3, when they are equal, the output of comparator A_2 goes HIGH, which resets Q_A to 0 and turns M_4 off. The value of V_{fm} at that time, V_a in Figure 7.15, is stored on C_a.

ϕ_4: V_{oenv} and V_{o1} are compared in A_2 and at the first point in ϕ_4 when they are equal, Q_B is reset to 0, M_5 turns off, and the value of V_{fm}, V_b in Figure 7.15, is stored on C_b. After that point, M_6 and M_7 are turned on and the charges of C_a, C_b and C_0 are shared. In the steady state, V_{fs} converges to the correct tuning voltage, $(V_a + V_b)/2$.

The relationship between V_{ol} and K (in Figure 7.12) was given in Eqn. (7.29), which shows a linear dependence of K on V_{ol}. It was demonstrated that the effect of K on the frequency tuning error is not significant for high values of Q. Thus, the value of V_{ol} is not critical for the tuning accuracy. Based on this conclusion, any offset voltage, V_{OFF}, of the comparator A_2 or the amplifier A_1 is unimportant since these offsets only change the effective level of V_{ol} by V_{OFF}. However, V_{OFF} should be less than the difference $(V_{oenv})_{max} - V_{ol}$ so that the peak of the envelope is above the effective value of the compared voltage level, V_{ol}. Since A_1 is used only in detecting the peak of V_{oenv}, it does not need to be fast. On the other hand, A_2 is used as a comparator, which might have a time delay that shifts the peak location of the output envelope and hence the tuning voltage. It has been shown [21] that the tuning error caused by this delay is negligible for any monotonic time waveform V_{fm}. Consequently, it can be concluded that frequency tuning is insensitive to comparator delays and offsets.

7.8.3 *Q*-tuning

As discussed before, Q should be fixed during the frequency tuning process, and it can only be updated during phase ϕ_1 (see Figure 7.14). Since the Q-tuning voltage is modified in the last phase of a four-phase cycle, the tuning process should start in ϕ_2 and end in ϕ_1. The starting point is not significant because of the continuous operation. The tuning principle is quite simple: if the output level V_{oenv} of the filter exceeds the peak input level V_{imax}, Q is reduced by decreasing the Q-control voltage by a small amount. If the output-envelope peak remains below the input peak, then the Q-tuning voltage and Q are increased. Since high-Q-tuning is very sensitive to comparator offset voltages an offset cancellation technique is used. The Q-tuning circuit is shown in Figure 7.17, and the process can be described as follows.

ϕ_2: The offset cancellation circuit goes into the reset mode. Capacitors C_{os} at the inputs of the differential comparator A_1 are connected to the reference input peak level V_{imax}, and the switches M_3 and M_4 are closed. This allows the offset voltage to be stored differentially on the capacitors C_{os}. Since the cancellation is differential, accuracy is limited only by the matching of the capacitors C_{os} and the switches. The delay element D, which consists of two inverters, serves to minimise charge injection effects [36].

ϕ_3: The capacitor C_c is pre-charged to $(V_{qs} + \Delta V)$, where V_{qs} is the current value of the Q-tuning voltage and $\Delta V = I_d R.V_{qs}$ is buffered through the unity-gain amplifier A_3, and the voltage difference ΔV is obtained using the resistors R and the current I_d.

ϕ_4: The reference level V_{imax} (from the input envelope detector) and the filter output envelope, V_{oenv}, are compared in A_1, A_2. If V_{oenv} remains below V_{imax} (Q is less than the desired value), V_P is low and the voltage on capacitor C_c remains at the preset value $(V_{qs} + \Delta V)$. If at some point V_{oenv} exceeds V_{imax} (Q is higher than desired), the output of comparator A_2 goes HIGH, V_P is set to V_{DD}, and the capacitor C_c is discharged to $(V_{qs} - \Delta V)$. Once set, V_P holds its value at V_{DD} until the end of phase ϕ_4 and then is reset to 0 for the next cycle. The clocking scheme in this phase is illustrated in Figure 7.18. Since the comparator output voltage (V_{A2}) during ϕ_3 is not

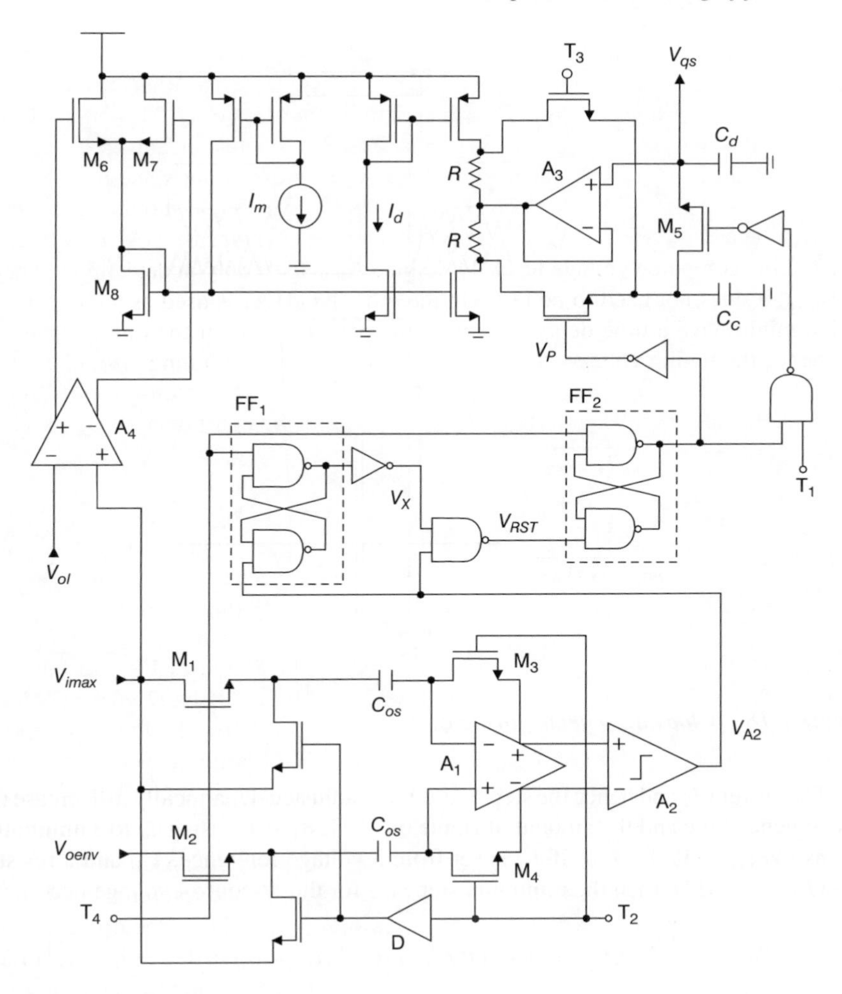

Figure 7.17 Q-tuning circuit

well defined, it can take either LOW or HIGH logic levels. If V_{A2} is high at the end of ϕ_3 due to finite comparator delay (t_C) it will settle to zero after T_4 becomes HIGH. Another flip-flop (FF_1 in Figure 7.17) is used to block the ambiguous comparator output. The inverted output of FF_1, V_X, narrows the phase ϕ_4 so that the actual comparison begins after V_{A2} settles to zero. The reset signal for the second flip-flop is now $V_{RST} = \overline{V_X \cdot V_{A2}}$, which is the correct signal.

ϕ_1: The transistor M_5 is turned on, and the charges of C_c and C_d are shared so as to generate the net change in V_{qs} as

$$\left(V_{qs}\right)_{n+1} - \left(V_{qs}\right)_n = \pm\Delta V \left[C_c/\left(C_c + C_d\right)\right] \tag{7.31}$$

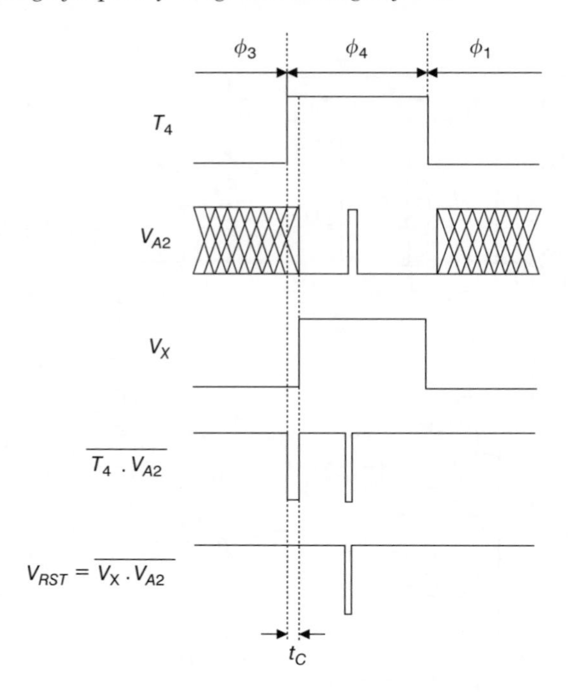

Figure 7.18 Clocking scheme during ϕ_4

The current I_d and hence the step size ΔV are adjusted dynamically to decrease the convergence time and the amount of ripple on V_{qs} [28], decreasing I_d to a minimum, I_m, as $(V_{imax} - V_{o1}) \to 0$. If I_m is set from a voltage reference, V_r, and a resistor R_m ($I_m = V_r/R_m$), then the minimum step size for the Q-control voltage becomes

$$\Delta V_{qs} = \pm V_r \left(R/R_m\right) \left[C_c/\left(C_c + C_d\right)\right] \tag{7.32}$$

which can be controlled accurately. By choosing $C_c \ll C_d$, ΔV_{qs} can be varied by a very small amount. However, there will always be a residual ripple on the voltage V_{qs} since the Q-tuning voltage in each cycle is either increased or decreased. The effect of this is the AM modulation of the signal at the filter output by this ripple. Note that the dynamic curve for I_d sets the step size but the sign of the step (direction of the change) comes from the offset-compensated comparator circuit, A_1 and A_2 in Figure 7.17.

7.8.4 Experimental results

A few experimental results will illustrate the performance. In Figure 7.19, the middle trace shows the filter's time-domain response for a 110 MHz reference signal whereas the bottom trace is the response for a 120 MHz signal. The top trace is the

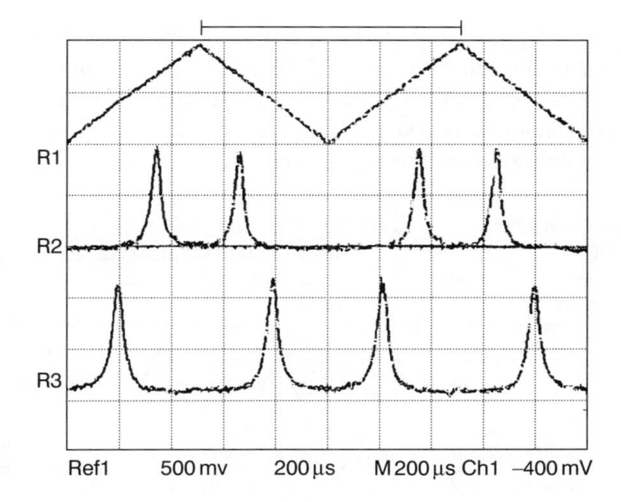

Figure 7.19 *The swept frequency tuning voltage V_{fm} (top trace, 500 mV/div) and the time-domain responses for 110 MHz (middle trace, 50 mV/div) and 120 MHz (bottom trace, 50 mV/div)*

Table 7.1 *Frequency-tuning measurements with fixed V_{qs}*

f_{ref} (MHz)	V_{fs} detected	f_0 (MHz) measured	Q	Frequency error(%)
105	0.816	105.30	32	0.29
110	0.720	110.25	46	0.23
115	0.608	115.25	77	0.22
120	0.474	120.05	83	0.04

swept frequency tuning voltage, V_{fm}, a triangular 1 V, 1 kHz wave, applied to the filter's frequency control input. The convergence speed of the tuning circuit is directly proportional to the frequency of V_{fm}. However, the sweep time must be long enough for the filter's parameters to be set and to converge at the chosen rate of V_{fm}, i.e. the filter's time-domain responsivity to V_{fm} should be taken into consideration when determining the sweep frequency.

Table 7.1 shows the frequency-tuning measurements[14] for fixed V_{qs}, i.e. the Q-tuning circuit is not active and the value of Q varies from 32 to 83. Note that the frequency tuning error is less than 0.3%. Both $\omega_0 = g_m/C$ and $Q = kg_m/g_0$ increase with g_m but Q increases faster due to Q-enhancement caused by the

[14] The circuit was implemented in a standard CMOS process, operating on a single 3 V power supply. Because of a need for external components such as buffers and the resulting matching problems, only measurements of the off-line tuning approach were performed.

frequency-dependent phase shift in the filter's feedback loop. Magnitude plots for these measurements are shown in Figure 7.20(a). The tuning range is limited by the transconductance range, $g_m(V_{fs})$; however, the extreme (minimum and maximum) frequency values cannot be used so that the peaking in the time-domain response that is required in the detection scheme is achieved.

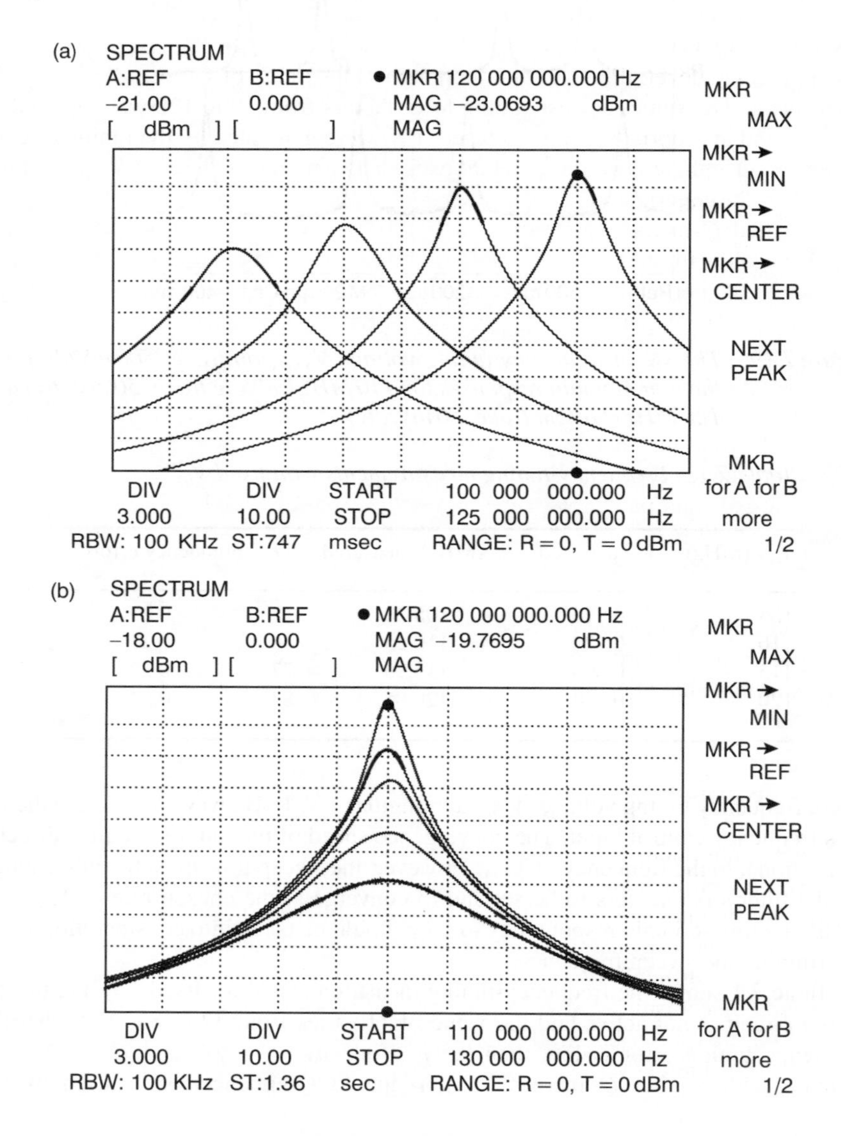

Figure 7.20 (a) Magnitude plots for the bandpass filter with f_0 = 105 MHz, 110 MHz, 115 MHz and 120 MHz; (b) 120 MHz bandpass filter for Q = 20, 33, 52, 71 and 120

Part of the tuning error is caused by clock feedthrough, which can be minimised using minimum-size devices for the switches. Partial cancellation of clock feedthrough is possible by using dummy transistors or complementary switches. Also, increasing the size of the output capacitors, C_o in Figure 7.16 and C_d in Figure 7.17, reduces the parasitic discharge and the ripple due to clock feedthrough.

To test the Q-tuning circuit, a DC voltage corresponding to the input signal level was directly applied to the input V_{imax} in Figure 7.17.[15] When changing the value of this applied DC level, the Q-tuning loop adjusts the value of Q accordingly. In the experiments, DC values corresponding to Q-factors from 10 to 120 were applied to the filter and appropriate tuning voltages, V_{qs}, were generated by the tuning circuit. However, since there is no direct relationship between the DC level at V_{imax} and the value of Q, the experimental Q-tuning error could not be determined. Table 7.2 shows the measured Q-factor for several V_{imax} values at a 120 MHz reference frequency. Magnitude plots for these measurements are shown in Figure 7.20(b). Note the stable well defined experimental transfer functions that are observed even for very large values of Q. The experimental frequency-tuning errors were found to be less than 0.3 per cent, making the method useful for integrated filters with quality factors in the range of a few hundred. The highest frequency of operation is determined by two envelope detectors, which do not need to be accurate as long as they are matched. Beyond the envelope detectors, the tuning operation uses only DC values and low-frequency logic signals. Note that the tuning circuitry is small and simple so that the area and power overhead required for tuning is minimised, and that the tuning operation is insensitive to comparator (or amplifier) offsets and delays. The envelope detectors are the most critical components; they extract the pole frequency and quality factor information. Better high-frequency envelope detectors with less noise will improve the tuning accuracy. Besides the crosstalk between blocks, filter operation is directly affected by the ripples on the control voltages, which may limit the maximum value of Q that can be attained from the circuit.

7.9 Conclusions

Building a continuous-time filter on an IC chip, compatible with any desired technology, is a well understood and relatively easy process. However, although many different techniques are available for designing filters with arbitrary transmission requirements, the task of tuning a filter in an economical and commercially acceptable way tends to be difficult. It is not easy to bring a filter to its prescribed specifications in the face of parasitics and poorly controlled process parameters, and neither is maintaining the prescribed transfer function under changing operating conditions. Generally, the tuning overhead (circuitry, power consumption, noise) becomes more expensive when filter parameters must be tuned over wide ranges or to tight tolerances. As a rule, frequency parameters are tuned by locking the pertinent RC products or

[15] This step was necessary because of the absence of a matching input envelope detector.

Table 7.2 Q-tuning measurements with a 120 MHz reference signal

V_{imax} (V)	V_{qs} (V) generated	Q measured
1.812	0.836	120
1.772	0.831	71
1.739	0.826	52
1.686	0.817	33
1.642	0.800	20

C/g_m ratios to an accurate crystal-controlled frequency derived from a clock or a sinusoidal oscillator. The locking scheme most often employed is a phase-locked or a frequency-locked loop. In some circumstances, such as when accurate phase measurements are difficult to make, a magnitude-locked loop may prove advantageous. For ease of measurement, quality factors are most often tuned by making use of the proportionality between gain and Q. This method assumes, of course, that such a simple relationship exists, which was correct for the circuit in Figure 7.4, but is by no means true for all filter circuits. Q-tuning is performed either by adjusting the transconductor phase or by compensating parasitic losses with negative elements. Tuning a quality factor accurately tends to lead to complex circuitry, especially for narrowband filters when Q is large. The designer is encouraged, therefore, to think through the system requirements and use the filter that is least demanding (lowest order, lowest quality factors) and consequently most economical to lay out, build and tune.

We have in this chapter presented the main ideas that have been proposed in the literature for the tuning of integrated continuous-time filters. To minimise interference problems, master-slave schemes are used most often. Although first-order master stages have been employed, typically, second-order filters or oscillators are used because they can be matched more reliably to second-order slave filter stages. Even for that case, matching of much better than 1 per cent between master and slaves on a chip is difficult to maintain. This problem increases at higher frequencies when parasitic components become more important. Whether the master-slave technique or a direct tuning method is used in cases where the filter has periodic inactive periods, it is helpful for better matching and tracking that a filter architecture is chosen that has many identical circuit blocks. The tuning algorithm will be simpler and the control circuitry smaller, less noisy and less power-hungry. DFT is a well understood and accepted requirement in integrated circuits, i.e. an IC is 'designed for test'. Analogously, it is crucial for success that an integrated analogue filter is 'designed for tuning'. The filter topologies must be tunable, must be insensitive to parasitics, and be constructed with well matched circuit blocks. The unavoidable overhead penalty for tuning will then be minimised.

7.10 References

1 TSIVIDIS, Y. and VOORMAN, J. A. (Eds.): *Integrated Continuous-Time Filters: Principles, Design and Implementations*, IEEE Press, New York, 1993.

2 SCHAUMANN, R., GHAUSI, M. S. and LAKER, K. R.: *Design of Analog Filters: Passive, Active RC and Switched Capacitor*, Prentice-Hall, Englewood Cliffs, 1990.

3 VANPETEGHEM, P. M. and SONG, R.: 'Tuning Strategies in High-Frequency Integrated Continuous-Time Filters', *IEEE Trans. Circuits and Systems*, Vol. CAS-36, pp. 136–139, January 1989.

4 PARK, C. S. and SCHAUMANN, R.: 'Design of a 4-MHz Analog Integrated CMOS Transconductance-*C* Bandpass Filter', *IEEE Journal of Solid-State Circuits*, Vol. SC-23, pp. 987–996, August 1988.

5 MIURA, K., OKADA, Y., SHIOMI, M., MASUDA, M., FUNAKI, E., OKADA, Y. and OGURA, S.: 'VCR Signal Processing LSIs with Self-adjusted Integrated Filters', *Proc. Bipolar Circuits and Technology Meeting (BCTM)*, pp. 85–86, 1986.

6 KOZMA, K. A., JOHNS, D. A. and SEDRA, A. S.: 'Automatic Tuning of Continuous-Time Filters Using an Adaptive Tuning Technique', *IEEE Trans. Circuits and Systems*, Vol. CAS-38, pp. 1241–1248, 1991.

7 KOZMA, K. A., JOHNS, D. A. and SEDRA, A. S.: 'Tuning of Continuous-Time Filters in Presence of Parasitic Poles', *IEEE Trans. Circuits and Systems*, I, Vol. CAS-40, pp. 13–20, 1993.

8 YAMAMOTO, T., KAMOSHIDA, I., KOGA, K., SAKAI, T. and SAWA, S.: 'FM Audio IC for VHS VCR Using New Signal Processing', *IEEE Trans. Consumer Electronics*, Vol. CE-35, pp. 723–731, November 1989.

9 SILVA-MARTINEZ, J., STEYAERT, M. and SANSON, W.: 'A Novel Approach for the Automatic Tuning of Continuous-Time Filters', *Proc. IEEE Int. Symp. Circuits and Systems* (ISCAS), pp. 1452–1455, 1991.

10 YAMAZAKI, H., OISHI, K. and GOTOH, K.: 'An Accurate Center Frequency Tuning Scheme for 450-kHz G_m-*C* Bandpass Filters', *IEEE Journal of Solid-State Circuits*, Vol. 34, No. 12, pp. 1691–1697, December 1999.

11 PENNOCK, J., FRITH, P. and BARKER, R. G.: 'CMOS Triode Transconductor Continuous-Time Filters', *IEEE Custom Integrated Circuits Conference* (CICC), pp. 378–381, 1986.

12 HURTIG, G.: 'Voltage-Tunable Multipole Bandpass Active Filters', *Proc. IEEE Int. Symp. Circuits and Systems* (ISCAS), pp. 569–572, 1974.

13 KRUMMENACHER, F.: 'Design Considerations in High-Frequency CMOS Transcon-ductance Amplifier Capacitor (TAC) Filters', *Proc. IEEE Int. Symp. on Circuits and Systems* (ISCAS), pp. 100–105, May 1989.

14 SCHAUMANN, R., BRAND, J. R. and LAKER, K. R.: 'Effects of Excess Phase in Multiple Feedback Active Filters', *IEEE Trans. Circuits and Systems*, Vol. CAS-27, No. 10, pp. 967–970, October 1980.

15 CHIANG, D. H. and SCHAUMANN, R.: 'A CMOS Fully-balanced Continuous-time IFLF Filter Design for Read/Write Channels', *Proc. IEEE Int. Symp. Circuits and Systems* (ISCAS), pp. I.167–I.170, May 1996.

16 CHIANG, D. H. and SCHAUMANN, R.: 'Performance Comparison of High-Order Analog Integrated Lowpass Filters Based on IFLF and Cascade Topologies', *IEE Proceedings, Circuits, Devices and Systems* (Special Issue on High-Frequency Analogue Filters), Vol. 147, No. 1, pp. 19–27, 2000.

17 WYSZYNSKI, A. and SCHAUMANN, R.: 'Frequency and Phase Tuning of Continuous-Time Integrated Filters Using Common-Mode Signals', *Proc. IEEE Int. Symp. Circuits and Systems* (ISCAS), pp. 5.269–5.272, June 1994.

18 WYSZYNSKI, A.: 'Design of Common-Mode Self-Tuned Fully Integrated Continuous-Time Transconductance-*C* Filters', PhD Thesis, Portland State University, 1998.

19 KUHN, W. B., ELSHABINI-RIAD, A. and STEPHENSON, F. W.: 'A New Tuning Technique for Implementing Very High Q, Continuous-Time, Bandpass Filters in Radio Receiver Applications', *Proc. IEEE Int. Symp. Circuits and Systems* (ISCAS), pp. 5.257–5.260, June 1994.

20 KRUMMENACHER, F. and JOEL, N.: 'A 4-MHz CMOS Continuous-Time Filter with On-Chip Automatic Tuning', *IEEE Journal of Solid-State Circuits*, Vol. SC-23, pp. 750–758, June 1988.

21 KARSILAYAN, A. I.: 'Automatic Tuning of High-*Q* Filters Based on Envelope Detection', PhD Thesis, Portland State University, 2000.

22 TSIVIDIS, Y.: 'Self-Tuned Filters', *Electronics Letters*, Vol. 17, No. 12, pp. 406–407, June 1981.

23 NGUYGEN, N. M. and MEYER, R. G.: 'Si IC-Compatible Inductors and *LC* Passive Filters', *IEEE Journal of Solid-State Circuits*, Vol. SC-25, pp. 1028–1031, 1990.

24 PIPILOS, S., TSIVIDIS, Y. P., FENK, J. and PAPANANOS, Y.: 'A Si 1.8 GHz RLC Filter with Tunable Center Frequency and Quality Factor', *IEEE Journal of Solid-State Circuits*, Vol. 31, No. 10, pp. 1517–1525, October 1996.

25 TSIVIDIS, Y.: *Operation and Modeling of the MOS Transistor*, McGraw-Hill, New York, 1987.

26 SCHAUMANN, R.: 'Simulating Lossless Ladders With Transconductance-C Circuits', *IEEE Trans. Circuits and Systems*, II, Vol. 45, No. 3, pp. 407–410, March 1998.

27 SZCZEPANSKI, S., SCHAUMANN, R. and PANKIEWICZ, B.: 'A CMOS Transconductance Element with Improved DC Gain and Wide Bandwidth for VHF Filters', *Journal of Analog Integrated Circuits and Signal Processing*, Vol. 10, No. 3, pp. 143–156, August 1996.

28 KARSILAYAN, A. I. and SCHAUMANN, R.: 'A Mixed-Mode Automatic Tuning Scheme for High-*Q* Continuous-Time Filters', *IEE Proceedings, Circuits, Devices and Systems* (Special Issue on High-Frequency Analogue Filters), Vol. 147, No. 1, pp. 57–64, 2000.

29 GOPINATHAN, V., TSIVIDIS, Y. P., TAN, K.-S. and HESTER, R. K.: 'Design Considerations for High-Frequency Continuous-Time Filters and Implementation

of an Antialiasing Filter for Digital Video', *IEEE Journal of Solid-State Circuits*, Vol. SC-25, pp. 1368–1378, December 1990.

30 KHOURY, J. M.: 'Design of a 15-MHz CMOS Continuous-Time Filter with On-Chip Tuning', *IEEE Journal of Solid-State Circuits*, Vol. SC-26, pp. 1988–1997, December 1991.

31 SEDRA, A. S. and SMITH, K. C.: *Microelectronic Circuits*, Oxford University Press, New York, 1998.

32 PARKERS, J. F. and CURRENT, K. W.: 'A CMOS Continuous-Time Band-pass Filter with Peak-Detection-Based Automatic Tuning', *Int. J. of Electronics*, Vol. 821, No. 5, pp. 551–564, 1996.

33 KHORRAMABADI, H. and GRAY, P. R.: 'High-Frequency CMOS Continuous-Time Filters', *IEEE Journal of Solid-State Circuits*, Vol. SC-19, pp. 939–948, December 1984.

34 BULT, K. and WALLINGA, H.: 'A CMOS Analog Continuous-Time Delay Line with Adaptive Delay-Time Control', *IEEE Journal of Solid-State Circuits*, Vol. SC-23, pp. 759–766, June 1988.

35 STEVENSON, J.-M. and SANCHEZ-SINENCIO, E.: 'An Accurate Quality Factor Tuning Scheme for IF and High-Q Continuous-Time Filters', *IEEE Journal of Solid-State Circuits*, Vol. 33, pp. 1970–1998, December 1998.

36 JOHNS, D. and MARTIN, K.: *Analog Integrated Circuit Design*, Wiley, New York, 1997.

Concluding remarks

Yichuang Sun

No matter to what extent digital techniques are used, analogue filters will always be needed for interfacing the real analogue world. Also, in recent years, the renewed interest in analogue, mixed-signal and RF circuits due to system-on-chip design has led to a new peak of research of high frequency integrated analogue filters. In fact, the high frequency integrated analogue filter has become a key component in achieving ubiquitous communication and computing. This book has been concerned with this fast growing area. Seven advanced topics have been included, namely OTA/g_m-C filters, MOSFET-C filters, active LC filters, log domain filters, switched current filters, analogue adaptive filters, and on-chip automatic filter tuning.

Whilst traditional active RC filters utilise the high gain operational amplifier in closed loop form, OTA/g_m-C filters dominating high frequency applications use the tunable transconductance amplifier in open loop form and do not use resistors. Chapter 1 has addressed the architectures, performance, design, tuning and implementation of OTA-C filters. Two integrator loop, LC ladder-based and multiple loop feedback g_m-C filters have been introduced, together with current-mode g_m-C filters. The diversity of design methods has been emphasised and guidance on choice of design for particular applications has been given, bearing in mind that each method has its own merits. Design examples using CMOS technology have also been given and automatic tuning issues have been discussed.

MOSFET-C filters result directly from active RC filters with the resistor being replaced by the tunable MOSFET. The MOSFET-C filter concept and practical design considerations have been presented in Chapter 2. Low power design and dynamic range issues have been discussed. MOSFET-C filters are also popular for high frequency applications. Use of balanced structures is particularly important for MOSFET-C filter design. CMOS has been almost exclusively used for MOSFET-C filters in the literature. This chapter has concluded that with MOSFET resistors and bipolar op-amps, BiCMOS is also an attractive technology for high-frequency high dynamic range MOSFET-C filters.

In the design of integrated filters, active simulation of the inductor has been utilised due to the difficulty in integrating the passive inductor on silicon. However, active inductor simulation cannot meet the requirements of RF wireless communications and bulky discrete inductors have had to be used in many applications. Attempts to put inductors on silicon have been more actively pursued recently due to the exploding mobile market. An excellent overview on active LC filters using integrated inductors has been given in Chapter 3. Basic principles, the state of the art, and practical design issues have been addressed. In particular, Q-enhancement methods, negative resistor implementations, dynamic range analysis, filter design and automatic tuning have been discussed.

Log domain filters are based on the exponential function of bipolar transistors. Rather than linearising the transistor, this method directly uses the non-linear characteristic to synthesise linear filters. Chapter 4 has thoroughly reviewed this new and exciting area. Basic ideas and concepts of log domain filters have been introduced. Synthesis methods for log domain filters have been summarised, which include the g_m-C based method and the translinear technique. Various practical performance and design issues have also been addressed. Log domain filters are perhaps the most revolutionary development in the recent rapid development of filter theory.

The filters studied in Chapters 1 to 4 are all continuous-time filters. Chapter 5 has addressed switched-current (SI) filters which are sampled-data analogue filters. Comparisons of SI filters and traditional switched capacitor filters have been made, showing the advantages of the SI filters. Different error performance criteria have been analysed for both class A and class AB SI filters in both single-ended and balanced form. Other considerations (such as low power and neutralisation) and simulation examples have also been given. SI filters have been successfully used for video frequency signal processing. Further increase in the operating frequency may be difficult due to the sampling requirement. Continuous-time input anti-aliasing and output smoothing filters may be needed in SI systems.

In many applications filter parameters must track some changes to achieve the best performance. Digital adaptive filters are only suitable for low frequency applications. In high frequency applications, analogue adaptive filters are the choice. Chapter 6 has overviewed the development of analogue adaptive filters. Adaptive algorithms and suitable filter structures have been presented. The chapter has covered both continuous-time and sampled-data adaptive filters. Interesting application areas have been discussed which include wired communications and magnetic storage systems.

In modern analogue filter design two key techniques have been widely used: the balanced structure and on-chip automatic tuning. Throughout the book, balanced architectures have been utilised for all kinds of filter. On-chip automatic tuning has also been mentioned in several chapters. Chapter 7 is exclusively focused on the difficult and important topic of on-chip automatic tuning. A detailed treatment has been given and basic principles and various tuning methods have been presented together with a new tuning scheme using envelope detectors and simple mixed-mode circuitry. Both frequency-tuning and Q-tuning have been addressed.

Throughout the book design methods and filter architectures have been stressed. In most cases they are general in the sense that they may be implemented in any IC technologies. Although CMOS, BiCMOS and bipolar have been more widely used for high frequency integrated analogue filter implementations, GaAs and SiGe have also been used for RF and microwave applications. In the last fifteen years or so, tremendous progress has been made in analogue filter research and development; however, many challenges remain ahead for us. For example, lower supply voltage and power consumption, GHz frequency operation and larger dynamic range are required for wireless communication systems. High-Q RF bandpass filter design will be a specific challenge for single chip transceivers. Linear phase filters/equalisers at higher frequencies for mass storage systems such as computer hard disk drives remain an active topic. Consumer electronic systems also often require very high performance analogue filters. Hopefully, this book has provided the reader with sufficient knowledge and skills to meet these challenges and to tackle any new filtering problems that may appear in the future.

Index